Atomically-Precise Methods for Synthesis of Solid Catalysts

RSC Catalysis Series

Editor-in-Chief:
Professor James J Spivey, *Louisiana State University, Baton Rouge, USA*

Series Editors:
Professor Chris Hardacre, *Queen's University Belfast, Northern Ireland*
Professor Zinfer Ismagilov, *Boreskov Institute of Catalysis, Novosibirsk, Russia*
Professor Umit Ozkan, *Ohio State University, USA*

Titles in the Series:
 1: Carbons and Carbon Supported Catalysts in Hydroprocessing
 2: Chiral Sulfur Ligands: Asymmetric Catalysis
 3: Recent Developments in Asymmetric Organocatalysis
 4: Catalysis in the Refining of Fischer–Tropsch Syncrude
 5: Organocatalytic Enantioselective Conjugate Addition Reactions: A Powerful Tool for the Stereocontrolled Synthesis of Complex Molecules
 6: *N*-Heterocyclic Carbenes: From Laboratory Curiosities to Efficient Synthetic Tools
 7: *P*-Stereogenic Ligands in Enantioselective Catalysis
 8: Chemistry of the Morita–Baylis–Hillman Reaction
 9: Proton-Coupled Electron Transfer: A Carrefour of Chemical Reactivity Traditions
10: Asymmetric Domino Reactions
11: C–H and C–X Bond Functionalization: Transition Metal Mediation
12: Metal Organic Frameworks as Heterogeneous Catalysts
13: Environmental Catalysis Over Gold-Based Materials
14: Computational Catalysis
15: Catalysis in Ionic Liquids: From Catalyst Synthesis to Application
16: Economic Synthesis of Heterocycles: Zinc, Iron, Copper, Cobalt, Manganese and Nickel Catalysts
17: Metal Nanoparticles for Catalysis: Advances and Applications
18: Heterogeneous Gold Catalysts and Catalysis
19: Conjugated Linoleic Acids and Conjugated Vegetable Oils
20: Enantioselective Multicatalysed Tandem Reactions
21: New Trends in Cross-Coupling: Theory and Applications
22: Atomically-Precise Methods for Synthesis of Solid Catalysts

How to obtain future titles on publication:
A standing order plan is available for this series. A standing order will bring delivery of each new volume immediately on publication.

For further information please contact:
Book Sales Department, Royal Society of Chemistry, Thomas Graham House, Science Park, Milton Road, Cambridge, CB4 0WF, UK
Telephone: +44 (0)1223 420066, Fax: +44 (0)1223 420247
Email: booksales@rsc.org

Visit our website at www.rsc.org/books

Atomically-Precise Methods for Synthesis of Solid Catalysts

Edited by

Sophie Hermans
Institut IMCN - Pole MOST, Universite catholique de Louvain,
Louvain-la-Neuve, Belgium
Email: sophie.hermans@uclouvain.be

Thierry Visart de Bocarmé
Service de Chimie Physique des Materiaux, Universite Libre de
Bruxelles, Brussels, Belgium
Email: Thierry.Visart.de.Bocarme@ulb.ac.be

THE QUEEN'S AWARDS
FOR ENTERPRISE:
INTERNATIONAL TRADE
2013

RSC Catalysis Series No. 22

Print ISBN: 978-1-84973-829-3
PDF eISBN: 978-1-78262-843-9
ISSN: 1757-6725

A catalogue record for this book is available from the British Library

© The Royal Society of Chemistry 2015

All rights reserved

Apart from fair dealing for the purposes of research for non-commercial purposes or for private study, criticism or review, as permitted under the Copyright, Designs and Patents Act 1988 and the Copyright and Related Rights Regulations 2003, this publication may not be reproduced, stored or transmitted, in any form or by any means, without the prior permission in writing of The Royal Society of Chemistry or the copyright owner, or in the case of reproduction in accordance with the terms of licences issued by the Copyright Licensing Agency in the UK, or in accordance with the terms of the licences issued by the appropriate Reproduction Rights Organization outside the UK. Enquiries concerning reproduction outside the terms stated here should be sent to The Royal Society of Chemistry at the address printed on this page.

The RSC is not responsible for individual opinions expressed in this work.

Published by The Royal Society of Chemistry,
Thomas Graham House, Science Park, Milton Road,
Cambridge CB4 0WF, UK

Registered Charity Number 207890

For further information see our web site at www.rsc.org

Printed and bound by CPI Group (UK) Ltd, Croydon, CR0 4YY

Preface

The idea of this book came up 2 years ago as an additional member of the RSC's 'Catalysis' series. When we were approached to edit it, it was not completely clear whether atomic-scale control of heterogeneous catalyst synthesis could indeed already be achieved with certainty. Is it a dream or reality? Recent developments in the preparation of catalysts, and especially multi-metallic nanoparticles, open up great perspectives to finely tune the catalytic properties of solids, either by their size and/or their composition that can now be adjusted and characterized with an atomic resolution.

We are proud to offer you this book with successful examples of catalysts controlled at the atomic scale. Rather than trying to find a unifying view on this subject matter, we have structured our book into self-contained chapters that each tackle atomic control using very different experimental methods. Each chapter represents one illustrative example of methodology that allows the synthesis of catalysts with a high degree of control. However, interestingly, there is a common ground: being more 'physical' or 'chemical' does not matter; each synthesis strategy presented is a 'bottom-up' approach, which aims at organizing matter with a high degree of complexity, rather than starting from bulk compounds and trying to downscale them.

The first three chapters use organometallic chemistry as the synthesis tool. The first chapter describes surface organometallic chemistry in order to build on solid supports single sites that mimic active species found in homogeneous catalysis. The second chapter immobilizes molecular metal complex catalysts within zeolites, to obtain uniform structures that facilitate understanding of factors governing catalytic activity. The third chapter presents the use of heteronuclear complexes or clusters as precursors of supported bimetallic nanoparticles with control of the composition within each individual nanoparticle. The fourth and fifth chapters disclose the use of atomically precise nanoclusters in catalytic applications to unravel the fundamental origins of structure–activity and structure–selectivity relationships. The next two chapters report more 'physical' methods with Chapter 6 covering monolayer electro-deposition techniques and Chapter 7 covering atomic layer deposition (ALD) in the vapor phase for the preparation of solid catalysts. Chapter 8 presents model catalysts (single crystals, films and fibers) that allow an understanding of reaction mechanisms at an advanced molecular level. Chapter 9 uses modern knowledge of nanoparticles (in terms of size, shape and compositional control) to prepare advanced electrocatalysts, presenting much improved activity and durability for electrochemical energy applications. Finally, Chapter 10 reports atom probe microscopies (APM) as ideal techniques to probe nano-structured catalysts at the atomic scale. We have deliberately chosen not to devote a whole chapter to a single characterization technique amenable to obtaining data at the atomic scale, even if these exist nowadays. One could think of HR-TEM and also *in situ* EXAFS or XRD techniques, for instance. However, throughout the examples treated in each chapter, the power (and limitations) of many such techniques appears clearly. The reader is referred to many other textbooks on modern characterization techniques to gather basic operating information.

Atomically precise methods for the synthesis of solid catalysts do exist. With its ten chapters emanating from different catalysis fields, this book proves it. Of course, it is not comprehensive and another collection of chapters could have made up the point. The preparation methods presented here seem to reach the ultimate level of control attainable, coming from bulk to microstructure, down to the nanoscale and eventually the atomic level. Is this the ultimate step of sophistication for synthesis methods that have been developed in heterogeneous catalysis over decades? Certainly not, as researchers always find new challenges to tackle. We are convinced that mastering the synthesis of nanostructured catalysts by means of atomically precise methods unravels a bright future for heterogeneous catalysis, its developments and applications.

Sophie Hermans and Thierry Visart de Bocarmé

Contents

Chapter 1 Synthesis of Well-defined Solid Catalysts by Surface Organometallic Chemistry   1
Frédéric Lefebvre

    1.1 Introduction   1
    1.2 Grafting Sites of the Support   3
    1.3 Formation of Grafted Organometallic Complexes by Reaction with One Hydroxyl Group   6
        1.3.1 Reaction of Metal Alkyl Complexes   6
        1.3.2 Reactivity of Metal Alkylidenes or Alkylidynes   8
    1.4 Further Reactions of Surface Species Formed by Reaction of an Organometallic Compound with One Hydroxyl Group   10
        1.4.1 Further Reactions with Hydroxyl Groups   10
        1.4.2 α-H, β-H and γ-H Abstraction   11
        1.4.3 Reaction with Hydrogen   12
        1.4.4 Reaction with Other Molecules   14
    1.5 Formation of Cationic Complexes on the Surface   15
    1.6 Applications in Catalysis   17
        1.6.1 Hydrogenolysis of Alkanes   17
        1.6.2 Alkane Metathesis   18
        1.6.3 Ethylene to Propene by Tungsten Hydride   18
        1.6.4 Olefin Metathesis   19
        1.6.5 Trimerization of Ethylene   19
        1.6.6 Epoxidation and Deperoxidation Reactions   20
    1.7 Activation of Ammonia and Nitrogen   21
    1.8 Conclusion   22
    References   22

Chapter 2	**Zeolite-supported Molecular Metal Complex Catalysts** *Isao Ogino*		**27**
	2.1 Introduction		27
	2.2 Synthesis		29
		2.2.1 Synthesis by an Ion-exchange Method	29
		2.2.2 Synthesis by a Chemical Vapor Deposition Method	30
		2.2.3 Synthesis by the Surface Organometallic Chemistry Approach	31
	2.3 Characterization Techniques		31
	2.4 Guide for the Synthesis and Characterization of Supported Molecular Metal Complexes with a High Degree of Structural Uniformity		33
		2.4.1 Types of Zeolite for Support Materials	33
		2.4.2 Precursor Metal Complexes	38
		2.4.3 Spectroscopic Characterization Techniques to Examine Structural Uniformity	41
	2.5 Molecular Chemistry of Zeolite-supported Metal Complex Catalysts		41
		2.5.1 Reactivity of Supported Metal Complexes	41
		2.5.2 Catalytic Cycle	46
	2.6 Conclusions and Future Directions		48
	References		50
Chapter 3	**Bimetallic Supported Catalysts from Single-source Precursors** *Sophie Hermans*		**55**
	3.1 Introduction		55
	3.2 Heteronuclear Species as Heterometallic Catalysts Precursors		57
		3.2.1 Complexes with Heterometallic M–M' Metal–Metal Bond	57
		3.2.2 Ions Pairs	58
		3.2.3 Heterometallic Complexes with Bridging Ligands	59
		3.2.4 Other Bimetallic Combinations	61
		3.2.5 Mixed-metal Precursors of Bulk Phases	61
	3.3 Clusters as Precursors		64
		3.3.1 Definition	64
		3.3.2 Synthesis	65

		3.3.3	Clusters in Catalysis	65
		3.3.4	Clusters as Precursors of Supported Bimetallic Nanoparticle Catalysts	66
	3.4	Conclusions and Future Challenges		82
	References			83

Chapter 4 Atomically Precise Gold Catalysis 87
Katla Sai Krishna, Jing Liu, Pilarisetty Tarakeshwar, Vladimiro Mujica, James J. Spivey and Challa S. S. R. Kumar

	4.1	Introduction		87
	4.2	Overview of Synthesis and Catalysis		88
		4.2.1	Synthesis of Atomically Precise Gold Nanoclusters	88
		4.2.2	Catalysis	91
	4.3	Overview of Electronic and Magnetic Structure of Catalysts		100
		4.3.1	Electronic Structure	102
		4.3.2	Magnetic Structure	108
	4.4	Correlation of Electronic Structure–Catalysis Relationship		113
	4.5	Correlation of Magnetic Structure–Catalysis Relationship		114
	4.6	Conclusions and Future Perspective		116
	Acknowledgements			117
	References			117

Chapter 5 Atomically Precise Gold Nanoclusters: Synthesis and Catalytic Application 123
Gao Li and Rongchao Jin

	5.1	Introduction		123
	5.2	Synthesis of Atomically Precise Gold Nanoclusters: Size-focusing Method		124
		5.2.1	The Case of $Au_{25}(SR)_{18}$ Nanoclusters	125
		5.2.2	The Case of $Au_{38}(SR)_{24}$ Nanocluster	126
	5.3	Crystal Structures of Gold Nanoclusters		128
	5.4	Thermal Stability of $Au_n(SR)_m$ Nanoclusters		129
	5.5	Reactivity and Catalytic Properties of $Au_n(SR)_m$ Nanoclusters		130
		5.5.1	Reversible Conversion Between $[Au_{25}(SR)_{18}]^0$ and $[Au_{25}(SR)_{18}]^-$	130
		5.5.2	Catalytic Oxidation	131
		5.5.3	Catalytic Selective Hydrogenation	136
		5.5.4	Catalytic Carbon–Carbon Coupling Reaction	138

5.6	Summary	141
	Acknowledgements	141
	References	141

Chapter 6 Electrochemical Atomic-level Controlled Syntheses of Electrocatalysts for the Oxygen Reduction Reaction — 144
Stoyan Bliznakov, Miomir Vukmirovic and Radoslav Adzic

6.1	Background	144
6.2	Electrochemical Deposition of Monolayers of Precious Metals on Different Transition Metal Supports	147
	6.2.1 Adsorption-driven Surface-limited Reactions for Atomic Monolayer Deposition	148
	6.2.2 Displacement-driven Surface-limited Reactions for Atomic Monolayer Deposition	149
6.3	Platinum Monolayer on Electro-deposited Mono- and Bi-metallic (Pd, PdAu, PdIr, NiW) Nanostructures: Highly Efficient Electrocatalysts for the ORR	156
	6.3.1 Electro-deposited Pd Nanostructures and Bimetallic Pd Alloys on Functionalized Carbon Substrates: Advanced Cores for Pt ML Shell Electrocatalysts	157
	6.3.2 Pt ML on Electro-deposited Pd/WNi Refractory Alloys: Advanced Fuel Cell Electrocatalysts	161
6.4	Conclusions	163
	Acknowledgements	164
	References	164

Chapter 7 Atomic Layer Deposition in Nanoporous Catalyst Materials — 167
Jolien Dendooven

7.1	Introduction	167
	7.1.1 Atomic Layer Deposition	168
	7.1.2 Conformality of ALD in Nanoporous Materials	171
	7.1.3 Opportunities of ALD in Supported Catalyst Preparation	177
7.2	ALD for Catalysis – a Literature Overview	179
	7.2.1 Early Work on Catalyst Preparation by ALD	179
	7.2.2 Supported Noble Metal Catalysts by ALD	181
	7.2.3 ALD for Photocatalysis	184

Contents   xi

		7.2.4	Synthesis of Catalytic Membranes by ALD	184
		7.2.5	Use of Ordered Mesoporous Supports in Catalyst Preparation by ALD	185
	7.3	ALD for Catalysis: Case Studies		186
		7.3.1	Case 1: Introducing Acid Sites in Ordered Mesoporous Materials	186
		7.3.2	Case 2: Introducing Photo-Active Nanoparticles in Mesoporous Films	189
	7.4	Conclusions		191
	Acknowledgements			192
	References			192

Chapter 8 Preparation and Characterization of Model Catalysts for the HCl Oxidation Reaction — 198
Christian Kanzler, Herbert Over, Bernd M. Smarsly and Claas Wessel

8.1	The Deacon Process: Oxidation of HCl	198
8.2	Why Model Catalysis?	199
8.3	Synthesis of Single Crystalline RuO$_2$ Films for Gaining Molecular Information on Stability and Activity	202
8.4	Atomic-Scale Properties of RuO$_2$(110)	204
8.5	What can be Learnt from Single Crystalline RuO$_2$ Model Catalyst?	205
8.6	Synthesis of Metal Oxide Fibers *via* Electrospinning	208
8.7	Electrospun RuO$_2$-based Fibers as Model Catalysts in the HCl Oxidation Reaction	211
8.8	CeO$_2$-based Catalysts: Alternatives to RuO$_2$-based Deacon Catalysts	213
8.9	Conclusions and Outlook	216
Acknowledgements		217
References		218

Chapter 9 Controllable Synthesis of Metal Nanoparticles for Electrocatalytic Activity Enhancement — 225
Qing Li, Wenlei Zhu and Shouheng Sun

9.1	Introduction		225
9.2	Synthesis of Monodisperse Metal NPs		226
	9.2.1	General Concept on NP Formation	226
	9.2.2	NP Size Control	228
	9.2.3	NP Shape Control	229
9.3	NP Activation for Catalysis		232

9.4	Applications of Metal NPs in Electrocatalysis		233
	9.4.1 NP Catalysis for ORR		233
	9.4.2 NP Catalysts for FAOR		239
	9.4.3 Metal NPs as Catalysts for Electrochemical Reduction of CO_2		242
9.5	Summary and Perspectives		243
Acknowledgements			244
References			244

Chapter 10 Investigating Nano-structured Catalysts at the Atomic scale by Field Ion Microscopy and Atom Probe Tomography 248
Cédric Barroo, Paul A. J. Bagot, George D. W. Smith and Thierry Visart DE Bocarmé

10.1	Introduction	248
10.2	Imaging and Local Chemical Analysis of Nanosized Crystals Before, During and After Catalysis	252
	10.2.1 Sample Preparation	252
	10.2.2 Field Ion Microscopy	254
	10.2.3 Atom Probe Tomography (APT)	256
10.3	Case Studies	259
	10.3.1 Catalytic Reactions and Surface Reconstructions on Metals: FIM Studies	259
	10.3.2 Surface Enrichment of Platinum Group Metal-based Alloys: APT Studies	267
	10.3.3 APT Studies of Catalytic Nanoparticles	280
10.4	Conclusions	289
Acknowledgements		290
References		291

Subject Index   **296**

CHAPTER 1

Synthesis of Well-defined Solid Catalysts by Surface Organometallic Chemistry

FRÉDÉRIC LEFEBVRE

Université Lyon 1, CPE Lyon, CNRS, UMR C2P2, LCOMS, Bâtiment CPE Curien, 43 Boulevard du 11 Novembre 1918, F-69616 Villeurbanne, France
Email: lefebvre@cpe.fr

1.1 Introduction

The knowledge in homogeneous catalysis is very high, due to the conceptual advance of molecular organometallic chemistry. Typically, reports in homogeneous catalysis provide not only information on the catalytic performances (activity, selectivity and life time), but also, in most cases, a detailed mechanistic understanding of the catalytic system. The actual elementary steps of the reaction, directly derived from the principles and the investigations of organometallic chemistry, are usually described. This knowledge allows a predictive approach of these systems, mainly based on the fact that it is possible to have only one well-defined catalytic species in the system. Unfortunately, from an industrial point of view, homogeneous catalysis suffers from many disadvantages and very often heterogeneous systems are preferred even if they are ill-defined and less active. The development of better catalysts in heterogeneous catalysis has always relied on empirical considerations since it is difficult to characterize the really active sites on the surfaces, as the so-called 'active sites' are usually in small number(s). At the present time, the number of accepted 'elementary steps' is still limited to a few examples mostly demonstrated by means of surface science, and the predictive approach, based on molecular concepts, is rare. The concept of surface organometallic chemistry has been developed as a possible answer to this problem. Its main objective is the creation on a support (which can be an oxide, a clay, a polymer *etc.*), of organometallic fragments which will be well-defined and uniform along the entire surface. These species will be characterized by all available physico-chemical methods in order to have a description of the coordination sphere around the metal as precise as possible, as for homogeneous complexes. This strategy, initially proposed by the group of J. M. Basset, has also been developed by other groups (see Table 1.1 for some examples) and has been the subject of numerous reviews and books.[1–9] In most cases the support was silica (flame

Table 1.1 Some examples of grafting reactions of organometallic complexes on various supports by use of surface organometallic chemistry.

Metal	Organometallic species/catalytic reaction	Ref.
Aluminium	Al(OiPr)$_3$ on mesoporous silicas	13
Calcium	Ca[N(SiMe$_3$)$_2$]$_2$ · 2THF on silica	14
Vanadium	V(=N*t*Bu)(CH$_2$*t*Bu)$_3$ on silica	15
Chromium	CrO$_2$Cl$_2$ on silica/model of Phillips polymerization catalyst	16
	Cr(CH$_2$*t*Bu)$_4$ on silica	17
Cobalt	Co[N(SiMe$_3$)$_2$]$_2$ on mesoporous silicas	13
Nickel	Ni(MeCN)$_6$(BF$_4$)$_2$ on silica and MCM-41/propene dimerization; ethylene oligomerization	18–20
Zinc	Zn[(*S,S*)-iPr-pybox](Et)$_2$ on silica	21
	Zn[N(SiMe$_3$)$_2$]$_2$ on mesoporous silicas	13
Zirconium	Zr(CH$_2$*t*Bu)$_4$ on amino-modified SBA-15	10
Molybdenum	Mo(=N)(NR$_2$)(OR)$_2$(pyr) on silica	22
	Mo(=CHCMe$_2$Ph)(=NAr)(OR)$_2$ on silica	23
Ruthenium	Ru(COD)(COT) on SiH groups of modified silica (COT = cyclooctatetraene)	11
Tantalum	Ta(=N*t*Bu)(CH$_2$CMe$_2$Ph)$_3$ on silica/oxo/imido heterometathesis	24
Tungsten	WMe$_6$ on silica	25
	(ArO)$_2$W(=O)(=CH*t*Bu) on silica/alkene metathesis	26
	W$_2$(NMe$_2$)$_6$ on silica	27
Rhenium	CH$_3$ReO$_3$ on silica–alumina/olefin metathesis	28
	[Re(CO)$_3$OH]$_4$ on silica	29
Iridium	Ir$_4$(CO)$_{12}$ on silica	30
Platinum	Pt(COD) complexes on silica (COD = cyclooctadiene)	31
Lanthanum	La(CH(PPh$_2$NSiMe$_3$)$_2$)(N(SiHMe$_2$)$_2$)$_2$ on silica	32,33
Neodymium	Nd(BH$_4$)$_3$(THF)$_3$ on silica/MMA and butyrolactones polymerization	34
	Nd(NR$_2$)$_3$ on silica/isoprene and butyrolactone polymerization	35
Cerium	[Ce(OiPr)$_3$N(SiMe$_3$)$_2$]$_2$ on MCM-41	36
Rare earths (Y, La, Nd, Sm)	Ln[N(SiMe$_3$)$_2$]$_3$ on silica	37–39
Gadolinium	Gd[N(SiHMe$_2$)$_2$]$_3$(THF)$_2$ on mesoporous silica	40

silica, porous silica or mesoporous silica) but there are some examples using other oxides such as alumina or magnesia (Table 1.1).

The grafted organometallic complexes can then be modified by using the classical rules of organometallic chemistry leading to species which are potentially active in catalysis. In these compounds the support can be a mono-, di- or tripodal ligand. Recently, a new dimension was added to this chemistry by performing, prior the reaction with the organometallic complex, a reaction replacing the grafting sites by other species such as N–H,[10] Si–H,[11] or phenol groups.[12]

As a consequence, it is now possible to prepare, on a surface, many organometallic complexes where the electronic and steric effects can be tuned easily. This knowledge now allows researchers to have a relatively predictive approach where the starting point is not the organometallic complex but a catalytic reaction. First, a catalytic cycle is proposed on the basis of the classical rules of organometallic chemistry. Second, a grafted organometallic complex which is one species involved in the postulated catalytic cycle is prepared. Third, the catalytic reaction is performed. Depending on the results the ligands around the metal are modified or, if the reaction does not proceed by the proposed catalytic cycle, another mechanism has to be proposed and tested.

We will describe here this surface chemistry *via* some examples mostly from work done in our laboratory. Our purpose will be to give the rules which will govern the reactivity of the organometallic complexes with the surface and not to compile a complete list of what can be made. We will first describe the grafting sites on the surface as this point will govern the reactivity of organometallic compounds. We will then describe the reaction of these sites with organometallic complexes and how the resulting species can be transformed into active sites before giving some examples in catalysis.

1.2 Grafting Sites of the Support

Organometallic complexes can be deposited on many supports, such as metals, zeolites, oxides or carbons. Depending on the nature, density and homogeneity of the reactive sites on the surface of these materials, different behaviors will be observed, leading to sometimes completely different catalytic applications. As an example we will describe in more detail silica, as it will be extensively used in the following, as it is the simplest support (see Table 1.2 for a non-exhaustive list of grafting sites on other supports). Silica can be considered as the simplest support as it contains only SiO_4 tetrahedra linked by $\equiv$Si–O–Si$\equiv$ bridges. It can be found in various forms such as silica gel, flame silica or mesoporous silica (like MCM-41 or SBA-15). In all cases, at room temperature, the surface is covered by hydroxyl groups $\equiv$Si–OH and siloxane bridges $\equiv$Si–O–Si$\equiv$ in interaction with adsorbed water molecules. Upon heating under vacuum at *ca.* 150 °C all water molecules are desorbed and the infrared spectrum shows mainly, in the ν(O–H) domain, a very broad band between 3700 and 3500 cm^{-1} attributed to $\equiv$Si–OH groups linked *via*

Table 1.2 Grafting sites of some supports which can be used in surface organometallic chemistry.

Support	Grafting sites
Alumina	Hydroxyl groups (at least five different types); Al–O–Al bridges (where the bond is not covalent); Lewis acid sites
Silica–Alumina	Silanol groups; Si–O(H)–Al bridges
Magnesia	Hydroxyl groups; lacunar magnesium sites
Zeolites	Protons with different locations which can be more or less accessible; extra-framework aluminium sites; silanol groups on the external surface of the crystallites
Carbon	All chemical functions can be found: alcohols, amines, ethers, thiols, ketones, aldehydes, carboxylic acid, *etc.*
Clays	Depending on the distance between the layers the possible grafting (or exchange) sites can be more or less accessible
Polymers	All non-inert chemical functions can be anchorage points for organometallic complexes

hydrogen bonds. Upon heating at higher temperature, condensation between two neighboring hydroxyl groups occurs, leading to the evolution of water molecules and formation of $\equiv$Si-O-Si$\equiv$ bridges. As a consequence, the intensity of the broad infrared band decreases and a new sharp band, attributed to isolated silanol groups, appears at *ca.* 3750 cm^{-1}. After heating at 500 °C only isolated silanols are present. Their amount can be determined by chemical methods (such as by their reactivity with CH$_3$Li) or physical techniques (such as quantitative solid-state ^{1}H MAS NMR). The two values can be different as the first method will only quantify the hydroxyl groups accessible to the reagents while the second one will give an estimation of the total number of hydroxyl groups. In the case of microporous solids the difference can be very important. In the case of flame silica, which is non-porous, the two methods lead to an OH density of *ca.* 1.4 OH nm^{-2}.[41,42] Upon heating under vacuum at 700 °C the OH density decreases to *ca.* 0.7 OH nm^{-2}. For such low values, one can reasonably suppose that the hydroxyl groups are far away from each other and so well-defined grafted organometallic isolated species will be expected upon reaction with these hydroxyl groups. This is the key point of surface organometallic chemistry. However, this view is not fully realistic even if it is sufficient in most cases; sometimes a more precise description of the silica support is needed for explaining the experimental data. First of all, there are not only monohydroxyl $\equiv$Si–OH groups on the surface but also dihydroxyl ones =Si(OH)$_2$, as evidenced by solid-state ^{29}Si MAS NMR. These =Si(OH)$_2$ groups give infrared bands at the same position as the $\equiv$Si–OH ones and so they cannot be distinguished by this method. On the samples heated at high temperature the ^{29}Si MAS NMR spectra become broader, preventing the separation of the different Si(OSi)$_{4-x}$(OH)$_x$ ($x = 0-2$) groups and so the presence of some =Si(OH)$_2$ species cannot be excluded even if their amount is probably very low. Studies by ^{1}H double quanta MAS NMR are more informative. This method allows the observation of protons pairs which are separated by less than *ca.* 5 Å. As a consequence, not only pairs of protons involved in hydrogen bonds are seen but also protons at a slightly higher distance. An intense signal is observed for pairs of isolated protons even on silica treated at 700 °C (for which the OH density is 0.7 OH nm^{-2}), probing unambiguously that the repartition of the hydroxyl groups is not fully homogeneous.[43] Additional experiments by triple quanta ^{1}H MAS NMR or spin counting were performed.[43,44] These experiments showed that some hydroxyl groups are present as nests of three silanols located on a cycle containing six silicon atoms as those observed on the (111) face of cristobalite. In that case, the distance between the hydrogen atoms will be *ca.* 3 Å. One, two or three cycles can be adjacent.

Finally, upon heating at a very high temperature (*ca.* 1000 °C), highly strained cycles containing two silicon atoms and two oxygen atoms are formed by condensation between two adjacent silanols. These cycles are highly reactive even if their amount is low (0.14 nm^{-2} on silica dehydroxylated at 1000 °C while the amount of residual hydroxyl groups is 0.4 nm^{-2}).

Scheme 1.1 summarizes the variety of species which are present on dehydroxylated silica as deduced from these studies. Depending on the nature of the silica (non-porous flame silica, mesoporous silica, *etc.*) the relative amount of these species will be different, leading then in some cases to different reaction products.

As shown above, even if silica can be considered as the simplest support, there is not only one species on its surface. For other oxides such as alumina the situation is more complicated and more than five different types of hydroxyl groups can be observed, with all their combinations, without taking into account the Lewis acid sites. The most complex support is probably carbon, as its surface contains a lot of functional groups covering all the fields of organic chemistry. This complexity of the surface support will have a consequence on the number of species which will be obtained upon reaction with organometallic complexes as the strength of the bond between the metal and the surface will be more or less strong. It is for this reason that preliminary studies are always made on silica and mainly on silica dehydroxylated at relatively high temperature (500 or 700 °C).

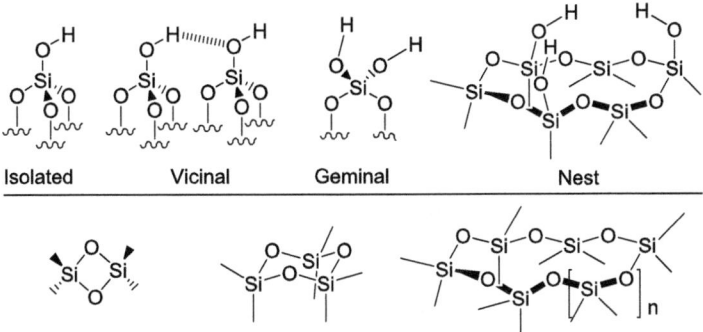

Scheme 1.1 Hydroxyl groups and siloxane cycles on the surface of silica.

1.3 Formation of Grafted Organometallic Complexes by Reaction with One Hydroxyl Group

As we have seen above, the active sites of the support are mainly (if carbon is excluded) hydroxyl groups; only their distribution and strength depend on the treatment and on the oxide under study. The formation of a chemical bond between the organometallic fragment and the solid will then pass, in most cases, through a reaction with these hydroxyl groups. We will describe here what will happen and by comparison of various supports and organometallic compounds how the reaction proceeds.

First of all, it is necessary to choose an organometallic compound with a M′–X bond for which the reaction M–OH + [M′]–X → M–O–[M′] + HX will be favored thermodynamically, M–OH being a hydroxyl group of the support. Many complexes can be chosen, such as chlorides or alkoxy derivatives. However, in these two cases the evolved hydrogen chloride or alcohols can further react with hydroxyl groups or M–O–M bridges of the support and so modify its properties. A typical example is the reaction at room temperature of tantalum methoxide Ta(OMe)$_5$ with silica dehydroxylated at high temperature.[45] ^{13}C CP-MAS NMR of the resulting material shows clearly two signals for methoxy species. One of them can be attributed to a methoxy group on tantalum as expected and a second to a methoxy group on the silica support (such species can be synthesized by treatment at relatively high temperature of silica with methanol). These silica methoxy groups are formed by reaction of evolved methanol with a siloxane bridge or a silanol group. As this reaction does not proceed at room temperature in the absence of the tantalum complex, the metal plays the role of catalyst for this reaction. The consequence is that the grafting reaction will not be clean and the starting treatment of the support for the creation of isolated grafting sites will not be efficient as new potential grafting sites will be created during the reaction. It is for this reason that the organometallic complexes which will be chosen must lead to inert X–H species. The best choice is to have evolution of alkanes which cannot be activated by these supports. For this purpose the ligands around the metal will be alkyl, alkylidene or alkylidyne groups.

1.3.1 Reaction of Metal Alkyl Complexes

Very often homoleptic organometallic complexes are chosen as they will lead to only one surface complex, all ligands being equivalent. One problem is that, kinetically, the reaction will be slow, compared for example with that achieved with alkoxy compounds, due to the fact that the first step, the physisorption on the support, will not be favored, the interaction between hydroxyl groups and alkyl groups not being strong. To overcome this problem the complex can be sublimed on the support, avoiding the use of a solvent, but this method can be used only when it has a sufficient vapor pressure and sometimes the sublimation is accompanied by a partial decomposition. In all cases the observed reaction can be simply written as:

$$\equiv Si-OH + [M]-R \rightarrow \equiv Si-O-[M] + R-H$$

When using homoleptic complexes and silica dehydroxylated at high temperature, well-defined species are obtained which are uniform over all the solid. This strategy has been applied to a lot of metals from the left (Ti, Zr, Hf) to the right (Sn, Ge) of the periodic table (see, for example, Basset et al.[46]). In all cases the same result was obtained but the reaction did not proceed at the same rate: For metals of the left, such as Ti or Zr, the grafting reaction occurred easily at room temperature while for metals of the right, such as Sn, it occurred only at high temperature (ca. 180 °C). This is related to a different reaction mechanism as evidenced by the use of other supports with a stronger acidity such as cloverite,[47] Y zeolite[48] or heteropolyacids.[49] Heteropolyacids such as $H_3PW_{12}O_{40}$ are molecular compounds more acidic than sulfuric acid. Surprisingly, they do not react with alkyl complexes of Ti or Zr while they react at room temperature (and even below) with tin complexes. These results can be understood as follows. For metal complexes of the left of the periodic table, which are highly electron deficient with empty d orbitals, the grafting reaction occurs via an attack of the M–C bond by the oxygen atom of the hydroxyl group followed by evolution of the alkane. So the first step is the formation of the bond with the surface (Scheme 1.2). For metal complexes of the right, which have a high electronic density, the first step is an attack by the proton, leading first to the evolution of alkane and then to the formation of the metal–support bond. As increasing the acidity decreases the oxygen nucleophilicity this explains the different reactivity as a function of the acidity of the support. A consequence is that it is not possible

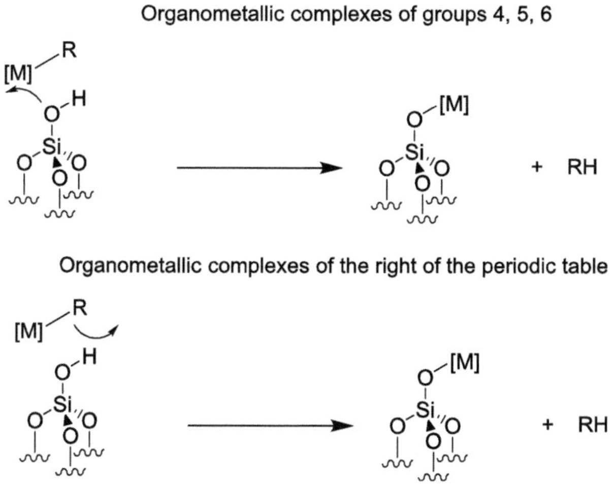

Scheme 1.2 First step of the grafting reaction of organometallic complexes on hydroxyl groups.

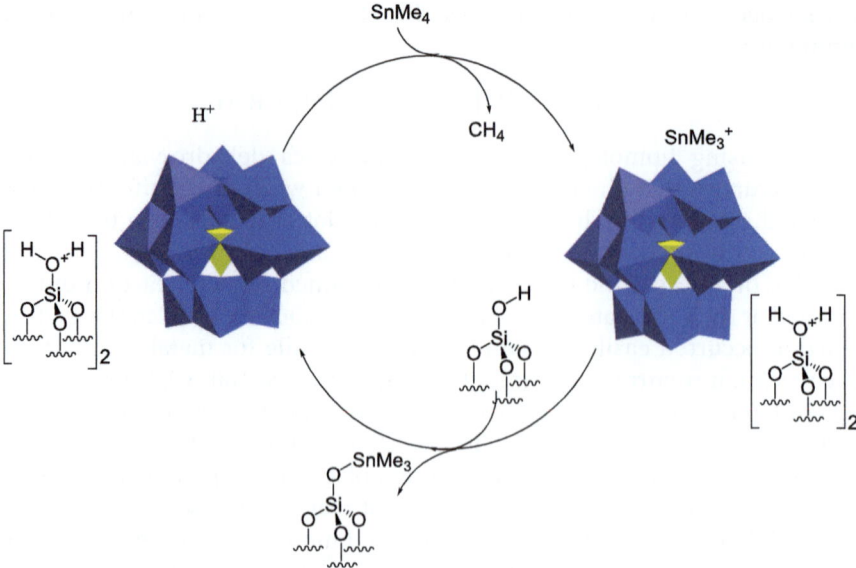

Scheme 1.3 Grafting reaction of tetramethyl tin on silica catalyzed by supported heteropolyacid.

to graft metal alkyl complexes of the left of the periodic table on highly acidic supports. Another consequence is that it is not possible to graft platinum methyl complexes on silica *via* breaking of the Pt–Me bond as these compounds are not stable thermally.

Another consequence of this mechanism is that it can be possible to graft metal complexes of the right of the periodic table by use of acidic species as catalysts: The grafting reaction of tetramethyl tin occurs at room temperature on $H_3PW_{12}O_{40}$ supported on silica but the amount of evolved methane exceeds by at least one order of magnitude the number of protons of the polyacid and can only be explained by a migration on the surface. The mechanism (Scheme 1.3) passes through an attack of the tin complex by the acidic proton of the heteropolyacid followed by a migration of the grafted tin species on the surface and restoration of the acidic proton.

1.3.2 Reactivity of Metal Alkylidenes or Alkylidynes

If for the elements of group 4 (Ti, Zr, Hf) it is possible to prepare metal complexes such as $M(-CH_2-C(CH_3)_3)_4$ without hydrogen atoms in β position of the metal (which can lead to side reactions, see below), it is not possible to obtain the corresponding species for those of groups 5 and 6. Indeed the steric hindrance around the metal is so high that α-H abstraction occurs, leading to the evolution of alkane and the formation of a metal alkylidene (for group 5) and even, after a second α-H abstraction, of a metal alkylidyne (or a dialkylidene). The resulting complexes are now heteroleptic as they

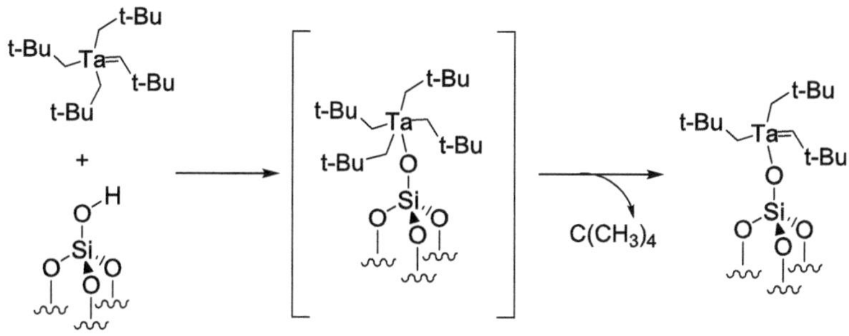

Scheme 1.4 Grafting reaction of trisneopentyl neopentylidene tantalum on a silica surface.

contain two types of ligands which could lead, a priori, to two types of surface complexes. However, only one surface species is obtained, for example ≡Si–O–Ta[CH$_2$–C(CH$_3$)$_3$]$_2$[=CH–C(CH$_3$)$_3$)] in the case of tantalum.[45,50] Two mechanisms can be proposed for the grafting reaction: (i) selective reaction of the metal alkyl moiety as for the alkyl complexes or (ii) addition of the oxygen atom on the carbene moiety followed by α-H abstraction and neopentane evolution (Scheme 1.4). The second mechanism was first proved by use of deuterated silica.[51] Indeed if the reaction proceeds *via* reaction with the alkyl moiety, all neopentane should be monodeuterated while in the second case only one fourth will be deuterated, as observed experimentally. Later the pentacoordinated intermediate was observed by solid-state ^{13}C NMR and its molecular analogue was synthesized confirming the above mechanism.[52]

In the case of alkylidynes complexes (for example W[CH$_2$–C(CH$_3$)$_3$]$_3$[≡CH–C(CH$_3$)$_3$)] quite the same reactivity is observed with the formation of surface alkylidynes species.[53] Another example is the rhenium complex Re[CH$_2$–C(CH$_3$)$_3$]$_2$[=CH–C(CH$_3$)$_3$)][≡CH–C(CH$_3$)$_3$)] which contains the three different ligands and whose reaction product with silica dehydroxylated at high temperature has been fully characterized by various physico-chemical methods and DFT calculations (Figure 1.1).[54] For this surface complex the presence of an agostic interaction between the metal and the hydrogen of the carbyne moiety could be proved by solid-state NMR. This illustrates the precision which can be attained in the characterization of a surface complex.

More recently, additional data on the nature of the bond between the metal and the silica support were obtained by ^{17}O solid-state NMR.[55] Not only the oxygen atom of the ≡Si–O–M bridge could be identified but also the presence of ≡Si–O–Si≡ bridges in small interaction with the metal was observed giving new insights into the structure of the grafted organometallic species and their interaction with the silica carrier.

This chemistry has then been extended to other complexes containing 'inert' ligands in combination with alkyl or alkylidene ones in view of the synthesis, for example, of olefin metathesis catalysts. This is the case of a tungsten imido complex based on the Schrock catalyst.[56,57] Another example

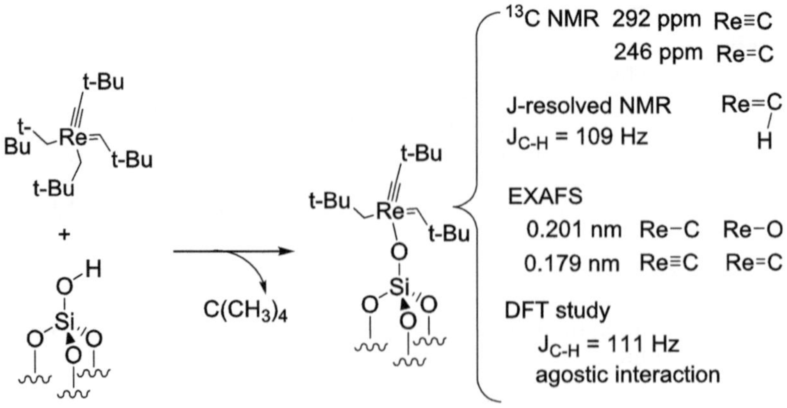

Figure 1.1 Reaction of the rhenium complex with silica and characterization of the reaction product.

is the synthesis of tungsten oxo species on the silica surface which can be considered as models of the industrial WO$_3$/SiO$_2$ metathesis catalyst.[58]

1.4 Further Reactions of Surface Species Formed by Reaction of an Organometallic Compound with One Hydroxyl Group

1.4.1 Further Reactions with Hydroxyl Groups

As described above, this reaction leads to the formation of a surface complex linked to the support by one bond. However, this species can further react with other hydroxyl groups if they are in close proximity. This can be observed in the case of silica dehydroxylated at moderate temperature (200 °C, for example) or in mesoporous silicas where the distribution of hydroxyl groups is not the same than in flame silica. Quite the same reaction as that described above will then occur, leading to a new evolution of alkane and formation of a digrafted species, one alkyl ligand being replaced by a ≡Si–O one. However, the question is: Is it possible to prepare cleanly a digrafted organometallic complex on the silica surface? Some people have claimed it to be so, but in most cases it is not possible as silica dehydroxylated at low temperature contains also isolated silanols as proved by infrared spectroscopy. A proof of this assumption was obtained in the case of the grafting reaction of the zirconium complex Zr[CH$_2$–C(CH$_3$)$_3$]$_4$ on silica dehydroxylated at various temperatures. For this purpose the reactivity of the surface complex with trimethylphosphine was studied (Scheme 1.5).[59] Indeed it had been reported that alkane could evolve, via α-H abstraction, upon reaction of alkyl complexes with a phosphine.[60] The monografted complex does not react with trimethylphosphine at room temperature probably due to a steric hindrance preventing the coordination of phosphorus on

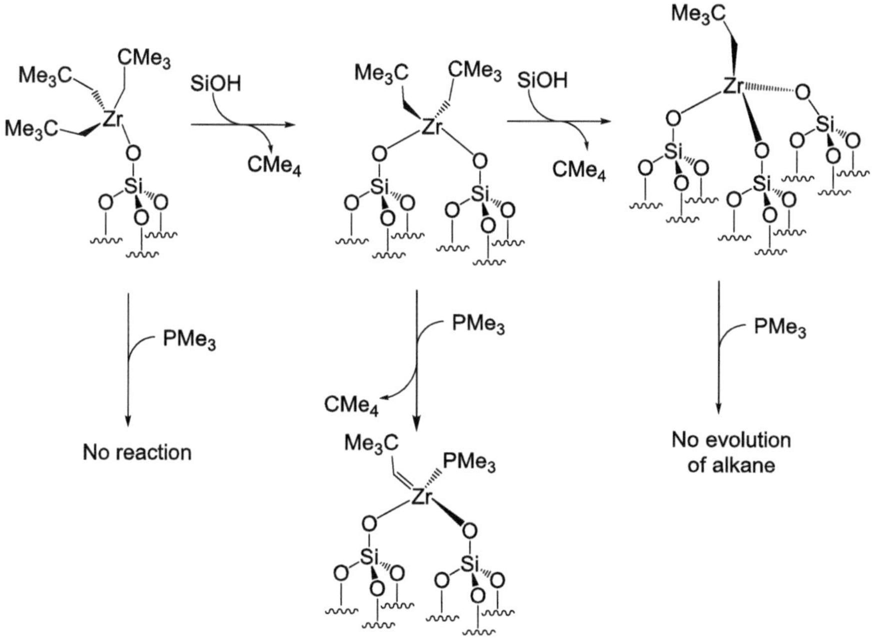

Scheme 1.5 Reactivity of the mono-grafted zirconium complex with silanol groups and trimethylphosphine.

zirconium. This point can be checked simply by studying the reactivity of the phosphine with the complex synthesized on highly dehydroxylated silica. In contrast the digrafted complex reacts with evolution of neopentane. As the tri-grafted complex has only one neopentyl ligand, it cannot undergo a α-H abstraction. By knowing the amount of neopentane evolved during this reaction and that during the grafting step, the relative proportions of the mono-, di- and tri-grafted species can be determined. The results are shown on Figure 1.2. Clearly it is not possible to prepare the digrafted complex alone, even if on silica dehydroxylated at 300 °C it represents more than 70% of the surface species. It is always present with the two other complexes. Only the monografted complex can be obtained as a pure species.

1.4.2 α-H, β-H and γ-H Abstraction

α-H abstraction can proceed in the case of sterically hindered complexes and leads to the evolution of alkane and the formation of a surface alkylidene which can be active in olefin metathesis. This reaction is also observed upon treatment of a digrafted zirconium with trimethyl phosphine (see above)[59] or by heating a digrafted chromium complex.[61]

Another reaction which can proceed is the β-H abstraction. This reaction is observed, for example, during the thermolysis of tin complexes supported on silica and leads to the evolution of alkene and the formation of a tin(II)

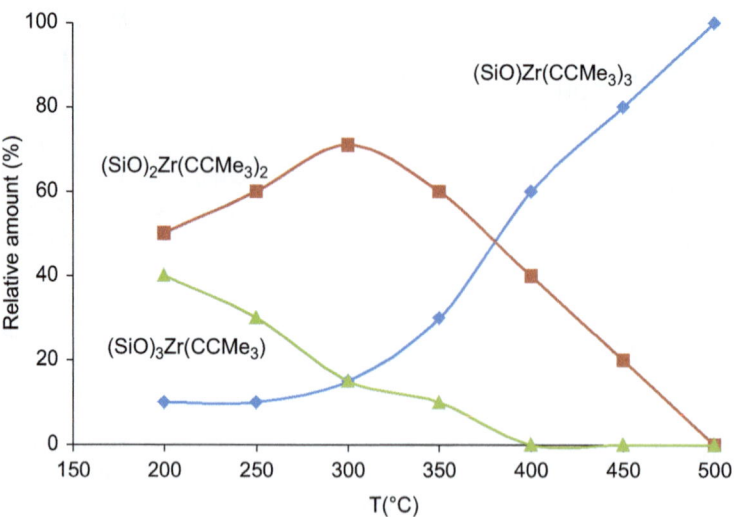

Figure 1.2 Relative proportions of $(\equiv SiO)_x Zr[CH_2C(CH_3)_3]_{4-x}$, where $x = 1-3$, after reaction of tetraneopentyl zirconium with silica dehydroxylated at various temperature.

species.[62] This reaction is also very important in catalysis, for depolymerization and hydrogenolysis or metathesis of alkanes. This point will be discussed in more details in the catalysis part.

γ-H abstraction has been proposed to occur during the thermolysis of some tetraneopentyl complexes of group 4.[63] This reaction leads to the evolution of neopentane and the formation of a metallacyclobutane. Even if it is relatively uncommon, it has been observed during the thermolysis of neopentyl hafnium complexes supported on silica.[64]

1.4.3 Reaction with Hydrogen

When looking at all above surface complexes only those containing the alkylidene ligand can be used in catalysis, for olefin metathesis. However, highly active species for the activation of alkanes can be prepared by hydrogenolysis. Treatment under hydrogen at moderate temperature (*ca.* 150 °C or below) of the alkyl, alkylidene and alkylidyne surface complexes leads to the formation of surface hydrides which are highly electron deficient and so potentially highly active catalysts. The expected reaction can simply be written by replacing one alkyl ligand by one hydrogen, one alkylidene one by two hydrogens and one alkylidyne one by three hydrogens with evolution of the corresponding alkane. However, the resulting species are so electron deficient and so reactive that they will react with neighboring siloxane bridges ≡Si–O–Si≡ leading to the formation of M–O–Si≡ bonds with the surface and the formation of a Si–H group. Thermodynamically this reaction is highly favored. The result is a digrafted metal hydride which can further

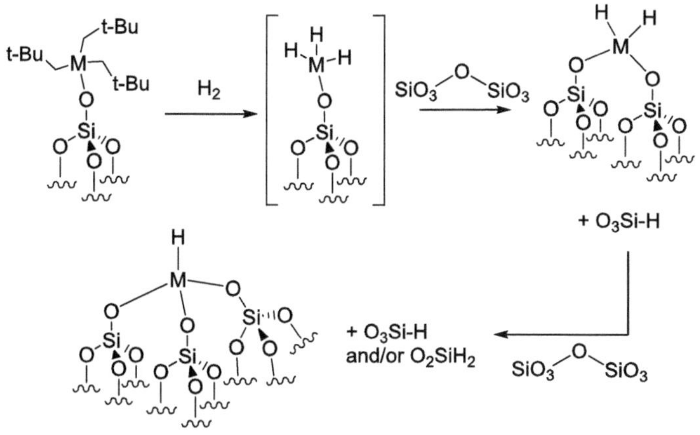

Synthesis of Well-defined Solid Catalysts by Surface Organometallic Chemistry 13

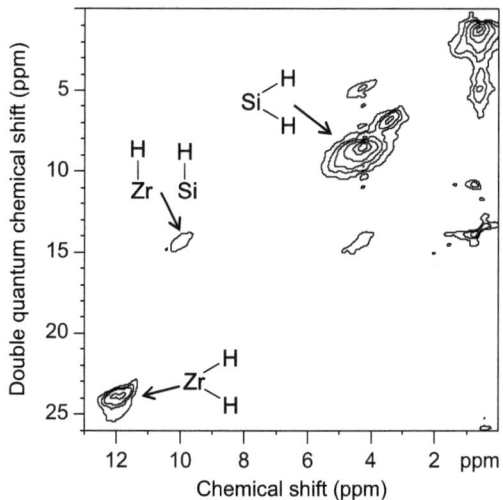

Scheme 1.6 Formation of group 4 hydrides by hydrogenolysis of the silica supported alkyl complexes.

Figure 1.3 2D double quanta ^{1}H MAS NMR spectrum of the reaction product of the hydrogenolysis of ≡Si–O–Zr[CH$_2$–C(CH$_3$)$_3$]$_3$ on silica.

react with another siloxane bridge leading to the formation of a tri-grafted metal hydride. This is the case for group 4 metal complexes which will lead to the formation of a mixture of mono- and dihydrides on the surface (Scheme 1.6).[65,66] The best characterization method for these species is two-dimensional (2D) double quanta ^{1}H MAS NMR which allows the observation of an autocorrelation peak for the dihydride (Figure 1.3). The formation of these two species is probably related to the heterogeneity of the silanol distribution on the solid as explained in the first part, the presence of nests preventing the formation of the third metal–surface bond. Indeed, the

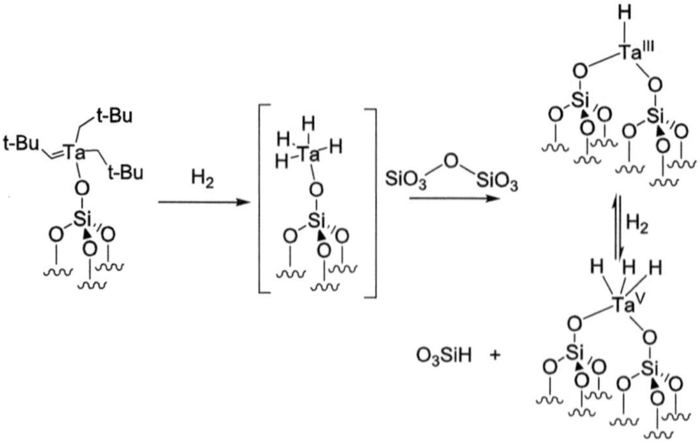

Scheme 1.7 Formation of mono- and dihydride tantalum surface complexes by hydrogenolysis of the surface alkyl alkylidene complex.

restructuration of the surface is so high that the hydride does not have a siloxane group in close proximity.

For elements of group 5 (mainly tantalum) the reaction is more complicated as this element can exist as tantalum(III) and tantalum(V). As a consequence a reductive elimination of hydrogen can be observed and the main species is a digrafted tantalum(III) monohydride which can also be in interaction with a siloxane bridge as shown by EXAFS.[67] In presence of hydrogen this tantalum(III) monohydride can coordinate one H_2 molecule *via* an oxidative addition, leading to a digrafted tantalum(V) trihydride in equilibrium with the monohydride (Scheme 1.7).

In the case of group 6 complexes, no hydride is obtained on silica and only the formation of large particles by sintering is observed. In contrast, when the alkylidyne complex of tungsten is grafted on alumina, a well-defined and highly active hydride is formed.[68] This is related to the strength of the bond between the metal and the surface and is well known in heterogeneous catalysis (platinum on alumina is used industrially instead of platinum on silica).

1.4.4 Reaction with Other Molecules

The supported alkyl (or hydride) complexes can also react with many molecules, such as alcohols or amines leading to the formation of various derivatives which are more stable than their precursors and can be used in reactions involving for example oxygen derivatives (epoxidation of olefins,[69] deperoxidation of cyclohexyl peroxide,[45] *etc.*). Depending on the amount of alcohol or amine which will be used and on the reaction conditions, different compounds can be obtained. In contrast to a direct grafting of the corresponding complexes, alkoxy derivatives for example, this method allows the obtention of well-defined and well-dispersed species on the surface, as the silica–metal bond had been created previously. One typical case is that of Ta(OMe)$_5$ which is a dimer. When treating the alkyl–alkylidene surface complex with methanol a monomer is obtained while reaction of the alkoxy derivative leads to the formation of a dimer, as shown by EXAFS. This can be of interest for applications in catalysis as in some cases dimers are expected to be the active sites. Depending on the molecule a lot of functions can be introduced around the metal. If the reaction is made on the complex prepared by reaction with a silica dehydroxylated at high temperature, this new species will have only one bond with the surface. If the starting silica had been dehydroxylated at low temperature, the main species will have two bonds with the surface and in the case of group 4 elements if the reaction is made with the hydride the main resulting species will have three bonds with the surface. However, and as for the direct grafting reaction of the alkoxy derivatives, alkoxy groups are also formed on the surface (see above). This point can be important in catalysis as it can modify the sorption properties of the solid and so can affect the kinetics of the reaction.

1.5 Formation of Cationic Complexes on the Surface

All the above synthesis methods allow only the preparation of neutral surface complexes linked to the surface by a covalent bond. However, in some cases, the catalytic species is a cation, for example metallocene species in the case of olefin polymerization or chromium cations for the trimerization of ethylene. Classically, this cation is formed by reaction of a neutral complex with a Lewis acid which must have a non-coordinating ability after its transformation into the corresponding anion. With regard to these conditions B(C$_6$F$_5$)$_3$ was found to be a very good candidate.

When applying this methodology to the design of single site heterogeneous catalysts on a surface such as silica, two different strategies may be used:

1. To graft first the Lewis acid co-catalyst directly to the surface in order to create a covalent bond with the surface and then to react this supported Lewis acid with the molecular complex in order to achieve the preparation of a 'floating' cationic complex.
2. To graft first a precursor complex directly with the silica surface in order to create a covalently bonded complex such as described above and in a second step react this surface complex with a Lewis co-catalyst which will abstract an alkyl group from the surface complex, leading to a cationic species covalently bonded to the surface.

These two methods will lead to different surface species the coordination sphere around the metal containing or not a covalent bond with the surface.

We will first describe one example of the first method with B(C$_6$F$_5$)$_3$ as a Lewis acid (Scheme 1.8) but this strategy can be applied to other compounds.

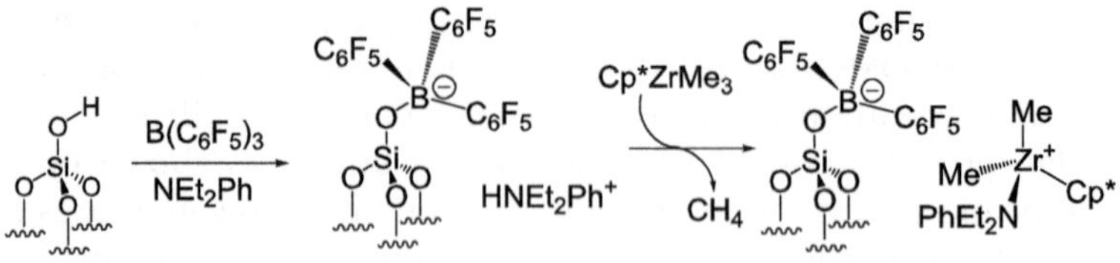

Scheme 1.8 Formation of a surface cationic zirconium complex by reaction of silica with a Lewis acid.

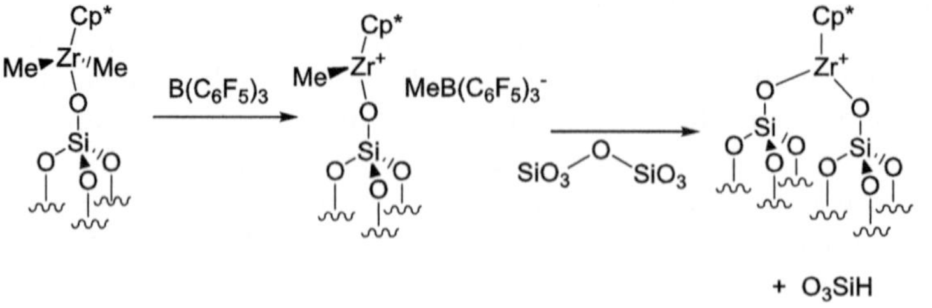

Scheme 1.9 Formation of a surface cationic complex by reaction of a surface neutral complex with a Lewis acid.

When B(C$_6$F$_5$)$_3$ is contacted with silica dehydroxylated at high temperature no reaction is observed. However, in presence of a tertiary amine like NEt$_2$Ph a grafting reaction occurs with formation of HNEt$_2$Ph$^+$ and the anionic grafted species [≡Si–O–B(C$_6$F$_5$)$_3$]$^-$ which has been fully characterized.[70] The tertiary amine acts as an activator of the inert silanol moiety and allows then the Lewis acid to coordinate the oxygen. Reaction of this surface ionic pair with a metallocene like Cp*ZrMe$_3$ leads to the evolution of one methane molecule and the formation of the surface species [≡Si–O–B(C$_6$F$_5$)$_3$]$^-$·[Cp*(Net$_2$Ph)ZrMe$_2$]$^+$ which has been fully characterized and is active in ethylene polymerization without the need of a cocatalyst such as methylaluminoxane (MAO).[71]

The other strategy (reaction of the organometallic complex with the silica support followed by abstraction of a methyl group by the borane compound) can also be illustrated by the same Cp*ZrMe$_3$ complex. Reaction of this complex with silica dehydroxylated at high temperature leads to the formation of the well-defined species ≡Si–O–ZrCp*Me$_2$ (Scheme 1.9) which can further react with B(C$_6$F$_5$)$_3$ without the need of a tertiary amine. The expected reaction (transfer of a methyl group from zirconium to boron) is observed but it is not the major species on the surface. The reason is that, as in the case of hydrogenolysis, this species is very reactive (note that in contrast to that prepared by the first strategy it does not have a coordinated amine). It can further react with a siloxane bridge *via* a methyl transfer to the surface leading to a digrafted cation which has loss all its methyl ligands and so is be inactive in catalysis.[72] The catalyst contains the two species, resulting in an activity lower than expected.

1.6 Applications in Catalysis

We will give some examples of reactions catalyzed by the above complexes with the aim to show the advantages of the synthesis of catalysts by surface organometallic chemistry.

1.6.1 Hydrogenolysis of Alkanes

As shown above the hydrides are prepared by treatment under hydrogen of the surface alkyl complexes. However, instead of the expected neopentane, only methane (and ethane in the case of group 4 elements) is observed in the gaseous phase. This reaction is also observed when an alkane is contacted with the hydride in presence of hydrogen. The key step is a β-alkyl transfer, quite similar to the β-hydrogen transfer but which occurs more rarely (Scheme 1.10). The driving force for the reaction is the hydrogenation of the olefin which will shift the equilibrium to the right. Indeed, in absence of hydrogen no reaction proceeds and the opposite reaction can be performed: the catalyst polymerizes ethylene *via* the classical mechanism of insertion in the Zr–C bond.

This reaction is of no interest for small alkanes but can become important when applied to polyolefins, which can be considered as alkanes with a very long chain. For example, it can be applied to the hydrogenolysis of waxes. Indeed Fischer–Tropsch synthesis leads to the formation of such side products which cannot be directly valorized industrially. Degradation by use of a group 4 hydride allows their direct transformation into diesel fuel and gasoline.[73,74]

Another interesting example is the transformation of polymers into more valuable compounds. For example, polystyrene can react with the zirconium hydride in presence of hydrogen. Three reactions are observed: (i) breaking of a carbon–carbon bond of the chain, (ii) breaking of a carbon–phenyl bond leading to the evolution of benzene, and (iii) hydrogenation of the phenyl group of the polymer. Depending on the conditions the last two reactions can be prominent, leading to the formation of a new styrene/ethylene/vinyl cyclohexane terpolymer with interesting physical properties.[75]

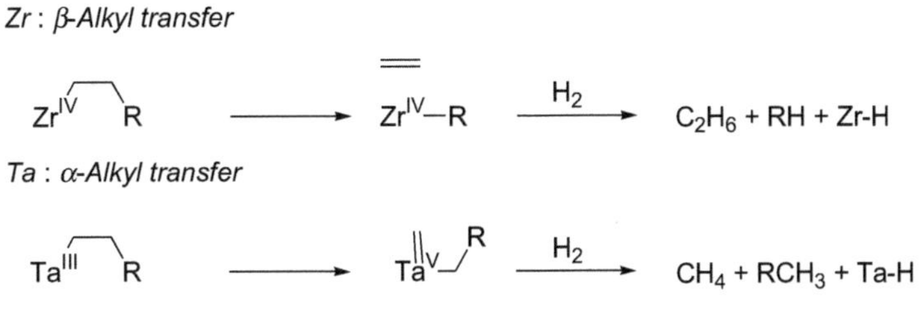

Scheme 1.10 Elementary reactions involved in the hydrogenolysis of alkanes by supported hydrides of groups 4 and 5.

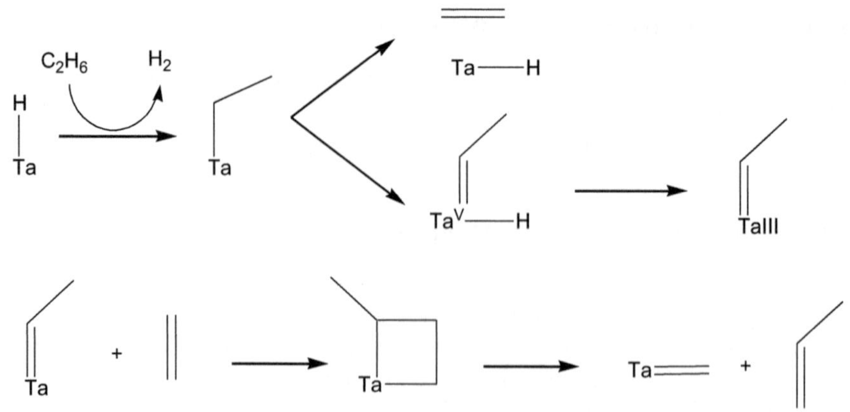

Scheme 1.11 Catalytic cycle proposed for the metathesis of alkanes.

1.6.2 Alkane Metathesis

In the case of tantalum and tungsten hydrides only methane is evolved during the hydrogenolysis of the corresponding surface alkyl complexes. The above mechanism involving a β-alkyl transfer is not sufficient to explain these results and it is necessary to take into account another mechanism for the cleavage of the C–C bond of ethane. This mechanism is an α-alkyl transfer, which will lead to the formation of a surface alkylidene (Scheme 1.10). Combination of the α- and β-alkyl transfer then allows an explanation of this cleavage, which occurs *via* the olefin metathesis reaction, evolving methane and propane from two ethane molecules (Scheme 1.11).[76] Propane is then transformed by the classical mechanism of hydrogenolysis into methane and ethane which can further react. Finally, only methane is obtained. The resulting reaction, transformation of two ethane molecules into one methane molecule and one propane molecule, is called alkane metathesis by analogy with what is observed for olefins. It can be decomposed into three steps: dehydrogenation of the alkane, olefin metathesis and, finally, hydrogenation of the resulting olefins.

1.6.3 Ethylene to Propene by Tungsten Hydride

As shown above, the tantalum and tungsten hydrides can react with alkanes leading to the formation of alkylidene species which are catalysts for olefin metathesis. However, the tungsten hydride can also perform the ethylene dimerization into butene. The consequence is that this system can directly transform ethylene into propene *via* the following mechanism (Scheme 1.12).[77] Ethylene is first dimerized into but-1-ene which is isomerized *via* a classical β-H abstraction mechanism. Then the cross-metathesis of ethylene and butane leads propene. These three different reactions occur on the same site.

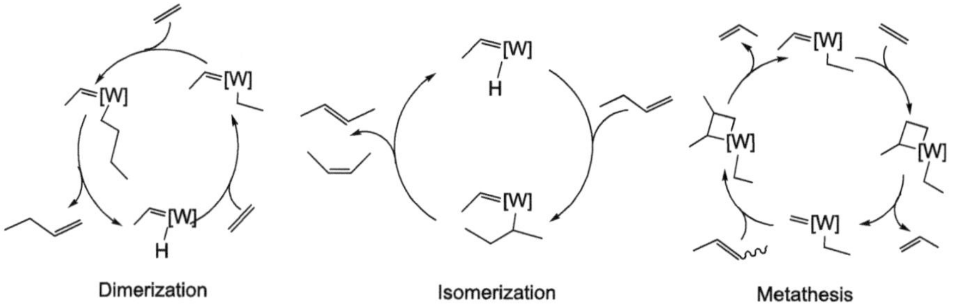

Synthesis of Well-defined Solid Catalysts by Surface Organometallic Chemistry 19

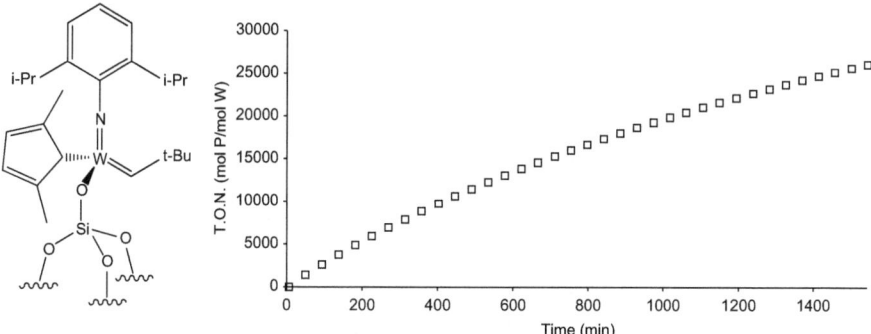

Scheme 1.12 Catalytic cycle proposed for the transformation of ethylene into propene on the supported tungsten hydride.

Figure 1.4 Propene metathesis (TON as a function of time) for the supported tungsten imido carbene shown on the left.

1.6.4 Olefin Metathesis

As shown above, olefin metathesis catalysts derived from those developed by Schrock and colleagues can be supported on silica.[57] The resulting materials can be very active for olefin metathesis. For example the system shown in Figure 1.4 can have turnover numbers higher than 20 000 after 20 h.

Recently, more active systems based on tungsten oxo complexes were developed and patented.[78] Typically these catalysts are based on the grafting reaction of W(=O)F[CH$_2$C(CH$_3$)$_3$]$_3$ on an oxide support and lead to very high activities in the metathesis of propene.

1.6.5 Trimerization of Ethylene

Selective trimerization of ethylene to 1-hexene has been one of the most exciting developments of olefin catalysis in the past decades. A supported catalyst for the ethylene oligomerization has been designed very recently by a molecular approach.[79] It operates efficiently without any co-catalyst and delivers 1-hexene in good selectivity. This system has been obtained by grafting reaction of TaCl$_2$Me$_3$ on dehydroxylated silica, leading to the

($\equiv$SiO)Ta^{V}Cl$_2$Me$_2$ precatalyst. During the initiation step ethylene reduces this complex, allowing then the generation of the key metallacycle intermediates and thus the selective production of the requested 1-hexene. DFT calculations have explained this unexpected reactivity and why other tantalum compounds are not active.[80]

1.6.6 Epoxidation and Deperoxidation Reactions

Such reactions involve oxygen or oxygenated molecules and so cannot be catalyzed by alkyl or hydride complexes such as those depicted above. However, by reaction of these species with alcohols it is possible to obtain new well-defined surface complexes which can be used for these reactions. Mono-, di- and tri-grafted titanium complexes can then be obtained which are active systems for the epoxidation of oct-1-ene. A study of these different systems and of their transformation during the catalytic reaction has allowed an understanding of the behavior of the industrial epoxidation catalyst developed by Shell, which deactivates rapidly at the beginning before becoming stable upon time.[81]

Another example is the asymmetric epoxidation of allyl alcohol. In solution this reaction is usually made by use of the Sharpless catalyst which is based on a titanium complex with a tartrate ligand. A system has been designed on the surface by taking into account that the coordination sphere of the metal must contain one bond with the surface, two bonds for coordination of the tartrate ligand and it must also accommodate the two reagents, allyl alcohol and the peroxide. As a consequence five bonds are needed around the metal and so a group 5 metal complex should be expected, such as tantalum. The coordination sphere around the metal can be built by (i) grafting reaction of the Ta[CH$_2$–C(CH$_3$)$_3$]$_3$[=CH–C(CH$_3$)$_3$] complex on silica dehydroxylated at high temperature (formation of the surface–metal bond); (ii) reaction of the grafted complex with ethanol in order to replace the alkyl and alkylidene ligands by ethoxy ones; (iii) addition of the asymmetric ligand, (+)-diisopropyl tartrate (Scheme 1.13). The resulting system is active for the asymmetric epoxidation of allyl alcohol with results quite comparable to those achieved in homogeneous catalysis.[69]

A third example is the deperoxidation of cyclohexyl hydroperoxide. An important way of synthesis of adipic acid developed by Rhodia passes through the oxidation of cyclohexane in three steps. In a first step cyclohexane is oxidized, by a classical Fenton mechanism, into a mixture of cyclohexyl peroxide, cyclohexanol and cyclohexanone. In a second step, this mixture is transformed into cyclohexanol and cyclohexanone by oxidation of the peroxide. Finally, the two oxygenated compounds are oxidized into adipic acid by nitric acid. Surface organometallic chemistry was used in order to find a heterogeneous system for the second step. Indeed this reaction is usually made in solution in presence of a chromium salt which is highly toxic. The aim of this study was mainly to obtain a purely heterogeneous system without any leaching even if it was not very active.

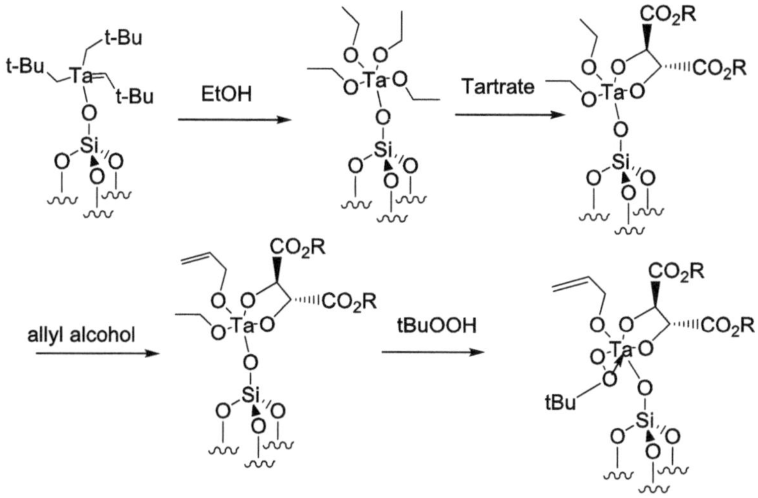

Scheme 1.13 Synthesis of a surface tantalum complex able to perform the asymmetric epoxidation of allyl alcohol.

A screening of all available systems showed that ethoxy complexes of tantalum grafted on silica, such as that depicted on Scheme 1.13, were active without any leaching for the deperoxidation of the industrial mixture. This system has been patented but unfortunately the activity was too low, preventing its industrial use.[45,82]

1.7 Activation of Ammonia and Nitrogen

Ammonia is a very important chemical as its production reaches *ca.* 10^8 tons per year. It is mainly produced by the Haber–Bosch process from nitrogen and hydrogen. The reaction proceeds at high pressure and relatively high temperature even if it is highly exothermic and so in conditions where the equilibrium is not shifted to the ammonia production. A lot of work has been done for a search of catalysts allowing this reaction to proceed in mild conditions and recently Schrock has proposed a catalytic cycle with a monometallic molybdenum complex.[83] However, up to now nothing had been made on a heterogeneous catalyst. It has been shown recently that the tantalum hydride complex can activate dinitrogen in presence of hydrogen leading to the formation of the amido imido complex of tantalum (≡SiO)$_2$Ta(=NH)(NH$_2$).[84] The reaction proceeds at moderate temperature and pressure (250 °C and 0.5 bar) and is completely shifted to the production of this surface complex. Recent DFT studies have shown that the reaction proceeds *via* successive heterolytic cleavages of hydrogen.[85] Unfortunately ammonia cannot be desorbed from this complex which is probably a stable species as it can also be obtained by the reaction of ammonia with the tantalum hydride.[86] However, it should probably be possible to have a catalytic cycle by adding a reaction of the amido imido complex with an organic moiety. Further studies in this direction are in progress.

1.8 Conclusion

In the course of this short review we have shown that it is possible to prepare well-defined organometallic fragments on an oxide surface. Only one species can be obtained on the surface when choosing the right conditions. This surface complex can further react with various molecules, leading to active species for a lot of reactions including the activation of alkanes, as species unknown in solution can be stabilized by the surface. The nature and the structure of the support on which the organometallic fragment will be grafted will have an effect on the conditions where only one species is observed and on the presence of different species after subsequent reactions of the grafted complex. For example, flame silica will lead to monografted complexes after dehydroxylation at a lower temperature than a mesoporous silica. The heterogeneity of the hydroxyl distribution has no effect on the grafting reaction but is important for the synthesis of surface hydrides by hydrogenolysis, as two species will be obtained. In all cases this approach allows the obtention of one or only few surface species with a well-defined coordination sphere around the metal, allowing structure–activity relationships to be performed. This method can also allow a predictive approach for the design of catalysts for a given reaction to be obtained.

References

1. *Modern Surface Organometallic Chemistry*, ed. J. M. Basset, R. Psaro, D. Roberto and R. Ugo, Wiley-VCH, 2009.
2. P. Sautet and F. Delbecq, *Chem. Rev.*, 2010, **110**, 1788–1806.
3. C. Dossi, A. Fusi, R. Psaro, D. Roberto and R. Ugo, *Mater. Chem. Phys.*, 1991, **29**, 191–199.
4. E. Cariati, D. Roberto, R. Ugo and E. Lucenti, *Chem. Rev.*, 2003, **103**, 3707–3732.
5. Y. Liang and R. Anwander, *Dalton Trans.*, 2013, **42**, 12521–12545.
6. R. Anwander, *Chem. Mater.*, 2001, **13**, 4419–4438.
7. S. I. Wolke and R. Buffon, *Quim. Nova*, 2002, **25**, 985–994.
8. C. Copéret, *Chem. Rev.*, 2010, **110**, 656–680.
9. N. Popoff, E. Mazoyer, J. Pelletier, R. M. Gauvin and M. Taoufik, *Chem. Soc. Rev.*, 2013, **42**, 9035–9054.
10. A. Bendjeriou-Sedjarari, J. M. Azzi, E. Abou-Hamad, D. H. Anjum, F. A. Pasha, K.-W. Huang, L. Emsley and J. M. Basset, *J. Am. Chem. Soc.*, 2013, **135**, 17943–17951.
11. F. Rascon, R. Berthoud, R. Wischert, W. Lukens and C. Coperet, *J. Phys. Chem. C*, 2011, **115**, 1150–1155.

12. N. Popoff, J. Espinas, J. Pelletier, K. C. Szeto, J. Thivolle-Cazat, L. Delevoye, R. M. Gauvin and M. Taoufik, *ChemCatChem*, 2013, **5**, 1971–1978.
13. Y. Liang, E. S. Erichsen and R. Anwander, *Dalton Trans.*, 2013, **42**, 6922–6935.
14. R. M. Gauvin, F. Buch, L. Delevoye and S. Harder, *Chem. - Eur. J.*, 2009, **15**, 4382–4393.
15. S. I. Wolke, R. Buffon and U. P. R. Filho, *J. Organomet. Chem.*, 2001, **625**, 101–107.
16. L. Zhong, M. Y. Lee, Z. Liu, Y. J. Wanglee, B. Liu and S. L. Scott, *J. Catal.*, 2012, **293**, 1–12.
17. S. L. Scott, A. Fu and L. A. MacAdams, *Inorg. Chim. Acta*, 2008, **361**, 3315–3321.
18. M. O. de Souza, R. F. de Souza, L. R. Rodrigues, H. O. Pastore, T. M. Gauvin, J. M. R. Gallo and C. Favero, *Catal. Commun.*, 2013, **32**, 32–35.
19. M. O. de Souza, L. R. Rodrigues, R. M. Gauvin, R. F. de Souza, H. O. Pastore, L. Gengembre, J. A. C. Ruiz, J. M. R. Gallo, T. S. Milanesi and M. A. Milani, *Catal. Commun.*, 2010, **11**, 597–600.
20. M. O. de Souza, L. R. Rodrigues, H. O. Pastore, J. A. C. Ruiz, L. Gengembre, R. M. Gauvin and R. F. de Souza, *Microporous Mesoporous Mater.*, 2006, **96**, 109–114.
21. J. Ternel, L. Delevoye, F. Agbossou-Niedercorn, T. Roisnel, R. M. Gauvin and C. M. Thomas, *Dalton Trans.*, 2010, **39**, 3802–3804.
22. M. Genelot, N. P. Cheval, M. Vitorino, E. Berrier, J. M. Weibel, P. Pale, A. Mortreux and R. M. Gauvin, *Chem. Sci.*, 2013, **4**, 2680–2685.
23. S. I. Wolke and R. Buffon, *J. Mol. Catal. A: Chem.*, 2000, **160**, 181–187.
24. P. V. Zhizhko, A. A. Zhizhin, O. A. Belyakova, Y. V. Zubavichus, Y. G. Kolyagin, D. N. Zarubin and N. A. Ustynyuk, *Organometallics*, 2013, **32**, 3611–3617.
25. M. K. Samantaray, E. Callens, E. Abou-Hamad, A. J. Rosini, C. M. Widdifield, R. Dey, L. Emsley and J. M. Basset, *J. Am. Chem. Soc.*, 2014, **136**, 1054–1061.
26. M. P. Conley, V. Mougel, D. V. Peryshkov, W. P. Forrest, D. Gajan, A. Lesage, L. Emsley, C. Coperet and R. R. Schrock, *J. Am. Chem. Soc.*, 2013, **135**, 19068–19070.
27. R. M. Gauvin, O. Coutelier, E. Berrier, A. Mortreux, L. Delevoye, J.-F. Paul, A.-S. Mamede and E. Payen, *Dalton Trans.*, 2007, 3127–3130.
28. A. W. Moses, C. Raab, R. C. Nelson, H. D. Leifeste, N. A. Ramsahye, S. Chattopadhyay, J. Eckert, B. F. Chmelka and S. L. Scott, *J. Am. Chem. Soc.*, 2007, **129**, 8912–8920.
29. D. Roberto, G. D'Alfonso, R. Ugo and M. Vailati, *Organometallics*, 2001, **20**, 4307–4311.
30. R. Psaro, C. Dossi, A. Fusi, R. D. Pergola, L. Garlaschelli, D. Roberto, L. Sordelli, R. Ugo and R. Zanoni, *J. Chem. Soc., Faraday Trans.*, 1992, **88**, 369–376.

31. P. Laurent, L. Veyre, C. Thieuleux, S. Donet and C. Coperet, *Dalton Trans.*, 2013, **42**, 238–248.
32. B. Revel, L. Delevoye, G. Tricot, M. Rastaetter, M. Kuzdrowska, P. W. Roesky and R. M. Gauvin, *Eur. J. Inorg. Chem.*, 2011, 1366–1369.
33. N. Ajellal, G. Durieux, L. Delevoye, G. Tricot, C. Dujardin, C. M. Thomas and R. M. Gauvin, *Chem. Commun.*, 2010, **46**, 1032–1034.
34. I. D. Rosal, M. J. L. Tschan, R. M. Gauvin, L. Maron and C. M. Thomas, *Polym. Chem.*, 2012, **3**, 1730–1739.
35. M. Terrier, E. Brulé, M. J. Vitorino, N. Ajellal, C. Robert, R. M. Gauvin and V. M. Thomas, *Macromol. Rapid Commun.*, 2011, **32**, 215–219.
36. A. R. Crozier, C. Schaedle, C. Maichle-Moessmer, K. W. Toernroos and R. Anwander, *Dalton Trans.*, 2013, **42**, 5491–5499.
37. R. M. Gauvin, T. Chenal, R. A. Hassan, A. Addad and A. Mortreux, *J. Mol. Catal. A: Chem.*, 2006, **257**, 31–40.
38. R. M. Gauvin and A. Mortreux, *Chem. Commun.*, 2005, 1146–1148.
39. R. M. Gauvin, L. Delevoye, R. A. Hassan, J. Keldenich and A. Mortreux, *Inorg. Chem.*, 2007, **46**, 1062–1070.
40. H. Skar, J. G. Seland, Y. Liang, N. A. Froeystein, K. W. Toernroos and R. Anwander, *Eur. J. Inorg. Chem.*, 2013, **2013**, 5969–5979.
41. N. Millot, C. C. Santini, F. Lefebvre and J. M. Basset, *C. R. Chim.*, 2004, **7**, 725–736.
42. M. E. Bartram, T. A. Michalske and J. W. Rogers Jr., *J. Phys. Chem.*, 1991, **95**, 4453–4463.
43. E. Grinenval, X. Rozanska, A. Baudouin, E. Berrier, F. Delbecq, P. Sautet, J. M. Basset and F. Lefebvre, *J. Phys. Chem. C*, 2010, **114**, 19024–19034.
44. B. C. Gerstein, M. Pruski and S.-J. Hwang, *Anal. Chim. Acta*, 1993, **283**, 1059–1079.
45. R. P. Saint-Arroman, PhD Thesis, Université Claude Bernard-Lyon 1, Lyon, 2002.
46. J.-M. Basset, A. Baudouin, F. Bayard, J.-P. Candy, C. Coperet, A. de Mallmann, G. Godard, E. Kuntz, F. Lefebvre, C. Lucas, S. Norsic, K. Pelzer, A. Quadrelli, C. Santini, D. Soulivong, F. Stoffelbach, M. Taoufik, C. Thieuleux, J. Thivolle-Cazat and L. Veyre, *Mod. Surf. Organomet. Chem.*, 2009, 23–73.
47. M. Adachi, J. Corker, H. Kessler, F. Lefebvre and J. M. Basset, *Microporous Mesoporous Mater.*, 1998, **21**, 81–90.
48. X. X. Wang, H. Zhao, F. Lefebvre and J. M. Basset, *Chem. Lett.*, 2000, 1164–1165.
49. N. Legagneux, A. de Mallmann, E. Grinenval, J. M. Basset and F. Lefebvre, *Inorg. Chem.*, 2009, **48**, 8718–8722.
50. E. L. Le Roux, M. Chabanas, A. Baudouin, A. de Mallmann, C. Coperet, E. A. Quadrelli, J. Thivolle-Cazat, J.-M. Basset, W. Lukens, A. Lesage, L. Emsley and G. J. Sunley, *J. Am. Chem. Soc.*, 2004, **126**, 13391–13399.
51. V. Dufaud, G. P. Niccolai, J. Thivolle-Cazat and J. M. Basset, *J. Am. Chem. Soc.*, 1995, **117**, 4288–4294.

52. M. Chabanas, E. A. Quadrelli, B. Fenet, C. Coperet, J. Thivolle-Cazat, J.-M. Basset, A. Lesage and L. Emsley, *Angew. Chem., Int. Ed.*, 2001, **40**, 4493–4496.
53. N. Merle, M. Taoufik, M. Nayer, A. Baudouin, E. Le Roux, R. M. Gauvin, F. Lefebvre, J. Thivolle-Cazat and J.-M. Basset, *J. Organomet. Chem.*, 2008, **693**, 1733–1737.
54. X. Solans-Monfort, J. S. Filhol, C. Coperet and O. Eisenstein, *New J. Chem.*, 2006, **30**, 842–850.
55. N. Merle, J. Trebosc, A. Baudouin, I. D. Rosal, L. Maron, K. Szeto, M. Genelot, A. Mortreux, M. Taoufik, L. Delevoye and R. M. Gauvin, *J. Am. Chem. Soc.*, 2012, **134**, 9263–9275.
56. B. Rhers, E. A. Quadrelli, A. Baudouin, M. Taoufik, C. Coperet, F. Lefebvre, J.-M. Basset, B. Fenet, A. Sinha and R. R. Schrock, *J. Organomet. Chem.*, 2006, **691**, 5448–5455.
57. B. Rhers, A. Salameh, A. Baudouin, E. A. Quadrelli, M. Taoufik, C. Coperet, F. Lefebvre, J.-M. Basset, X. Solans-Monfort, O. Eisenstein, W. W. Lukens, L. P. H. Lopez, A. Sinha and R. R. Schrock, *Organometallics*, 2006, **25**, 3554–3557.
58. E. Mazoyer, N. Merle, A. de Mallmann, J.-M. Basset, E. Berrier, L. Delevoye, J.-F. Paul, C. P. Nicholas, R. M. Gauvin and M. Taoufik, *Chem. Commun.*, 2010, **46**, 8944–8946.
59. V. Riollet, M. Taoufik, J.-M. Basset and F. Lefebvre, *J. Organomet. Chem.*, 2007, **692**, 4193–4195.
60. G. A. Rupprecht, L. W. Messerle, J. D. Fellmann and R. R. Schrock, *J. Am. Chem. Soc.*, 1980, **120**, 6236–6244.
61. J. A. N. Ajjou, S. L. Scott and V. Paquet, *J. Am. Chem. Soc.*, 1998, **120**, 415–416.
62. C. Nedez, F. Lefebvre, A. Choplin, J. M. Basset and E. Benazzi, *J. Am. Chem. Soc.*, 1994, **116**, 3039–3046.
63. Y. D. Wu, Z. H. Peng, K. W. K. Chan, X. Liu, A. A. Tuinman and Z. Xue, *Organometallics*, 1999, **18**, 2081–2090.
64. G. Tosin, C. C. Santini, M. Taoufik, A. de Mallmann and J.-M. Basset, *Organometallics*, 2006, **25**, 3324–3335.
65. C. Thieuleux, E. A. Quadrelli, J.-M. Basset, J. Doebler and J. Sauer, *Chem. Commun.*, 2004, 1729–1731.
66. J. Corker, F. Lefebvre, C. Lecuyer, V. Dufaud, F. Quignard, A. Choplin, J. Evans and J. M. Basset, *Science*, 1996, **271**, 966–969.
67. V. Vidal, A. Theolier, J. Thivolle-Cazat, J. M. Basset and J. Corker, *J. Am. Chem. Soc.*, 1996, **118**, 4595–4602.
68. E. Le Roux, M. Taoufik, C. Coperet, A. de Mallmann, J. Thivolle-Cazat, J.-M. Basset, B. M. Maunders and G. J. Sunley, *Angew. Chem., Int. Ed.*, 2005, **44**, 6755–6758.
69. D. Meunier, A. Piechaczyk, A. de Mallmann and J. M. Basset, *Angew. Chem. Int. Ed.*, 1999, **38**, 3540–3542.
70. N. Millot, A. Cox, C. C. Santini, Y. Molard and J. M. Basset, *Chem. - Eur. J.*, 2002, **8**, 1438–1442.

71. N. Millot, C. C. Santini, A. Baudouin and J. M. Basset, *Chem. Commun.*, 2003, 2034–2035.
72. N. Millot, S. Soignier, C. C. Santini, A. Baudouin and J. M. Basset, *J. Am. Chem. Soc.*, 2006, **128**, 9361–9370.
73. C. Larabi, N. Merle, S. Norsic, M. Taoufik, A. Baudouin, C. Lucas, J. Thivolle-Cazat, A. de Mallmann and J.-M. Basset, *Organometallics*, 2009, **28**, 5647–5655.
74. S. Norsic, C. Larabi, M. Delgado, A. Garron, A. de. Mallmann, C. Santini, K. Szeto, J. M. Basset and M. Taoufik, *Catal. Sci. Technol.*, 2012, **2**, 215–219.
75. J. M. Basset, E. Kuntz, R. Gauvin, D. Laurenti-Savoure, C. Jenny and P. Bres, *Fr. pat.*, FR2842202 A1, 2004.
76. J. M. Basset, C. Coperet, L. Lefort, B. M. Maunders, O. Maury, E. Le Roux, G. Saggio, S. Soignier, D. Soulivong, G. J. Sunley, M. Taoufik and J. Thivolle-Cazat, *J. Am. Chem. Soc.*, 2005, **127**, 8604–8605.
77. M. Taoufik, E. Le Roux, J. Thivolle-Cazat and J.-M. Basset, *Angew. Chem., Int. Ed.*, 2007, **46**, 7202–7205.
78. M. Taoufik, E. Le Roux, J. Thivolle-Cazat and J. M. Basset, *U.S. Pat. Appl. Pat.*, 20120316374 A1, 2012.
79. Y. Chen, E. Callens, E. Abou-Hamad, N. Merle, A. J. P. White, M. Taoufik, C. Coperet, E. L. Roux and J. M. Basset, *Angew. Chem., Int. Ed.*, 2012, **51**, 11886–11889.
80. Y. Chen, R. Credendino, E. Callens, M. Atiqullah, M. A. Al-Harthi, L. Cavallo and J. M. Basset, *ACS Catal.*, 2013, **3**, 1360–1364.
81. C. Rosier, PhD Thesis, Université Claude Bernard-Lyon 1, Lyon, 1999.
82. E. Fache, F. Igersheim, D. Bonnet, J. M. Basset, F. Lefebvre and R. P. Saint-Arroman, *PCT Int. Appl.*, WO 2002085826 A2, 2002.
83. D. V. Yandulov and R. R. Schrock, *Science*, 2003, **301**, 76–78.
84. P. Avenier, M. Taoufik, A. Lesage, X. Solans-Monfort, A. Baudouin, A. de Mallmann, L. Veyre, J. M. Basset, O. Eisenstein, L. Emsley and E. A. Quadrelli, *Science*, 2007, **317**, 1056–1060.
85. X. Solans-Monfort, C. Chow, E. Goure, Y. Kaya, J. M. Basset, M. Taoufik, E. A. Quadrelli and O. Eisenstein, *Inorg. Chem.*, 2012, **51**, 7237–7249.
86. P. Avenier, A. Lesage, M. Taoufik, A. Baudouin, A. de Mallmann, S. Fiddy, M. Vautier, L. Veyre, J.-M. Basset, L. Emsley and E. A. Quadrelli, *J. Am. Chem. Soc.*, 2007, **129**, 176–186.

CHAPTER 2

Zeolite-supported Molecular Metal Complex Catalysts

ISAO OGINO

Division of Chemical Process Engineering, Graduate School of Engineering, Hokkaido University, K13W8, Kita-Ku, Sapporo, Hokkaido 060-8628, Japan
Email: iogino@eng.hokudai.ac.jp

2.1 Introduction

Supported metal catalysts have found numerous applications in industry.[1,2] Many of the supported metal catalysts used in industry are non-uniform in the structure of active sites. Non-uniformity may help one to discover new catalytic reactions because a spectrum of different structures of active sites allows various catalytic reactions to be tested simultaneously. However, because a tiny fraction of supported species sometimes plays a dominant role in catalysis, it remains challenging to understand the structure–performance relationship of supported metal catalysts. If the structure of active site is highly uniform and simple enough, it allows incisive structural characterization by spectroscopic techniques. Furthermore, uniformity and simplicity of the structure of supported metals allow the elucidation of fundamental molecular chemistry by rigorous theoretical calculations at the density functional theory (DFT) level. These investigations will build a foundation for understanding the complex chemistry that takes place on industrial catalysts as well as the design of new selective catalysts, ultimately with more complex structures.

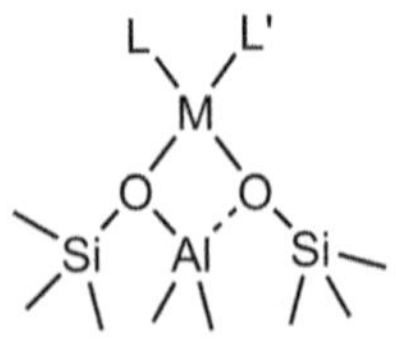

Figure 2.1 Schematic representation of a cationic metal complex (M(L)(L′)) anchored on a zeolite. M represents a metal. L and L′ represent ligands.

When supported metals are small and consist of a single atom or a few atoms, a support plays important roles as ligands to control their catalytic properties.[3,4] Thus, non-uniformity of a support leads to a spectrum of various catalytic properties of supported metals. Using crystalline supports such as zeolites is advantageous because the hindrance of surface non-uniformity may be largely overcome. Zeolites provide well-defined anchoring sites for cationic metal species because isomorphous substitution of Si^{4+} with Al^{3+} forms a negative charge in the zeolite framework, which is compensated by an exchangeable cation.[5] When a metal complex is anchored on a zeolite, the zeolite often acts as an anionic bidentate ligand as schematically shown in Figure 2.1.

Because the density of anchoring sites in a zeolite can be varied by changing its Si/Al ratio, the spatial distance between supported metal complexes can be controlled to some extent. In addition, zeolites offer unique environments for metals when they are anchored in molecular-sized and ordered micropores of zeolites, enabling some catalytic reactions that are difficult to be achieved using other catalysts.[5-7] Unique catalytic performance of zeolite-supported metal complex catalysts has been demonstrated in various reactions such as the oxidation of methane to methanol[8-10] and $deNO_x$ reaction.[11-13]

This chapter describes the synthesis and characterization of zeolite-supported molecular metal complex catalysts (supported mononuclear metal complexes that function essentially as molecules). These catalysts are in a subclass of supported metal catalysts. There are many excellent reviews that show detailed investigations of the local structures of zeolite-supported metal complexes and their relationships with catalytic performance.[14-17] The focus of this chapter is the synthesis method of structurally uniform supported species, characterization to demonstrate the uniformity of supported species, and investigations of structure–performance relationships using structurally uniform supported species.

Structural uniformity of the zeolite-supported metal complex is a key to determine their structures precisely and understand catalytic chemistry at the molecular level. Spectroscopic techniques such as extended X-ray absorption fine structure (EXAFS) and X-ray absorption near edge structure (XANES) spectroscopies provide exact structural information if the supported species have a high degree of uniformity. Structural uniformity offers the opportunity for investigations of catalytic cycles in detail as performed for molecular catalysts in solution; a combination of transient experiments and DFT calculations identifies reaction intermediates and spectators to decipher a whole catalytic cycle. Advantages of using zeolites as support materials and synthesizing structurally uniform supported species have been demonstrated by various researchers. Miessner et al. reported the synthesis and IR characterization of dealuminated zeolite Y-supported rhodium dicarbonyl complexes.[18] The sharpness of the IR bands (ν_{CO}) characterizing the supported rhodium species indicate their structural uniformity. Because the initially prepared species have high structural uniformity, their structure as well as structural changes under reactive atmosphere can be investigated in detail. Goellner et al. investigated the details of the structure of the dealuminated zeolite Y-supported rhodium dicarbonyl by IR, EXAFS, and DFT calculations as described in Section 2.4.2.[19] Miessner investigated the structural changes of supported rhodium dicarbonyls when they are treated in a flow of 10% H_2 in N_2 at 473–523 K and suggested the formation of supported rhodium monocarbonyls on the basis of IR spectroscopy data.[20] Vayssilov and Rösch used DFT calculations to examine the IR band assignments done by Miessner and suggested the presence of dihydrogen and hydride ligands in addition to carbonyl ligands.[21] More recently, Vityuk, Alexeev and Amiridis reported the synthesis and characterization of zeolite-supported rhodium complex with carbonyl and hydride ligands by the reaction of supported $Rh(CO)(C_2H_4)$ with H_2.[22] Liang et al. reported the synthesis of zeolite-supported rhodium diethene complexes (supported $Rh(C_2H_4)_2$) and demonstrate the dynamic uniformity of the supported species using variable-temperature solid state NMR spectroscopy as described in Section 2.4.3.[23,24] Liang et al. also demonstrated how the uniformity of supported species allows precise determination of their chemistry.[25] Kletnieks et al. showed how the structural uniformity of zeolite-supported molecular rhodium complexes allows one to investigate reactivity of supported species and their functions in a catalytic cycle at an unprecedented level as described in the Section 2.5.2.[26]

In the first part of this chapter (Sections 2.2 and 2.3), the general synthesis methods and characterization techniques of zeolite-supported metal complex catalysts are described. In the second part (Section 2.4), several issues regarding the synthesis of zeolite-supported metal complexes with high structural uniformity as well as characterization techniques to examine structural uniformity are described. Then, in the third part, investigations on reactivity and catalytic performances of zeolite-supported molecular metal complex catalysts are presented using some examples reported in the literature. In the final part, future directions in addition to conclusions are stated.

2.2 Synthesis

2.2.1 Synthesis by an Ion-exchange Method

Zeolite-supported metal complex catalysts are often synthesized by an ion-exchange method.[9,27] In a typical synthesis, a zeolite is contacted with an aqueous solution containing metal salts at controlled pH to avoid precipitation of metal hydroxides in the solution. Although many parameters (*e.g.*, pH, temperature, type of metal salts, extent of ion-exchange and calcination conditions) affect the structures of the resultant supported metal complexes,[27] this method is still straightforward to implement, cost-effective and scalable for commercial applications. Therefore, a large number of zeolite-supported metal complexes have been synthesized this way. Furthermore, some of the zeolite-supported metal complexes synthesized by this method show prospective catalytic performances, which are enabled by the unique structures of supported metal complexes formed within micropores or cages of zeolites. Examples of catalytic reactions include the oxidation of methane to methanol catalyzed by $[Cu_2O]^+$ anchored on ZSM-5 zeolite[8–10] and deNO$_x$ reaction catalyzed by Cu-ZSM-5.[11]

Synthesis by an ion-exchange method often forms diverse species.[28] For example, when zinc cations are introduced into ZSM-5 zeolite by an ion-exchange method, multiple species having different nuclearity (mononuclear zinc complexes and binuclear clusters) could form.[29] Synthesis by an ion-exchange method often accompanies calcination of ion-exchanged catalysts at high temperatures to remove water and activate supported metal ions for catalytic reactions. Calcination induces migration of metals and results in the formation of diverse metal species. Because metal cations are stabilized by maximizing their coordination numbers, metal cations tend to migrate into small cages such as sodalite cages or hexagonal prisms that may not be accessible by reactants.[27] When mixtures are present, understanding surface chemistry becomes challenging. Therefore, if the goal of synthesis is to obtain uniform structure of supported metal complexes, this method may not be suitable.

2.2.2 Synthesis by a Chemical Vapor Deposition Method

Other synthesis methods of zeolite-supported metal complex catalysts include chemical vapor deposition (CVD) of volatile metal complex precursors.[12,30–32] For example, Sachtler and Chen synthesized Fe-containing ZSM-5 zeolite by subliming FeCl$_3$ vapor into the cavities of the proton form of ZSM-5.[12] Their CVD method avoids a potential oxidation of Fe^{2+} to Fe^{3+} and precipitation of FeOOH and/or Fe(OH)$_3$ that often form when the ion-exchange method is used. Furthermore, this method allows high iron loadings as high as a Fe/Al ratio of 1. Koningsberger *et al.* reported that the catalyst synthesized according to the above method contains binuclear oxo/hydroxo-complexes in Fe/ZSM-5 as evidenced by X-ray absorption spectroscopy (XAS).[30] In other examples of synthesis by this method, Iglesia *et al.* reported the synthesis of zeolite-supported Re^{7+}-oxo complex catalyst by sublimation of Re$_2$O$_7$ onto HZSM-5 zeolite.[31] Their data show the formation of site-isolated Re-oxo species at acidic sites of the zeolite that catalyzes the oxidation reaction of C_2H_5OH to ethyl acetate or acetal.

2.2.3 Synthesis by the Surface Organometallic Chemistry Approach

The surface organometallic chemistry approach merges disciplines of solution organometallic chemistry and solid surface chemistry.[33–40] Numerous examples of syntheses of oxide-supported metal complexes *via* this route have been reported. From the perspective of synthesis of zeolite-supported metal complexes with uniform structure, the surface organometallic chemistry approach provides several advantages over other methods. First, synthesis by ligand exchange of an organometallic precursor does not require high temperature treatments that may result in the formation of multiple supported species. Second, a wealth of information about organometallic chemistry provides a basis for investigating the structure and catalytic chemistry of supported metal complexes. Finally, organometallic precursors can be designed to have ligands that mimic zeolite surface and those that function as intermediates in a catalytic cycle.

Organometallic compounds having labile ligands such as η^3-allyl and alkyls have been used as metal complex precursors in the synthesis of oxide-supported metal complexes.[14,33–40] These organometallic compounds can be reacted with zeolite in an organic solvent or contacted with zeolite after being vaporized. Ligands in the initially prepared samples can be post-synthetically exchanged with other ligands as described later in this chapter.

Gates and his co-workers reported the syntheses of a family of zeolite-supported mononuclear metal complex catalysts with uniform structure using the surface organometallic approach. Some criteria of choices of organometallic precursors and zeolites to synthesize structurally uniform supported metal species are described in detail in the Section 2.4.

2.3 Characterization Techniques

Zeolite-supported molecular metal complex catalysts are characterized by multiple techniques such as EXAFS and XANES spectroscopies, transmission electron microscopy (TEM), solid state NMR, IR, UV-visible spectroscopies and X-ray photoelectron spectroscopy (XPS).[40–42] DFT calculations guide band assignments in spectroscopic characterization and aid in selecting a candidate structure model. Strengths and limitations for each characterization techniques have been summarized in several review papers and some of them are listed in Table 2.1.[16,42] An important point to note is that these techniques are complementary to each other and thus multiple techniques are required to characterize supported metal complex catalysts. In addition, it is important to understand the limitations of each technique. For example, EXAFS provides direct information about local structures around a particular atom regardless of the phase of the materials. This technique determines the inter-atomic distance accurately (± 0.02 Å).[43,44] However, it gives average information of the whole sample. Therefore, characterization of zeolite-supported metal complexes consisting of

Table 2.1 Characterization techniques to investigate zeolite-supported molecular metal complex catalysts.

Technique	Information	Limitations
EXAFS spectroscopy	Metal–ligands and metal–support interactions (inter-atomic distances between an absorber metal atom and backscatter atoms and number of backscatter atoms)	Average information; not suitable for high temperature experiments; cannot distinguish light backscatter atoms (such as C, O, and N *etc.*)
XANES spectroscopy	Oxidation state of supported metal complexes	Average information
TEM	Location of supported metal complexes and their nuclearity	Local information; electron damages to samples
IR spectroscopy	Ligands on a metal; species present on the surface; uniformity of supported metal complexes by CO as a probe molecule[18]	Assignments of bands may not be straightforward
NMR spectroscopy	Ligands on a metal; dynamical uniformity of supported metal complex (*e.g.*, supported rhodium diethene complex[24])	Assignments of bands may not be straightforward
UV-visible spectroscopy	Ligand field environment of supported metal complexes	Not specific to identify supported species
Density functional theory	Estimated spectral assignments; structural models of supported metal complexes; prediction of intermediate species in a catalytic cycle	Simplified model for a support

multiple species gives the average inter-atomic distance as described in Section 2.4.1 (the limited number of structural parameters are allowed to be optimized in analysis[45] and thus it is difficult to model several structures in one model). Determinations of the number of atoms surrounding a particular atom are accompanied by relatively large errors (± 20%). Rejection of one structural model over other models is not straightforward and requires careful error analysis[46] as well as information from other techniques. Unfortunately, detailed processes of EXAFS analysis are not provided in many papers. Characterization of metal-low-Z backscatter contributions (*e.g.*, metal–oxygen and metal–carbon) by EXAFS is more challenging than that of metal-high-Z backscatter contributions (*e.g.*, metal–metal). This difficulty is relieved when supported metal complexes have terminal CO ligands because multiple scattering effects of EXAFS provide a powerful analysis tool.[47] Structural uniformity facilitates characterization by EXAFS. Bare *et al.* used EXAFS and XRD to elucidate structures of active sites of highly crystalline Sn-β-zeolite.[48] This zeolite has tin atoms within the zeolite framework. Their work elucidated the structure at tin sites as well as their distribution in the zeolite framework. Their work also suggests that a similar technique may be applicable to characterization of zeolite-supported molecular metal complex catalysts if they have high structural uniformity as well as uniform distribution within a zeolite.

Some of the spectroscopic techniques listed in Table 2.1 can be carried out under reactive atmosphere to investigate catalysts under working conditions[49-54] and sometimes multiple techniques are used in concert with transient method to characterize their formation of active catalytic species on supports and to show how the structures evolve during catalysis.[55] Results are correlated with catalysis data obtained in separate experiments using a flow reactor. In those cases, flow patterns in sample cells used in spectroscopic measurements should match those used in the catalytic reaction experiments so that mass transfer does not disguise results and direct comparisons can be made.

Structural uniformity of zeolite-supported metal complexes has been investigated by IR spectroscopy using CO as a probe molecule. Solid state ^{13}C MAS NMR spectroscopy was used to investigate dynamic structural uniformity of zeolite-supported rhodium diethene complex. These examples are described in detail in Section 2.4.3.

2.4 Guide for the Synthesis and Characterization of Supported Molecular Metal Complexes with a High Degree of Structural Uniformity

2.4.1 Types of Zeolite for Support Materials

There are several criteria for choosing the type of a zeolite as a support material. The sites where cationic metal complexes are anchored are acid sites (near framework aluminium sites) of zeolites (Figure 2.1). Therefore, to ensure site isolation of supported metal complexes, a high silica–alumina ratio is preferred. At the same time, acidic sites should be accessible to the metal precursors. Ideally, these anchoring sites are crystallographically equivalent so that these sites function as identical ligands.

Choice of the type of a zeolite is limited by the relative pore aperture of a zeolite.[56] FAU zeolite is often used because of the largest pore aperture among commercially available zeolites.[18-26,32] Ogino et al. investigated molecular sieving effects in the synthesis of zeolite-supported rhodium and ruthenium complexes.[56] Zeolites tested in these investigations were HSSZ-42, Hβ, HMOR, and HZSM-5. Critical molecular dimensions of the precursors (Rh(η^2-C$_2$H$_4$)$_2$(acac) (acac = acetylacetonate, C$_5$H$_7$O$_2^-$, see Section 2.4.2 for more details) (**I**), Ir(η^2-C$_2$H$_4$)$_2$(acac) (**II**), Ru(η^2-C$_2$H$_4$)$_2$(acac)$_2$ (**III**), and Rh(CO)$_2$(acac) (**IV**)) and the narrowest pore apertures of the zeolites are shown in Figure 2.2.

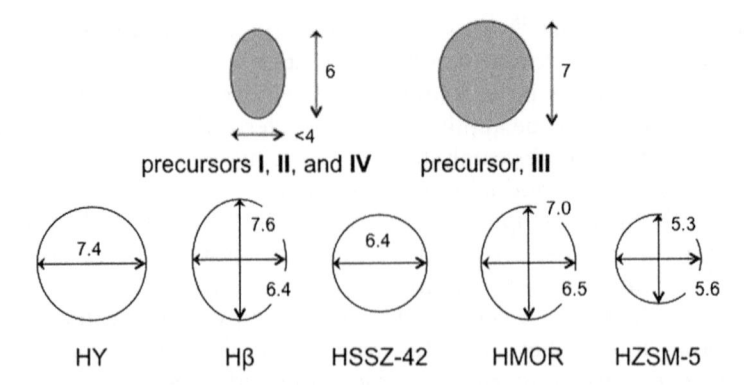

Figure 2.2 Critical molecular dimensions of the precursors (Rh(η^2-C$_2$H$_4$)$_2$(acac) (**I**), Ir(η^2-C$_2$H$_4$)$_2$(acac) (**II**), Ru(η^2-C$_2$H$_4$)$_2$(acac)$_2$ (**III**), and Rh(CO)$_2$(acac) (**IV**)) and the narrowest pore apertures of the zeolites. The dimensions are in Å. Adapted from ref. 56.

Using IR and EXAFS spectroscopies, approximate ratios of chemisorbed and physisorbed species were obtained (Table 2.2). CO ligands in Rh(CO)$_2$(acac) and those in the supported Rh(CO)$_2$ show different ν_{CO} frequencies; IR data characterizing Rh(CO)$_2$(acac) in a THF solution gives two sharp ν_{CO} bands (FWHM < 5 cm^{-1}) at 2010 cm^{-1} (symmetric band) and 2081 cm^{-1} (antisymmetric band) because of C_{2v} symmetry of this complex. Because electron density of oxygen atoms of acac is less than those of acid sites of these zeolites (*i.e.*, acidity of zeolite is stronger than Hacac whose pK_a is 8.99 in water at 298 K), rhodium atoms in chemisorbed rhodium complexes are more electron deficient. Thus, rhodium atoms in the chemisorbed complexes back-donate less electron density to an antibonding orbital of CO, leading to a blue shift of CO stretching band. Structural uniformity of supported rhodium dicarbonyls leads to the sharp CO stretching bands that are similar to those in a solution, and therefore it allows unambiguous assignments of the bands for chemisorbed or physisorbed species. Furthermore, the IR data were corroborated by EXAFS spectroscopy data.

When the precursor Rh(CO)$_2$(acac) was adsorbed in large-pore zeolite Hβ, all of the precursor was chemisorbed; IR data show two ν_{CO} bands that are blue-shifted by >35 cm^{-1} as compared with those characterizing the precursor in a THF solution and EXAFS data characterizing the zeolite-supported sample show longer Rh–O distances (2.12 Å) (Table 2.2). However, when HSSZ-42, which has smaller pore apertures, was used as a support, only a fraction of the metal complexes underwent slow ligand exchange and the rest became physisorbed in the zeolite. IR data show four ν_{CO} bands (two for chemisorbed rhodium dicarbonyl and the other for physisorbed Rh(CO)$_2$(acac)). Consequently, EXAFS data provide only the average bond distance of Rh–O (2.08 Å, Table 2.2).

Comparison of sizes of metal complexes with pore apertures of zeolites shows that when the pore diameters of the zeolites are approximately 1 Å greater than the critical diameter of the precursor (Table 2.3), all of the acac

Table 2.2 Rh–O distances characterizing Rh(CO)$_2$(acac) and supported metal complexes.

Zeolite support	Rh–O distance (Å)	Schematic drawing of molecular or supported metal complex	Characterization technique	Remarks
None	2.040		XRD	Precursor Rh(CO)$_2$(acac)
Hβ	2.12	Rh–Al: 3.08 Å	EXAFS	Chemisorbed species
HY	2.15	Rh–Al: 2.73 Å	EXAFS	Chemisorbed species
HSSZ-42	2.08 (average value)	Mixture	EXAFS	Chemisorbed + physisorbed species

Adapted from ref. 56.

ligands are readily removed from the precursor. When the transport restriction is significant, however, the supported metal complexes are inferred to be concentrated near the pore mouths of the zeolites.

Because commercially available zeolites have the limitation in the pore aperture of 0.73 nm, synthesis of new zeolites having larger apertures is desired. Numerous advances have been made in syntheses of new zeolites in the past few decades and there are some excellent reviews and books available.[5,57–70] Syntheses of extra-large pore zeolites (>12 MR),[57] layer expanded zeolites with large side pockets within layers,[58] and two-dimensional zeolite nanosheets[59–70] with large side pockets[64,66,67,70] are suitable for accommodating large metal complex precursors. Synthesis of two-dimensional zeolites is an emerging area in zeolite synthesis.[59,62] Two-dimensional zeolites are interesting particularly if they have large side pockets wherein metal complexes are anchored.

Lu et al. anchored the same iridium complex, Ir(η^2-C$_2$H$_4$)$_2$(acac), within one-dimensional channels of HSSZ-53 zeolite.[71] SSZ-53 zeolite is a 14-ring extra-large pore zeolite with pore dimensions of 6.4×8.7 Å, which is large

Table 2.3 Zeolite-supported metal complexes: summary of synthesis results.

Zeolite support	Pore dimensionality and pore structure	Si/Al ratio (atomic)	Precursor	Approximate critical diameter of precursor (Å)	Approximate fraction of acac ligands removed from metal atoms
HY	3D 12-ring, 7.4 Å	30	Rh(η^2-C$_2$H$_4$)$_2$(acac)	6	1
			Ir(η^2-C$_2$H$_4$)$_2$(acac)	6	1
			cis-Ru(η^2-C$_2$H$_4$)$_2$(acac)$_2$	7	0.5
Hβ	3D 12-ring, 7.6×6.4 Å 5.5 Å	19	Rh(CO)$_2$(acac)	6	1
			Rh(η^2-C$_2$H$_4$)$_2$(acac)	6	1
			cis-Ru(η^2-C$_2$H$_4$)$_2$(acac)$_2$	7	0.7
HSSZ-42	Undulating 1D 12-ring, 6.4 Å	15	Rh(CO)$_2$(acac)	6	0.8
			cis-Ru(η^2-C$_2$H$_4$)$_2$(acac)$_2$	7	0.1
HMOR	2D 12-ring, 6.5×7.0 Å 8-ring, 5.7×2.6 Å	19	Rh(CO)$_2$(acac)	6	0.6
			cis-Ru(η^2-C$_2$H$_4$)$_2$(acac)$_2$	7	0.1
HZSM-5	3D 10-rings, 5.3×5.6 Å 5.1×5.3 Å	15	cis-Ru(η^2-C$_2$H$_4$)$_2$(acac)$_2$	7	0.1

Table adapted from ref. 56.

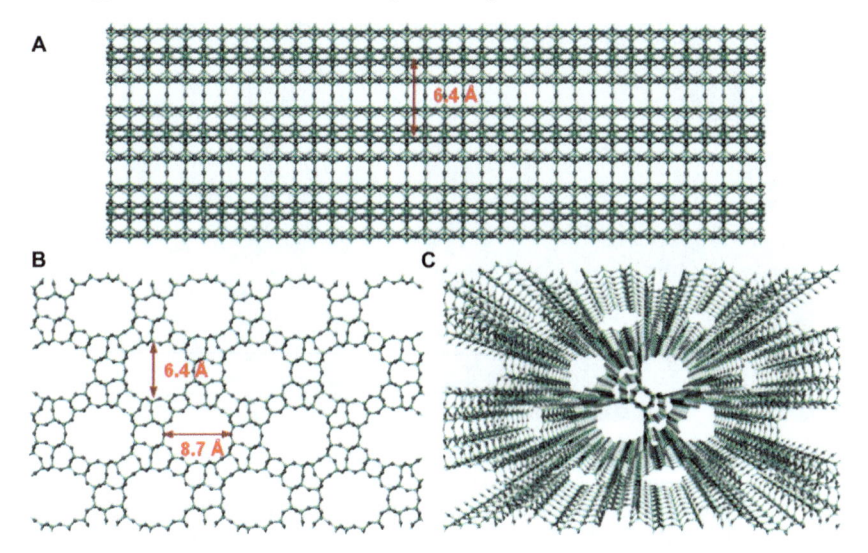

Figure 2.3 Models of zeolite HSSZ-53 framework: views from the [010] (A) and [100] (B) projections, and perspective view from the [100] projection (C). The double-headed arrows indicate the diameters of a HSSZ-53 channel in each projection.
Reproduced from ref. 53.

enough to accommodate this iridium complex (Figure 2.3).[72] HSSZ-53 used in this investigation is highly crystalline in the aluminosilicate form, which is suitable for anchoring a structurally uniform supported iridium complex.

Reaction of Ir(η^2-C$_2$H$_4$)$_2$(acac) with HSSZ-53 proceeds *via* exchange of acac ligands with bidentate zeolite ligands at acid sites with retaining two ethene ligands intact, yielding Ir(C$_2$H$_4$)$_2$ species (see Section 2.4.2 for more details). Highly crystalline one-dimensional HSSZ-53 zeolite facilitates characterization by aberration-corrected STEM in the high-angle annular dark-field (HAADF) mode high-resolution electron microscopy because of simpler structure and less damage by the electron beams than those for conventional zeolites such as three-dimensional zeolite Y. Furthermore, the synthesized catalyst allows detailed characterization by aberration-corrected STEM in the high-angle annular dark-field (HAADF) mode because of a high contrast of iridium ($Z=77$) and other light elements such as silicon ($Z=14$), aluminium ($Z=13$), oxygen ($Z=8$) and the proton ($Z=1$). Using this technique, Lu *et al.* provided direct evidence of mononuclear iridium complexes anchored within the 14-ring channels (Figure 2.4).

EXAFS spectroscopy data show that mononuclear Ir(η^2-C$_2$H$_4$)$_2$ binds to the surface of the zeolite *via* two Ir–O bonds with an Ir–O distance of 2.10 Å, which is a typical value for chemisorbed metal complexes. IR data using CO as a probe molecule show CO stretching bands that indicate the presence of exclusively chemisorbed iridium species.

38 Chapter 2

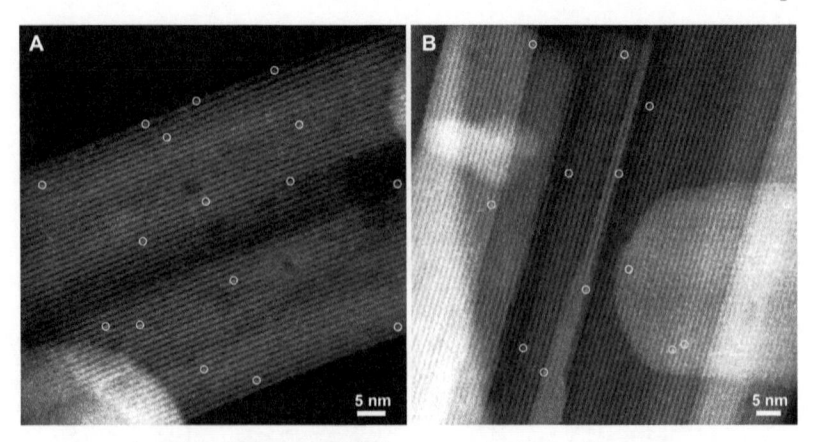

Figure 2.4 HAADF-STEM (Z-contrast) images characterizing the catalyst prepared by the reaction of Ir(η^2-C$_2$H$_4$)$_2$(acac) with zeolite HSSZ-53: (A) the initially prepared catalyst and (B) the catalyst after it had been used for ethene hydrogenation. The images show individual Ir atoms (some of them indicated by white circles) well dispersed inside the one-dimensional channels of the zeolite and in the absence of detectable iridium clusters. Reproduced from ref. 53.

Control of the distribution of acidic sites of zeolites provides another benefit to the synthesis of structurally uniform supported metal complexes. Synthesis of a zeolite with specific tetrahedral aluminium sites is challenging particularly when zeolites consist of various crystallographically different T atoms. However, a recent report shows that the synthesis of zeolites by a particular combination of SDAs induces a specific distribution of aluminium sites.[73]

2.4.2 Precursor Metal Complexes

A variety of organometallic precursors can be used in the synthesis of oxide-supported metal complexes.[33–35] Although various organometallic precursors (see examples in ref. 33–35) can be also used in the synthesis of zeolite-supported metal complexes, metal complex precursors are preferred to have ligands that mimic the surface of zeolite ligands. Acetylacetonate (C$_5$H$_7$O$_2^-$, abbreviated as acac hereinafter) ligands schematically shown in Figure 2.5 act as anionic bidentate ligands and coordinate to a cationic metal ion *via* two oxygen atoms. Because acac ligands mimic oxygen atoms at an acidic site of a zeolite (Figures 2.1 and 2.5), the ligand exchange of acac ligands with those from a zeolite is expected to cause a minimal structural exchange (slight changes in the metal–oxygen bond distances are often observed by ligand exchange because of the differences in the electron density of oxygen atoms in acac and those in a zeolite, see Section 2.4.1).

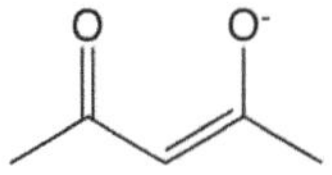

Zeolite-supported Molecular Metal Complex Catalysts 39

Figure 2.5 Acetylacetonate (anionic form of pentane-2,4-dione, Hacac).

LM(acac) + 2HO(O)Al$_{zeolite}$ → LM{O$_2$Al} + Hacac-HO(O)Al

Scheme 2.1 Schematic representation of reaction of an acetylacetonate metal complex, LM(acac), with an acid site of a zeolite. The brackets indicate the surface of a support.

Gates and his co-workers used a series of metal acac complexes to synthesize zeolite-supported metal complex catalysts. Examples of the metal complex precursors include Au(CH$_3$)$_2$(acac),[74] Rh(CO)$_2$(acac),[19,55] Rh(η^2-C$_2$H$_4$)$_2$(acac),[23,46,47,75,76] Ir(η^2-C$_2$H$_4$)$_2$(acac)[71,77–80] and Ru(η^2-C$_2$H$_4$)$_2$(acac)$_2$.[53–63] Metal atoms in these complexes are cationic and have formal charges of +1 (Au, Rh, Ir) or +2 (Ru). Syntheses of these samples are typically conducted in Schlenk flask using a standard air-exclusion technique. A calcined zeolite in the proton form is reacted with a metal precursor complex in a dry n-pentane solvent under an inert atmosphere. Upon coming into contact with zeolites, acac ligands in metal acac complexes undergo ligand exchange with oxygen atoms of the zeolite. After the ligand exchange completes, the solvent is removed in dynamic vacuum. Supported catalysts synthesized by this method are readily used for catalytic reactions without pretreatments at elevated temperatures. Using metal complexes having other ligands such as halides complicates structural characterization[19,76] and may cause undesired catalytic consequences.

Ligand exchange of acac ligands with bidentate ligands of a zeolite is triggered presumably by the protonation of an acac ligand by a proton of a zeolite and dissociated Hacac (protonated form of acac) becomes adsorbed on the surface of a zeolite support (Scheme 2.1).[82,83] Protonated acac ligands are removed from the metal center and become adsorbed on a zeolite as characterized by IR spectroscopy.[23,77,82,83] Interaction energy of protonated acac (Hacac) with a silanol group via hydrogen bonding was estimated as 6 kcal mol^{-1} by DFT calculations.[83] Furthermore, this value could easily be 10 kcal mol^{-1} depending on the exact form of the hydroxyl group. Thus, the interaction of Hacac with a zeolite surface provides an additional driving force for the facile ligand exchange. The ligand exchange reaction appears to be stoichiometric if a zeolite has enough number of accessible acid sites as compared with the amount of the metal precursors[81,84] and pore aperture of a zeolite is large enough to allow the access of metal complexes to the acid sites of the zeolite.[56]

Goellner et al. synthesized a zeolite-supported rhodium complex by the reaction of Rh(CO)$_2$(acac) with dealuminated zeolite Y (denoted as HY).[19] Rh(CO)$_2$(acac) is a well-known metal complex precursor used in

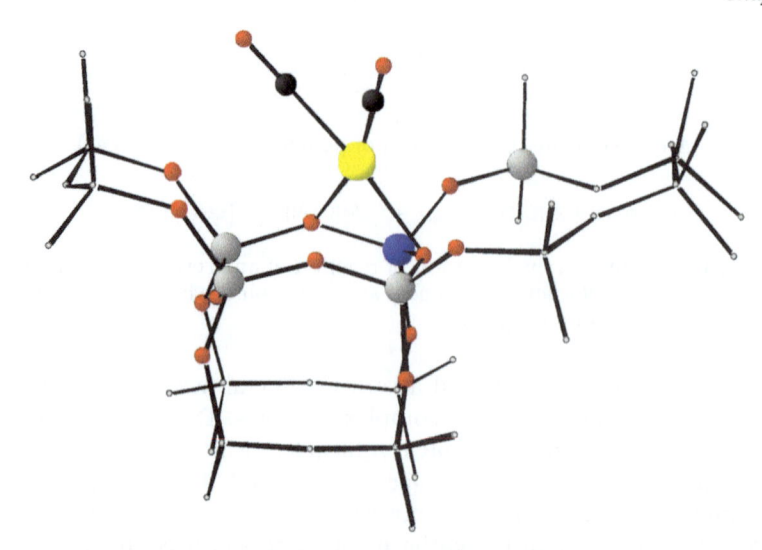

Figure 2.6 Location of rhodium dicarbonyl complex at a four-ring of faujasite: rhodium (yellow), carbon (black), and oxygen (red). The atoms included in the isolated cluster model T4 are shown as circles: aluminium (blue), silicon (light gray), oxygen (red). The dangling bonds of the cluster model are capped by hydrogen atoms.
Reproduced from ref. 19 and color coded.

organometallic chemistry in solution. Cuny and Buchwald used this complex to synthesize a rhodium complex catalyst for hydroformylation.[85] In their example, the acac ligand is replaced with a bis-organobisphosphite ligand *in situ*. Goellner *et al.* showed that the reaction of Rh(CO)$_2$(acac) with zeolite HY forms supported Rh(CO)$_2$ (note that this species is cationic and the formal charge of Rh is +1) at an acid site near Al in a four-ring of the zeolite (see Figure 2.6).[19] This supported rhodium complex is formally 16e species (8e for Rh(I), 4e for two CO ligands, 4e for the zeolite). The uniformity of the supported rhodium complexes allows incisive characterization; their EXAFS spectroscopy and DFT calculation data show that Rh(CO)$_2$ is bonded *via* two Rh–O bonds with a Rh–O distance of 2.15–2.20 Å and retain a (pseudo-square) planar symmetry (Figure 2.6). Rh–O bond lengths are slightly elongated from 2.040 Å determined for the precursor (by structural analysis by single crystal X-ray diffraction of the precursor Rh(CO)$_2$(acac) in a solid state) to 2.15–2.20 Å for the supported species by the ligand exchange. The longer Rh–O distances for the supported species are caused by higher electron density of oxygen atoms of the zeolite than those of acac (see Section 2.4.1). Structural uniformity of the supported rhodium complexes allows the selection of the best structural model for the supported rhodium molecular complex among several candidate models; neither of models corresponding Rh(CO)$_2$ in a three-hollow site nor the same species in a six-ring is adequate to account for the EXAFS data.

When reactive ethene ligands are used in acac metal complex precursors (*e.g.*, Rh(η^2-C$_2$H$_4$)$_2$(acac),[23] Ir(η^2-C$_2$H$_4$)$_2$(acac)[77] and Ru(η^2-C$_2$H$_4$)$_2$(acac)$_2$[81,84]), these ligands act as important roles in catalytic cycles of hydrogenation[77,23] and dimerization of olefins.[76,81]

2.4.3 Spectroscopic Characterization Techniques to Examine Structural Uniformity

IR spectroscopy using CO as a probe molecule is a powerful technique to characterize supported metals.[86–89] Miessner *et al.* were the first to demonstrate the uniformity of zeolite-supported rhodium(I) dicarbonyls by this technique.[18] They infer the high uniformity of the supported rhodium dicarbonyls by the sharpness of CO stretching bands (FWHM < 5 cm^{-1}). The degree of the narrowness of the ν_{CO} bands is similar to those observed for Rh(CO)$_2$(acac) dissolved in hexane.[90] Similar degree of narrowness of ν_{CO} bands were observed in the characterization of various zeolite-supported mononuclear metal complexes formed from the reactions of Rh(η^2-C$_2$H$_4$)$_2$(acac),[23,91] Ir(η^2-C$_2$H$_4$)$_2$(acac),[71,77] or Ru(η^2-C$_2$H$_4$)$_2$(acac)$_2$[81,84] with zeolite HY by this technique. However, when Rh(η^2-C$_2$H$_4$)$_2$(acac) was chemisorbed on MgO that is crystalline but has vastly different anchoring sites, IR data show much broader peaks (FWHM > 20 cm^{-1}) because of inhomogeneous structures of supported Rh(η^2-C$_2$H$_4$)$_2$.[92]

Ehresmann *et al.* demonstrate dynamical uniformity of zeolite-supported Rh(η^2-C$_2$H$_4$)$_2$ complexes (structure of the supported rhodium complex is shown in Figure 2.7) by variable-temperature ^{13}C CP-MAS NMR spectroscopy.[24] ^{13}C resonance of π-bonded ethene ligands is broadened as a result of a conflict of ^{1}H–^{13}C dipolar decoupling and random anisotropic reorientation of the ^{1}H–^{13}C bond vector. No signal was observed at 293 K because these two processes occur on the same timescale for all spin-pairs in the sample (Figure 2.8). This result indicates that the ethene ligands in this sample have a common rotational barrier, which indicates that supported Rh(η^2-C$_2$H$_4$)$_2$ complexes have common electronic and steric properties. The important point to note is that this phenomena is also observed for pure Rh(η^2-C$_2$H$_4$)$_2$(acac) crystals.[93] These data clearly show the uniformity of the zeolite-supported Rh(η^2-C$_2$H$_4$)$_2$ complexes.

2.5 Molecular Chemistry of Zeolite-supported Metal Complex Catalysts

2.5.1 Reactivity of Supported Metal Complexes

When metals are atomically dispersed, as in the case of zeolite-supported metal complex catalysts, zeolite supports play important roles as ligands to control catalytic properties of metals. Thus, subtle structural changes caused

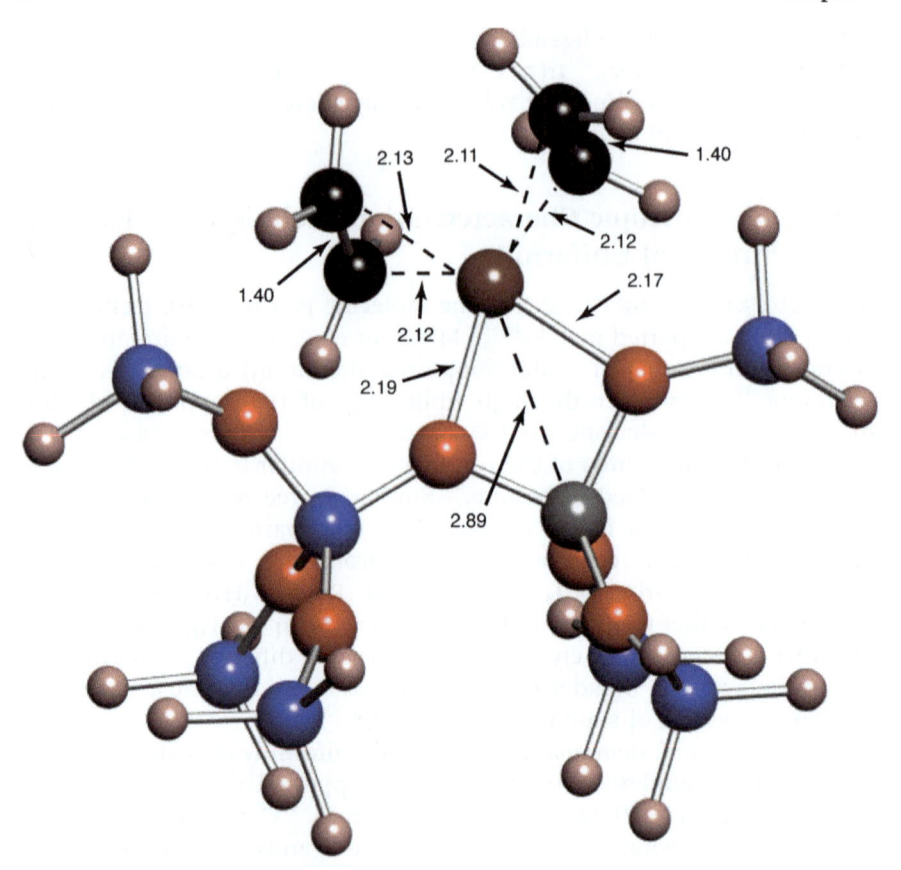

Figure 2.7 Geometry of {Rh(η^2-C$_2$H$_4$)$_2$} bonded to a zeolite anion cluster, optimized at the B3LYP level.
Reproduced from ref. 24.

by the ligand exchange of acac with oxygen atoms of a zeolite have a significant catalytic consequence. Gates and Ogino reported a systematic variations of chemisorbed and physisorbed mononuclear ruthenium complexes by reacting Ru(η^2-C$_2$H$_4$)$_2$(acac)$_2$ with zeolite HY.[84] When Ru loadings were varied from 1 to 3 wt% (corresponding Al/Ru atomic ratios of 6–2), only the sample with a Ru loading of 1 wt% underwent complete ligand exchange of acac ligands with oxygen atoms of the zeolite; samples with higher loadings had a mixture of chemisorbed and physisorbed species. The variations of the fractions of chemisorbed species were confirmed by various results; chemisorbed species show different color from that for the precursor but the color of the samples became similar to the original metal precursor complex as Ru loading was increased. Imaginary part of the EXAFS data, which does not suffer from nonlinear interference of coordination shells that might appear in the modulus, shows systematic variations of coordination shells

Zeolite-supported Molecular Metal Complex Catalysts

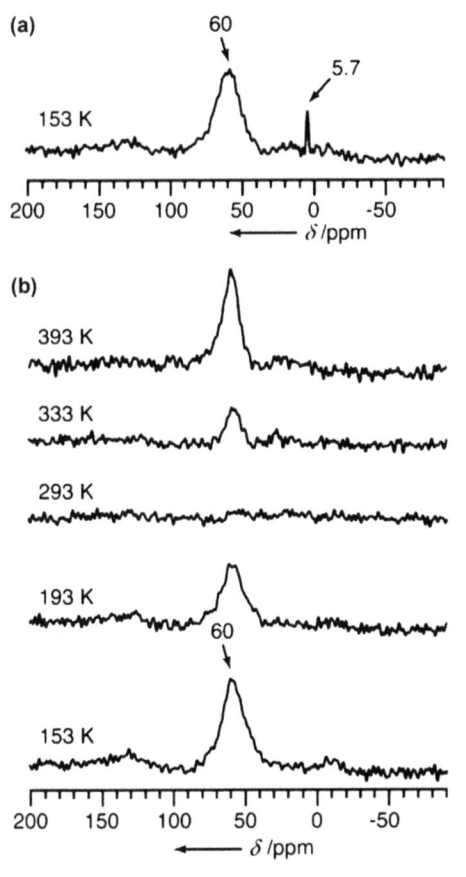

Figure 2.8 ^{13}C CP-MAS NMR spectra (75.4 MHz) of the sample formed from Rh(η^2-C$_2$H$_4$)$_2$(acac) and zeolite HY and then exchanged with ^{13}C-labeled ethene. (a) Variable-temperature spectra showing the temperature dependence of the signal intensity. (b) Spectrum obtained at low temperature after heating and subsequent cooling of the sample, which demonstrates the results of a reverse-spillover reaction. The signal at δ = 5.7 ppm corresponds to ethane.
Reproduced from ref. 24.

from the sample containing only chemisorbed species to the sample containing largely physisorbed species.

When the zeolite-supported ruthenium complexes were tested in ethene dimerization reaction, the data show that only the chemisorbed species are active for this reaction. The turnover frequencies expressed with respect to the amount of chemisorbed species are virtually the same among the samples at steady state. Furthermore, butene selectivity, which is influenced by the residence time of reactant ethene with respect to the number of 'catalytically active species (chemisorbed species)' in a flow reactor, shows the same value for the sample containing only chemisorbed and a mixture of chemisorbed and physisorbed species at steady state. Because a short residence time yields a higher 1-butene yield (longer residence times allows 1-butene to isomerize to other isomers), their data suggest that ethene spends nearly the same time in a flow reactor with respect to the number of 'chemisorbed species' and thus the data show that chemisorbed species are the catalytically active species for this dimerization reaction. These data demonstrate the important role of the zeolite as ligands;[4,84] coordination of oxygen atoms of the zeolite to the ruthenium atoms is essential.

Lu *et al.* investigated roles of ligands using $Ir(\eta^2\text{-}C_2H_4)_2(acac)$ supported on various zeolites (HSSZ-53, Hβ, and HY), γ-Al_2O_3, and MgO. Supported iridium species in these samples are isostructural $(Ir(\eta^2\text{-}C_2H_4)_2)$. Ethene ligands in all catalysts undergo facile ligand exchange with CO by the treatment of 10% CO at 300 K at 1 bar, forming supported *gem*-dicarbonyl iridium complexes, $Ir(CO)_2$. However, the supported iridium complexes exhibit markedly different catalytic activities in ethene hydrogenation. When the electron density of iridium atoms is characterized by IR spectroscopy (CO as a probe molecule and XANES spectroscopy (comparison of the white line intensities), the results show a systematic variation of the electron density; electron density of iridium atoms becomes higher in the order of MgO < Al_2O_3 « HSSZ-53 < zeolite Hβ < zeolite HY (Figure 2.9).

The subtle differences of the frequencies of ν_{CO} among the three zeolites are resolved by the sharpness of ν_{CO} bands that is enabled by the uniformity of supported iridium species. For example, slight red-shifts in ν_{CO} (≈ 4 cm^{-1}) was also observed for zeolite Hβ-supported rhodium complexes as compared with zeolite HY.[25,91] The IR data are complemented by XANES data (Figure 2.8). In addition to the variation of the electron density of iridium atoms in these catalysts, Lu *et al.* show a correlation between the electron density and the easiness of ligand exchange of carbonyl ligands of the supported $Ir(CO)_2$ with C_2H_4. Their data show that CO ligands coordinated to electron-deficient iridium atoms are more easily exchanged with ethene ligands. This result may be explained by weaker bonds for Ir–CO (caused by less back-donation from the iridium to CO) as well as easier coordination of ethene to a vacant orbital for electron-deficient iridium atoms as in the case of supported rhodium complex (*vide infra*). The results highlight the analogy between roles of supports in catalysis by anchored metal complexes and those of ligands in organometallic solution catalysis. The mechanism for the ligand exchange of CO with C_2H_4 (*e.g.*, associative *versus* dissociative) has not been fully investigated yet and requires further investigations. The classical examples in organometallic complexes with 16e, d^8 square symmetry (*e.g.*, rhodium(I) complexes) is known to proceed *via* the associative mechanism with an 18e intermediate having a trigonal bipyramid with the incoming ligand (C_2H_4 in the example shown above) in the equatorial plane.[94] In this mechanism, good π-acid ligands are known to facilitate leaving of a ligand in the trans position. Furthermore, because organometallic 16e complexes including rhodium(I) often undergo ligand exchange *via* the associative mechanism (see Scheme 2.2),[94] a similar

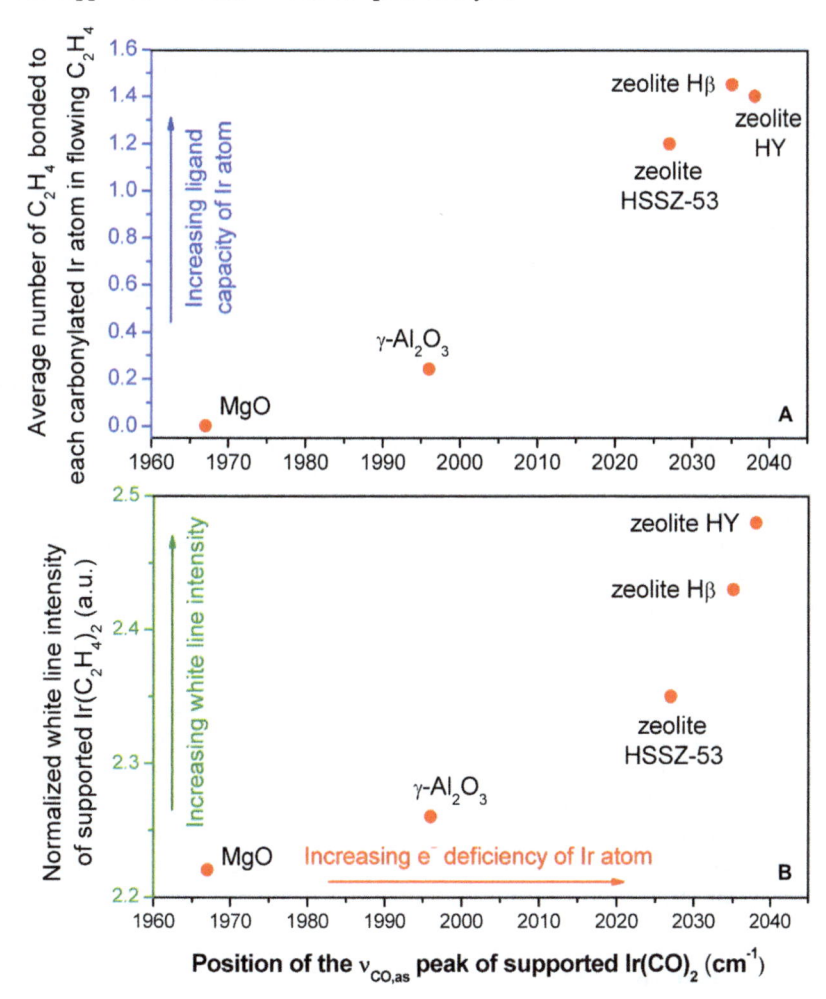

Zeolite-supported Molecular Metal Complex Catalysts 45

Figure 2.9 (A) Correlation between the average number of C_2H_4 groups bonded to iridium when $Ir(CO)_2$ complexes were treated in flowing C_2H_4 at 298 K and 1 bar and (B) normalized white-line intensities of the initially prepared $Ir(CO)_2$ complexes characterizing the complexes on various supports and the frequency of ν_{CO} bands of the $Ir(CO)_2$ complexes (a higher ν_{CO} frequency indicates that the iridium complex is more electron-deficient). The $Ir(CO)_2$ complexes were prepared by treating $Ir(C_2H_4)_2$ complexes with a pulse of CO in helium at 298 K. Reproduced from ref. 80.

$$L_nM + L' \xrightarrow{k_1} L_nM\text{-}L' \xrightarrow{fast} L_{n-1}M\text{-}L' + L$$

$$\text{Rate} = k_1[L'][L_nM]$$

Scheme 2.2 Associative ligand exchange of L with L' for the metal complex L_nM.

mechanism may be applied to the supported Ir(CO)$_2$ system. Another issue that remains to be elucidated is whether a support zeolite plays any important role in the ligand exchange.[81] For metal complexes catalysts in solution, reactivity of ligands is often influenced by polarity and nature of solvents.

Lu et al. also reported a correlation between the electron density on supported diethene iridium complexes and the rate of ethene hydrogenation. The rate of ethene hydrogenation increases as the Ir atom becomes more electron-deficient. They reported that H–D exchange rate became significantly higher when switching the support from the electron-donating MgO to the electron-withdrawing HSSZ-53.

Liang et al. investigated details of ligand exchange process of zeolite-supported Rh(η^2-C$_2$H$_4$)$_2$. They show that π-bonded C$_2$H$_4$ ligands readily undergo ligand exchange with C$_2$H$_4$ molecules present in the gas phase over the sample via the formation of an intermediate, ethyl ligands that is assisted by reverse spillover.[25] Ethene ligands in the supported Rh(η^2-C$_2$H$_4$)$_2$ are readily exchanged with CO ligands, forming supported Rh(CO)$_2$, consistent with calculated average bond energy of R–L (53 kJ mol^{-1} for L = CO as compared with 40 kJ mol^{-1} for L = C$_2$H$_4$). Upon exposure of the supported Rh(CO)$_2$ to a C$_2$H$_4$ flow, one of two CO ligands is exchanged by C$_2$H$_4$, forming supported Rh(CO)(η^2-C$_2$H$_4$). This exchange occurs because the experiment was conducted in flow conditions far from equilibrium. When supported Rh(η^2-C$_2$H$_4$)$_2$ complex is contacted with H$_2$ in helium, it forms supported rhodium monohydride with C$_2$H$_4$. When the same supported Rh(η^2-C$_2$H$_4$)$_2$ complex is contacted with H$_2$ in a nitrogen flow at room temperature, it forms supported rhodium dinitrogen Rh(N$_2$)$_2$ via a rhodium nitrogen complex (Rh(C$_2$H$_5$)(N$_2$) or Rh(N$_2$)). These results suggest new catalysis for supported rhodium complexes to convert dinitrogen.

2.5.2 Catalytic Cycle

Structural uniformity of zeolite-supported metal complexes offers the unique opportunity to investigate catalytic cycles rigorously. Kletnieks et al. investigated the catalytic cycle of acetylene cyclotrimerization on zeolite-supported rhodium metal complex.[26] The complete cycle is shown in Figure 2.10. The initial structure of the supported rhodium complexes (denoted as the precursor, **PRE**) determined by a combination of IR, NMR and EXAFS spectroscopies aided by DFT calculations is site-isolated Rh(η^2-C$_2$H$_4$)$_2$ species bonded to the surface of the zeolite via two Rh–O bonds (similar to that shown in Figure 2.7). **PRE** has highly uniform structure as described in the preceding sections. When acetylene is passed over **PRE**, they readily replace the reactive ethene ligands and form a supported bis-acetylene complex (**SI$_1$**). **PRE** enters into a catalytic cycle of acetylene cyclotrimerization by being converted to the stable intermediate **SI$_1$**. The intermediate **SI$_1$** was detected by solid state ^{13}C MAS NMR spectroscopy (Figure 2.11) and assignments of chemical shifts were aided by DFT

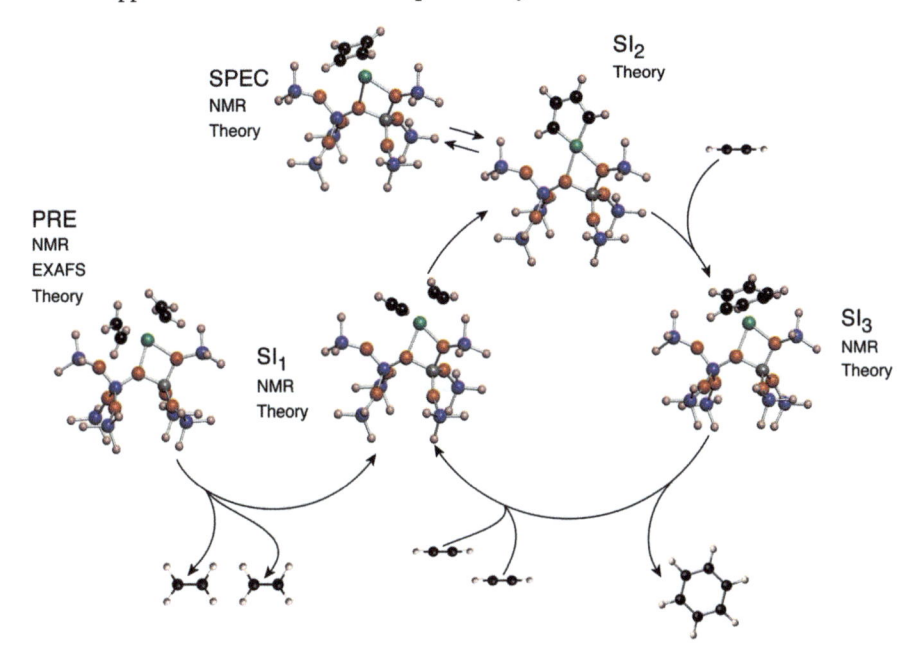

Figure 2.10 Catalytic cycle for cyclotrimerization of acetylene catalyzed by a single-site Rh$^+$ center bonded to a zeolite support by two Rh–O bonds. The reactive organic groups bonded to Rh in the as-synthesized catalyst (at left, **PRE**), two ethene ligands, allow facile entry into a catalytic cycle. **PRE** has been characterized by EXAFS and ^{13}C NMR spectroscopies as well as density functional theory. When the sample is brought in contact with gas-phase acetylene, the ethene is replaced by acetylene, to form the stable intermediate **SI$_1$**, initiating the catalytic cycle. This intermediate undergoes ring closure *via* transition state **TS$_{1\to 2}$**, forming the metallacycle **SI$_2$**, which equilibrates with the more stable cyclobutadiene complex (**SPEC**, a spectator species). The pool consisting of **SI$_2$** and **SPEC** crosses *via* transition state **TS$_{2\to 3}$** into the stable benzene complex **SI$_3$**. Dissociation of product benzene into the gas phase, compensated by adsorption of reactant acetylene, closes the cycle. Reproduced from ref. 26.

calculations. **SI$_1$** is converted to supported benzene complex **SI$_3$** as characterized by ^{13}C MAS NMR spectroscopy. DFT calculations suggest another intermediate **SI$_2$** (supported metallacycle complex), which is in equilibrium with a spectator (**SPEC**, supported cyclobutadiene complex). **SPEC** is also characterized by NMR spectroscopy. The benzene ligand in **SI$_3$** is replaced with two acetylene molecules, forming benzene as a product in the gas phase. The ligand exchange of benzene with acetylene completes the catalytic cycle. This work illustrates that the uniformity of **PRE** sets the starting point of the structure of the supported rhodium complex in the reaction and allows rigorous characterization of the structures of intermediates in the catalytic cycle.

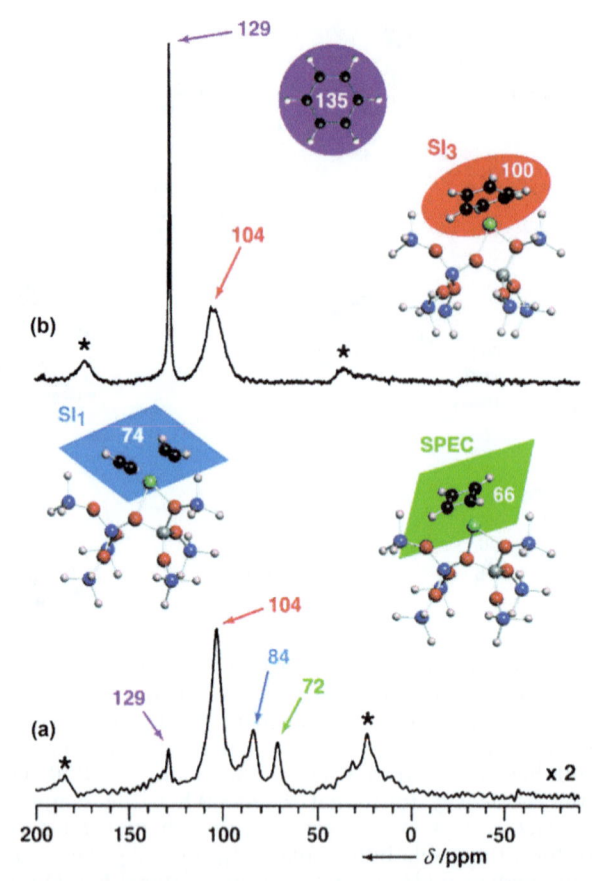

Figure 2.11 ^{13}C MAS NMR spectra of the zeolite-supported Rh$^+$ ion sampling two distributions of the species over the catalytic cycle of Figure 2.9. (a) Entering the cycle by exchanging [^{13}C$_2$]acetylene for the unlabeled ethene ligands in **PRE** leads to **SI$_1$** (84 vs. 74 ppm theoretical), **SPEC** (72 vs. 66 ppm theoretical), **SI$_3$** (104 vs. 100 ppm theoretical), and product, free benzene (129 vs. 135 ppm theoretical). (b) Entering the cycle by displacing unlabeled ethene ligands with excess [^{13}C$_6$]benzene leads to **SI$_3$** directly, and decyclotrimerization to preceding parts of the cycle is essentially precluded by thermodynamics. The symbol * denotes rotational sidebands. Theoretical chemical shift values were computed by using the GIAO formalism with the B3LYP exchange-correlation functional with the Stuttgart effective core potential (ECP) basis set on Rh and a polarized triple zeta basis set on the other atoms. The calculated geometries are shown in the figure.
Reproduced from ref. 26.

2.6 Conclusions and Future Directions

Supported metal catalysts have brought various technological breakthroughs in industry. However, many of their structure–performance relationships have not been well understood. One of the difficulties is associated with structural non-uniformity of supported species. As a tiny fraction of supported metals sometimes plays a dominant role in catalysis, it remains challenging to understand catalytic chemistry of these catalysts. When metals are atomically dispersed, supports act as ligands to control the catalytic properties of metals. Consequently, non-uniformity of supports leads to the formation of multiple supported species, leading to diverse catalytic properties. Zeolites provide the opportunity to synthesize structurally uniform supported species because they are crystalline materials and have nearly uniform anchoring sites. Site-isolated cationic metal complexes can be anchored on these sites *via* the surface organometallic approach if zeolites have large apertures and low density of anchoring sites. One of the advantages of using the surface organometallic approach over other synthesis methods is that organometallic precursors can be designed to incorporate ligands that mimic zeolite surface and those that function as intermediates in a catalytic cycle. Among various organometallic precursors, metal acac complexes are preferred because ligand exchange of bidentate acac ligands with oxygen ligands of zeolites readily occurs *via* protonation of acac ligands by zeolites. Acac ligands mimic the surface of zeolites that also act as bidentate ligands so that the ligand exchange causes minimal structural changes. A series of zeolite-supported metal complexes have been synthesized using metal acac precursors such as Rh(CO)$_2$(acac), Rh(η^2-C$_2$H$_4$)$_2$(acac), Ir(η^2-C$_2$H$_4$)$_2$(acac), and Ru(η^2-C$_2$H$_4$)$_2$(acac)$_2$. Resultant supported metal complexes have a high degree of uniformity as proven by IR spectroscopy using CO as a probe molecule and/or variable-temperature ^{13}C MAS NMR spectroscopy. Uniformity of supported metal complexes in these catalysts allows incisive characterization by multiple spectroscopic techniques such as EXAFS and IR spectroscopies. Furthermore, these experimental results are corroborated by rigorous structural modeling by DFT calculations. Structural uniformity of supported species also allows detailed investigations on reactivity of ligands, genesis of catalytically active species, and a whole catalytic cycle by a combination of multiple characterization techniques with DFT calculations at an unprecedented level. These investigations show that supported metal complexes in these catalysts function essentially as molecules.

Remaining challenges in the synthesis of zeolite-supported molecular metal complex catalysts are (i) how to anchor metal complexes at crystallographically equivalent positions in a zeolite and (ii) how to anchor larger metal complexes. Although zeolites are crystalline and have nearly uniform anchoring sites, distribution of anchoring sites are often random and in some cases zeolites have multiple anchoring sites that are not equivalent crystallographically. If zeolites are synthesized with precise control of the distribution of anchoring sites (acid sites), that will allow one to synthesize zeolite-supported molecular metal complex catalysts more precisely, which ultimately enables structural analysis of supported metal complexes by single-crystal X-ray diffraction. Anchoring bulky metal complexes in zeolites is another challenge because of the limitation of small pore apertures for conventional zeolites. Emergence of two-dimensional zeolites as well as new extra-large pore zeolites will enable the use of bulky metal acac complexes. If these challenges are overcome and a wide range of zeolite-supported molecular metal complex catalysts are designed and synthesized with a high degree of precision, the investigations of catalytic chemistry of supported catalysts will be greatly facilitated. Ultimately, new selective catalysts that may have complex structures are synthesized with atomically-precise synthesis method.

References

1. B. C. Gates, *Catalytic Chemistry*, Wiley, New York, 1992.
2. G. Ertl, *Handbook of Heterogeneous Catalysis*, 2nd edn, Wiley-VCH, Weinheim Chichester, 2008.
3. A. M. Argo, J. F. Odzak, F. S. Lai and B. C. Gates, *Nature*, 2002, **415**, 623–626.
4. M. Flytzani-Stephanopoulos and B. C. Gates, *Annu. Rev. Chem. Biomol. Eng.*, 2012, **3**, 545–574.
5. J. Čejka, A. Corma and S. I. Zones, *Zeolite and Catalysis: Synthesis, Reactions and Applications*, Wiley-VCH, Weinheim Chichester, 2010.
6. D. E. De Vos, M. Dams, B. F. Sels and P. A. Jacobs, *Chem. Rev.*, 2002, **102**, 3615–3640.
7. D. E. De Vos, B. F. Sels and P. A. Jacobs, *CATTECH*, 2002, **6**, 14–29.
8. M. H. Groothaert, P. J. Smeets, B. F. Sels, P. A. Jacobs and R. A. Schoonheydt, *J. Am. Chem. Soc.*, 2005, **127**, 1394–1395.
9. P. J. Smeets, M. H. Groothaert and R. A. Schoonheydt, *Catal. Today*, 2005, **110**, 303–309.
10. J. S. Woertink, P. J. Smeets, M. H. Groothaert, M. A. Vance, B. F. Sels, R. A. Schoonheydt and E. I. Solomon, *Proc. Natl. Acad. Sci. U. S. A.*, 2009, **106**, 18908–18913.
11. H. Yahiro and M. Iwamoto, *Appl. Catal., A*, 2001, **222**, 163–181.
12. H.-Y. Chen and W. M. H. Sachtler, *Catal. Today*, 1998, **42**, 73–83.
13. J. Pérez-Ramírez, F. Kapteijn, G. Mul and J. A. Moulijn, *Chem. Commun.*, 2001, 693–694.
14. E. A. Pidko, E. J. M. Hensen and R. A. van Santen, *Proc. R. Soc. A*, 2012, **468**, 2070–2086.
15. P. J. Smeets, J. S. Woertink, B. F. Sels, E. I. Solomon and R. A. Schoonheydt, *Inorg. Chem.*, 2010, **49**, 3573–3583.
16. U. Deka, I. Lezcano-Gonzalez, B. M. Weckhuysen and A. M. Beale, *ACS Catal.*, 2013, **3**, 413–427.
17. P. Vanelderen, J. Vancauwenbergh, B. F. Sels and R. A. Schoonheydt, *Coord. Chem. Rev.*, 2013, **257**, 483–494.
18. H. Miessner, I. Burkhardt, D. Gutschick, A. Zecchina, C. Morterra and G. Spoto, *J. Chem. Soc., Faraday Trans. 1*, 1989, **85**, 2113–2126.

19. J. F. Goellner, B. C. Gates, G. N. Vayssilov and N. Rösch, *J. Am. Chem. Soc.*, 2000, **122**, 8056–8066.
20. H. Miessner, *J. Am. Chem. Soc.*, 1994, **116**, 11522–11530.
21. G. N. Vayssilov and N. Rösch, *J. Am. Chem. Soc.*, 2002, **124**, 3783–3786.
22. A. D. Vityuk, O. S. Alexeev and M. D. Amiridis, *J. Catal.*, 2014, **311**, 230–243.
23. A. J. Liang, V. A. Bhirud, J. O. Ehresmann, P. W. Kletnieks, J. F. Haw and B. C. Gates, *J. Phys. Chem. B*, 2005, **109**, 24236–24243.
24. J. O. Ehresmann, P. W. Kletnieks, A. Liang, V. A. Bhirud, O. P. Bagatchenko, E. J. Lee, M. Klaric, B. C. Gates and J. F. Haw, *Angew. Chem., Int. Ed.*, 2006, **45**, 574–576.
25. A. J. Liang, R. Craciun, M. Chen, T. G. Kelly, P. W. Kletnieks, J. F. Haw, D. A. Dixon and B. C. Gates, *J. Am. Chem. Soc.*, 2009, **131**, 8460–8473.
26. P. W. Kletnieks, A. J. Liang, R. Craciun, J. O. Ehresmann, D. M. Marcus, V. A. Bhirud, M. M. Klaric, M. J. Hayman, D. R. Guenther, O. P. Bagatchenko, D. A. Dixon, B. C. Gates and J. F. Haw, *Chem. - Eur. J.*, 2007, **13**, 7294–7304.
27. W. M. H. Sachtler and Z. Zhang, in *Advances in Catalysis*, ed. D. D. Eley, H. Pines and P. B. Weisz, Academic Press, New York, 1993, vol. 39, pp. 129–220.
28. A. Zecchina, M. Rivallan, G. Berlier, C. Lamberti and G. Ricchiardi, *Phys. Chem. Chem. Phys.*, 2007, **9**, 3483–3499.
29. S. M. T. Almutairi, B. Mezari, P. C. M. M. Magusin, E. A. Pidko and E. J. M. Hensen, *ACS Catal.*, 2011, **2**, 71–83.
30. A. A. Battiston, J. H. Bitter and D. C. Koningsberger, *Catal. Lett.*, 2000, **66**, 75–79.
31. H. S. Lacheen, P. J. Cordeiro and E. Iglesia, *J. Am. Chem. Soc.*, 2006, **128**, 15082–15083.
32. I. J. Drake, Y. Zhang, D. Briggs, B. Lim, T. Chau and A. T. Bell, *J. Phys. Chem. B*, 2006, **110**, 11654–11664.
33. J.-M. Basset, B. C. Gates, J. P. Candy, A. Choplin, M. Leconte, *et al.*, Eds. *Surface Organometallic Chemistry : Molecular Approaches to Surface Catalysis*, Springer, New York, 1988.
34. Y. Iwasawa, in *Advances in Catalysis*, ed. D. D. Eley, H. Pines and P. B. Weisz, Academic Press, 1987, vol. 35, pp. 187–264.
35. C. Copéret, M. Chabanas, R. Petroff Saint-Arroman and J.-M. Basset, *Angew. Chem., Int. Ed.*, 2003, **42**, 156–181.
36. J. Guzman and B. C. Gates, *Dalton Trans.*, 2003, 3303–3318.
37. J. M. Notestein, E. Iglesia and A. Katz, *J. Am. Chem. Soc.*, 2004, **126**, 16478–16486.
38. J. M. Notestein, L. R. Andrini, V. I. Kalchenko, F. G. Requejo, A. Katz and E. Iglesia, *J. Am. Chem. Soc.*, 2007, **129**, 1122–1131.
39. B. C. Gates and H. H. Lamb, *J. Mol. Catal.*, 1989, **52**, 1–18.
40. J. Long, X. Wang, G. Zhang, J. Dong, T. Yan, Z. Li and X. Fu, *Chem. - Eur. J.*, 2007, **13**, 7890–7899.

41. V. Ortalan, A. Uzun, B. C. Gates and N. D. Browning, *Nat. Nanotechnol.*, 2010, **5**, 506–510.
42. J. C. Fierro-Gonzalez, S. Kuba, Y. L. Hao and B. C. Gates, *J. Phys. Chem. B*, 2006, **110**, 13326–13351.
43. D. C. Koningsberger, B. L. Mojet, G. E. van Dorssen and D. E. Ramaker, *Top. Catal.*, 2000, **10**, 143–155.
44. O. Alexeev and B. C. Gates, *Top. Catal.*, 2000, **10**, 273–293.
45. E. A. Stern, *Phys. Rev. B*, 1993, **48**, 9825–9827.
46. Error Reporting Recommendations: A Report of the Standards and Criteria Committee, Adopted by the International XAFS Society, Standards and Criteria Committee, July 26, 2000, Ako, Japan. www.i-x-s.org/OLD/subcommittee_reports/sc/err-rep.pdf.
47. F. B. M. Vanzon, P. S. Kirlin, B. C. Gates and D. C. Koningsberger, *J. Phys. Chem.*, 1989, **93**, 2218–2222.
48. S. R. Bare, S. D. Kelly, W. Sinkler, J. J. Low, F. S. Modica, S. Valencia, A. Corma and L. T. Nemeth, *J. Am. Chem. Soc.*, 2005, **127**, 12924–12932.
49. G. T. Palomino, S. Bordiga, A. Zecchina, G. L. Marra and C. Lamberti, *J. Phys. Chem. B*, 2000, **104**, 8641–8651.
50. Y. Zhang, D. N. Briggs, E. de Smit and A. T. Bell, *J. Catal.*, 2007, **251**, 443–452.
51. M. H. Groothaert, J. A. van Bokhoven, A. A. Battiston, B. M. Weckhuysen and R. A. Schoonheydt, *J. Am. Chem. Soc.*, 2003, **125**, 7629–7640.
52. A. Yamaguchi, A. Suzuki, T. Shido, Y. Inada, K. Asakura, M. Nomura and Y. Iwasawa, *J. Phys. Chem. B*, 2002, **106**, 2415–2422.
53. K. K. Bando, T. Wada, T. Miyamoto, K. Miyazaki, S. Takakusagi, Y. Koike, Y. Inada, M. Nomura, A. Yamaguchi, T. Gott, S. Ted Oyama and K. Asakura, *J. Catal.*, 2012, **286**, 165–171.
54. H. Yamashita, M. Matsuoka, K. Tsuji, Y. Shioya, M. Anpo and M. Che, *J. Phys. Chem.*, 1996, **100**, 397–402.
55. I. Ogino and B. C. Gates, *J. Phys. Chem. C*, 2010, **114**, 8405–8413.
56. I. Ogino, C. Y. Chen and B. C. Gates, *Dalton Trans.*, 2010, **39**, 8423–8431.
57. M. E. Davis, *Nature*, 2002, **417**, 813–821.
58. P. Wu, J. F. Ruan, L. L. Wang, L. L. Wu, Y. Wang, Y. M. Liu, W. B. Fan, M. Y. He, O. Terasaki and T. Tatsumi, *J. Am. Chem. Soc.*, 2008, **130**, 8178–8187.
59. W. J. Roth and J. Čejka, *Catal. Sci. Technol.*, 2011, **1**, 43–53.
60. J. Čejka, G. Centi, J. Perez-Pariente and W. J. Roth, *Catal. Today*, 2012, **179**, 2–15.
61. G. Bellussi, A. Carati, C. Rizzo and R. Millini, *Catal. Sci. Technol.*, 2013, **3**, 833–857.
62. U. Díaz and A. Corma, *Dalton Trans.*, 2014, **43**, 10292–10316.
63. M. Choi, K. Na, J. Kim, Y. Sakamoto, O. Terasaki and R. Ryoo, *Nature*, 2009, **461**, 246–250.
64. A. Corma, V. Fornés, S. B. Pergher, T. L. M. Maesen and J. G. Buglass, *Nature*, 1998, **396**, 353–356.

65. A. Corma, U. Díaz, M. E. Domine and V. Fornés, *J. Am. Chem. Soc.*, 2000, **122**, 2804–2809.
66. S. Maheshwari, E. Jordan, S. Kumar, F. S. Bates, R. L. Penn, D. F. Shantz and M. Tsapatsis, *J. Am. Chem. Soc.*, 2008, **130**, 1507–1516.
67. I. Ogino, M. M. Nigra, S.-J. Hwang, J.-M. Ha, T. Rea, S. I. Zones and A. Katz, *J. Am. Chem. Soc.*, 2011, **133**, 3288–3291.
68. E. A. Eilertsen, I. Ogino, S. J. Hwang, T. Rea, S. Yeh, S. I. Zones and A. Katz, *Chem. Mater.*, 2011, **23**, 5404–5408.
69. I. Ogino, E. A. Eilertsen, S.-J. Hwang, T. Rea, D. Xie, X. Ouyang, S. I. Zones and A. Katz, *Chem. Mater.*, 2013, **25**, 1502–1509.
70. X. Ouyang, S.-J. Hwang, R. C. Runnebaum, D. Xie, Y.-J. Wanglee, T. Rea, S. I. Zones and A. Katz, *J. Am. Chem. Soc.*, 2013, **136**, 1449–1461.
71. J. Lu, C. Aydin, A. J. Liang, C. Y. Chen, N. D. Browning and B. C. Gates, *ACS Catal.*, 2012, **2**, 1002–1012.
72. A. Burton, S. Elomari, C.-Y. Chen, R. C. Medrud, I. Y. Chan, L. M. Bull, C. Kibby, T. V. Harris, S. I. Zones and E. S. Vittoratos, *Chem. - Eur. J.*, 2003, **9**, 5737–5748.
73. A. B. Pinar, C. Márquez-Álvarez, M. Grande-Casas and J. Pérez-Pariente, *J. Catal.*, 2009, **263**, 258–265.
74. J. C. Fierro-Gonzalez, S. Kuba, Y. Hao and B. C. Gates, *J. Phys. Chem. B*, 2006, **110**, 13326–13351.
75. P. Serna and B. C. Gates, *J. Am. Chem. Soc.*, 2011, **133**, 4714–4717.
76. P. Serna and B. C. Gates, *Angew. Chem., Int. Ed.*, 2011, **50**, 5528–5531.
77. A. Uzun, V. A. Bhirud, P. W. Kletnieks, J. F. Haw and B. C. Gates, *J. Phys. Chem. C*, 2007, **111**, 15064–15073.
78. E. Bayram, J. Lu, C. Aydin, A. Uzun, N. D. Browning, B. C. Gates and R. G. Finke, *ACS Catal.*, 2012, **2**, 1947–1957.
79. J. Lu, P. Serna and B. C. Gates, *ACS Catal.*, 2011, **1**, 1549–1561.
80. J. Lu, C. Aydin, N. D. Browning and B. C. Gates, *Langmuir*, 2012, **28**, 12806–12815.
81. I. Ogino and B. C. Gates, *J. Am. Chem. Soc.*, 2008, **130**, 13338–13346.
82. J. C. Kenvin, M. G. White and M. B. Mitchell, *Langmuir*, 1991, **7**, 1198–1205.
83. I. Ogino, M. Y. Chen, J. Dyer, P. W. Kletnieks, J. F. Haw, D. A. Dixon and B. C. Gates, *Chem. - Eur. J.*, 2010, **16**, 7427–7436.
84. I. Ogino and B. C. Gates, *Chem. - Eur. J.*, 2009, **15**, 6827–6837.
85. G. D. Cuny and S. L. Buchwald, *J. Am. Chem. Soc.*, 1993, **115**, 2066–2068.
86. K. I. Hadjiivanov and G. N. Vayssilov, *Adv. Catal.*, 2002, **47**, 307–511.
87. H. A. Aleksandrov, V. R. Zdravkova, M. Y. Mihaylov, P. St Petkov, G. N. Vayssilov and K. I. Hadjiivanov, *J. Phys. Chem. C*, 2012, **116**, 22823–22831.
88. K. Hadjiivanov and H. Knözinger, *Surf. Sci.*, 2009, **603**, 1629–1636.
89. D. Nachtigallová, O. Bludský, C. Otero Areán, R. Bulánek and P. Nachtigall, *Phys. Chem. Chem. Phys.*, 2006, **8**, 4849–4852.
90. S. Kawi, Z. Xu and B. C. Gates, *Inorg. Chem.*, 1994, **33**, 503–509.

91. I. Ogino and B. C. Gates, *J. Phys. Chem. C*, 2010, **114**, 2685–2693.
92. V. A. Bhirud, J. O. Ehresmann, P. W. Kletnieks, J. F. Haw and B. C. Gates, *Langmuir*, 2006, **22**, 490–496.
93. S. A. Vierköetter, C. E. Barnes, G. L. Garner and L. G. Butler, *J. Am. Chem. Soc.*, 1994, **116**, 7445–7446.
94. R. H. Crabtree, *The organometallic chemistry of the transition metals*, 3rd edn, John Wiley, New York, 2001.

CHAPTER 3

Bimetallic Supported Catalysts from Single-source Precursors

SOPHIE HERMANS

Université catholique de Louvain, Institut de la Matière Condensée et des Nanosciences (IMCN), Place Louis Pasteur 1/3, B-1348 Louvain-la-Neuve, Belgium
Email: sophie.hermans@uclouvain.be

3.1 Introduction

Since the famous works of Sinfelt and co-workers,[1] it is generally admitted that bimetallic catalysts are advantageous for many catalytic reactions. Bimetallic catalysts usually present synergistic effects, *i.e.*, catalytic activity, selectivity and/or stability that outperform the performance of each individual metal taken separately.[2] One of the two metals might even be completely inactive alone, in which case it is usually referred to as a promoter element. Many useful bimetallic associations have appeared over the years, which usually involve pairs of transition metals, but might also extend to main-group metals or even lanthanides. In catalytic active phases used in bulk forms (such as oxides), new heterometallic phases are usually identified which are responsible for the improvement (for example mixed oxides). In supported catalysts, an intimate contact between the two metals is also usually recognized as being vital for the bimetallic effect to

RSC Catalysis Series No. 22
Atomically-Precise Methods for Synthesis of Solid Catalysts
Edited by Sophie Hermans and Thierry Visart de Bocarmé
© The Royal Society of Chemistry 2015
Published by the Royal Society of Chemistry, www.rsc.org

take place. Bimetallic catalysts have found their way towards many industrial applications.[3] Industrial bimetallic supported catalysts that are commercialized on a large scale include, for example, Re–Pt, Sn–Pt and Ir–Pt formulations for naphtha reforming and Rh–Pt and related catalysts for car exhausts.

Bimetallic effects in heterogeneous catalysts might arise from several effects which historically were classified into two broad categories: geometric and electronic effects. The first term, also called the 'ensemble effect', has been used mainly for situations where the second metal is not necessarily active in the catalytic reaction but plays a role on the size of the catalytic islands of the active metal. This of course modifies the activity and/or selectivity of the catalyst as a whole. It also encompasses other situations were the second metal plays a dispersion effect (by presenting favorable interactions with the support for example) or blocks some specific adsorption sites which are responsible for lower selectivity. This might also increase resistance to deactivation. The second term involves some (at least partial) electron transfer from one metal to the other, affecting its redox properties and/or chemisorption ability. This is typically observed for alloy or core–shell bimetallic supported particles. However, bimetallic effects might be more subtle than these two classes and result from a combination of factors. The second metal could for example favor the adsorption of one of the reactants or even its activation (for instance H_2 or O_2 dissociative chemisorption), lowering the activation barrier. It is usually identified by synergistic effects, *i.e.*, the bimetallic catalyst being more active than both equivalent monometallic catalysts taken separately or even physically mixed in the reactor.

The conventional preparation procedures for bimetallic supported catalysts involve mainly co-impregnation or co-precipitation, which lead to inhomogeneous composition on the support surface. Each obtained supported particle might be bimetallic but enriched in one or the other metal or even monometallic. It is extremely difficult to form alloy nanoparticles of some metals that favor self-interactions. Moreover, these conventional preparations usually involve high-temperature activation treatments that lead to larger sintered particles.

This chapter will describe the use of heterometallic molecular precursors for the preparation of mixed-metal supported heterogeneous catalysts. Heterometallic precursors not only allow control on bimetallic composition but also higher metal dispersion. It will cover first heterometallic binuclear complexes taken as precursors of bimetallic catalysts. Then in a second section, the present chapter will describe the use of organometallic cluster compounds (*i.e.*, multinuclear complexes) in catalysis. The main concept is to use clusters as precursors of supported nanoparticles, in the hope of retaining size and composition of the cluster core in the final nanoparticles. The goal of this chapter is not to be comprehensive on the subject but to illustrate by some nice examples taken from the past literature the key concepts in this area of research.

3.2 Heteronuclear Species as Heterometallic Catalysts Precursors

3.2.1 Complexes with Heterometallic M–M' Metal–Metal Bond

As early as in the 1980s, the group of B. C. Gates investigated the use of organometallic compounds presenting formally a heterometallic M–M' bond as precursor of bimetallic catalysts.[4] The predicted advantages of such precursors is that the metals are already in zero-oxidation state and the ligands contain mainly C, H and O, which should favor alloy formation.

Given the industrial importance of the Pt/Ir/Sn system for naphtha reforming, a bimetallic catalyst was prepared on NaY zeolite by adsorption of the organometallic precursor $(COD)_2Ir–SnMe_3$.[5] It was compared to the analogous materials prepared by adsorption on NaY of either $Ir(CO)_2(acac)$ or $Ir(CO)_2(acac) + SnMe_3OH$ (in a 1:1 molar ratio). The three samples were activated by heating to 773 K under a stream of H_2. The sample prepared from the heterometallic precursor presented particles 1–10 nm in diameter while co-adsorption of monometallic precursors gave only 1 nm particles evenly distributed within the zeolite support. It seems that the single-source precursor is too bulky to enter the super-cages of the NaY zeolite and reacts quickly on its external surface as a result of the exchangeable COD ligands, which explains larger particles formation upon activation. EDX analysis showed that the larger metal particles (>5 nm) of Ir–Sn/NaY were approximately 1:1 Ir/Sn alloys, while the smaller particles (1 nm) present in Ir + Sn/NaY prepared from monometallic precursors appeared to be Ir-rich. The Ir/Sn/NaY catalysts were examined for the dehydrogenation of propane, propane hydrogenolysis, and the temperature-programmed reaction of CO. The Ir/Sn catalyst derived from a single-source precursor, Ir–Sn/NaY, had a significantly lower hydrogenolysis activity, ascribed to electronic modification of Ir by Sn as well as by dilution of active Ir ensembles, as well as lower dispersion.

Binuclear oxomethoxide Re and Mo complexes $Re_2O_3(OMe)_6$, $ReMoO_2(MeO)_7$ (Figure 3.1), and $Mo_2O_2(OMe)_8$ were encapsulated inside the microporous Y zeolite and mesoporous SiO_2 and Al_2O_3 by incipient wetness.[6] The authors pointed out three major advantages of such alkoxides precursors: (i) high solubility in organic polar solvents, (ii) their oxidation resulting in the

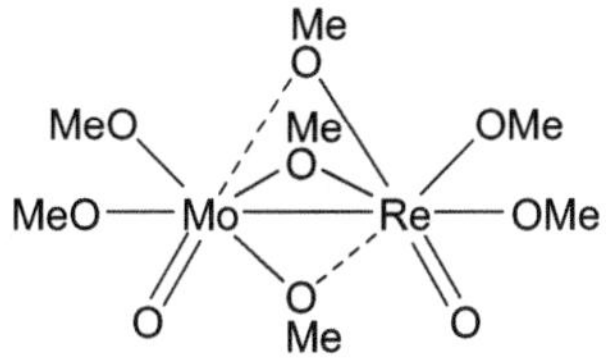

Figure 3.1 Molecular structure of $ReMoO_2(MeO)_7$.

corresponding bimetallic oxides occurs at low temperature, and (iii) the molecular dimensions of these complexes can be adjusted by varying the size of alkoxide ligands. It was found that the binuclear Re and Mo complexes lose their ligands when entering the micropores of NaY zeolite, so that just metal sub-oxide cores remain. The opposite situation was observed in the case of silica-supported samples, where infrared spectroscopy revealed characteristic bands attributable to the organic groups of the oxomethoxide complexes. The supported complexes were calcined to the corresponding oxides at 450 °C for 4 h in air. The heterobimetallic MoReO$_6$/NaY catalyst exhibited the highest acidity.

3.2.2 Ions Pairs

A family of heterometallic species that has emerged in the literature as precursors for bimetallic catalysts is metal salts in which the cation contains one metal while the anion contains the other. In these species, there is no formal heterometallic M–M' bond, but the ionic pair provides a close proximity between the two metallic elements. This ensures more intimate mixing than by successive or co-impregnation procedures.

One such example involved the use of a platinum–tin compound [Pt(NH$_3$)$_4$][SnCl$_6$], which presents a NaCl type lattice in the solid state with Pt cations and Sn anions but without direct Pt–Sn interaction.[7] A Pt-Sn/Al$_2$O$_3$ catalyst prepared from this compound was compared with a more traditional one prepared by co-impregnation of H$_2$PtCl$_6$ and SnCl$_4 \cdot$ 5H$_2$O. TPR demonstrated a deeper tin reduction for the sample prepared with the bimetallic precursor. Moreover, it allowed a higher Pt dispersion and more pronounced electronic interaction between Pt and Sn.

Another example with Pt in the anion involves the [(bipy)$_3$Ni](PtCl$_6$) species.[8] This was used as precursor for PtNi nanoparticles supported on carbon, following an original preparation procedure: ball milling of carbon and precursor followed by thermal treatment. It gave better catalytic activity than a commercial E-TEK Pt/C catalyst for the electro-oxidation of methanol. The same authors also tested the molecular precursor [(bipy)$_3$Ru](PtCl$_6$) for the preparation of direct methanol fuel cell catalyst: after ball milling the precursor with Vulcan Carbon XC 72, the homogenized mixture was heated in 5% H$_2$/95% Ar at 550 °C for 2 h to yield carbon-supported PtRu alloy particles by reductive decomposition.[9] TEM showed that the PtRu nanoparticles were larger than materials prepared from separated chlorine metal salts but the activity was higher and also better than the commercial E-TEK (20 wt%) Pt catalyst. Interestingly, the authors recognize that the catalyst composition may not be easily adjusted when using this procedure.

The group of Ali Reza Rezvani reported recently further application of this idea. First they used [Zn(H$_2$O)$_6$]$_2$[Ni(NCS)$_6$] to prepare a Zn–Ni/SiO$_2$ catalyst that was compared to references prepared by impregnation and co-precipitation methods.[10] It was shown by molar conductivity measurements that the mixed compound forms a 2 : 1 electrolyte in water and dissociates to give the cationic $[Zn(H_2O)_6]^{2+}$ and the anionic $[Ni(NCS)_6]^{4-}$ species. Upon calcination of the silica-supported compound, ZnO and NiO particles form that are of smaller sizes hence more active in water gas shift reaction than the reference material. Another cationic–anionic complex, $[Cu(H_2O)_6]$-$[Ni(dipic)_2]$, was prepared from the reaction of $(NH_4)_2[Ni(dipic)_2]$ with $Cu(NO_3)_2 \cdot 3H_2O$ and also supported on silica and calcined.[11] In this case, a mixed $Ni_{0.75}Cu_{0.25}O$ phase was observed in addition to CuO and NiO. Again, it displayed higher activity than the commercial Sud-Chemie MDC-7 catalyst and the reference catalyst prepared by a conventional impregnation method.

3.2.3 Heterometallic Complexes with Bridging Ligands

Heterometallic complexes with no formal M–M' bond but stabilized through bridging ligands have also been used as precursors of bimetallic catalysts. For example, the group of C. M. Lukehart used a chloro-bridged Pt–Ru complex (Figure 3.2a) for the preparation of Pt_1Ru_1/C nanocomposite.[12] The complex immobilized on Vulcan carbon was activated by rapid microwave heating and gave supported nanoparticles presenting the nominal ratio of

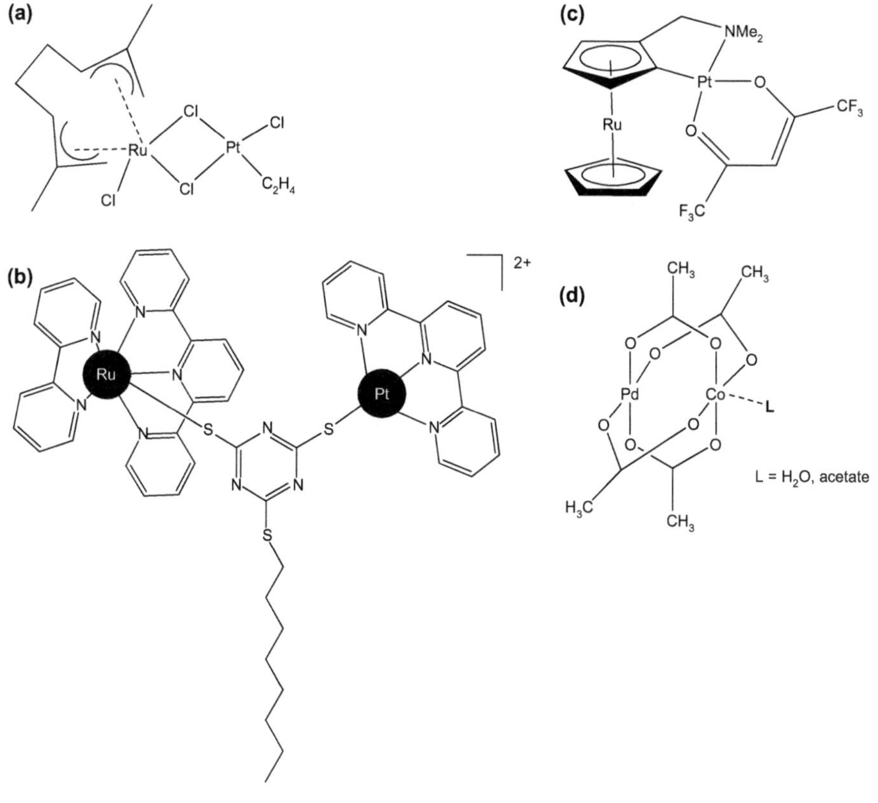

Figure 3.2 Example of hetero-bimetallic complexes exempt of M–M' bond presenting bridging ligands.

the two metals. The material obtained was tested as anode in direct methanol fuel cell (DMFC) and gave superior performance thanks to the ideal composition at the nanoscale. On-particle EDXS confirmed the homogeneity of composition. The same precursor was tested as precursor of DMFC anode catalysts supported on a range of different carbon nanotubes and nanofibers supports.[13] It was found that the best system was supported on narrow tubular herringbone graphitic carbon nanofibers, even if the nanoparticles obtained are enriched in Pt and mixed with Ru-only particles. The same group explored the preparation of Pt–Re/Vulcan carbon nanocomposites using a non-cluster Pt–Re (1:1) bimetallic precursor:[14] characterization by TEM, on-particle EDXS and powder XRD confirmed the alloy formation at the nanoscale but the obtained performance as DMFC anode catalysts did not outperform that of the Pt–Ru/C systems.

Another group reported the use of ligand-bridged heteronuclear complex for the formation of PtRu alloy (Figure 3.2b).[15] It displayed a higher electrocatalytic activity and durability in the methanol oxidation reaction than PtRu alloy prepared by pyrolysis of a mixture of K_2PtCl_4 and $RuCl_3$. This was tentatively explained from XAFS measurements that suggested the presence of weakly coordinated oxygen species in the PtRu alloy prepared from the Pt–Ru complex, which could become an active oxidant, *i.e.*, surface Ru–OH, under electrochemical conditions.

By reacting a ruthenocene derivative with a $Pt(DMSO)_2Cl_2$ complex followed by ligand exchange of the Cl and DMSO for the hfac fluorinated ligand, another Pt–Ru heterometallic complex presenting a bridging ligand was synthesized (Figure 3.2c).[16] As a result of its volatility conferred by the hfac ligand, it could be used as precursor for a preparation method involving chemical vapour deposition (CVD) of the mixed-metal complex as a thin film on a 'carbon paper' support under a slow flow of O_2 carrier gas. It was tested for electrochemical methanol oxidation: the film deposited at 300 °C gave results consistent with the presence of Pt–Ru binary alloy. When the film was deposited at 400 °C, characterization data indicated phase segregation into Pt-rich alloy and RuO_2 and the electrochemical results degraded.

Other binuclear combinations have been explored similarly. The ligand-bridged dinuclear precursor $[PdPtCl_2(\mu\text{-dppm})_2]$ where dppm = $Ph_2PCH_2PPh_2$ was deposited on Zr-doped mesoporous silica by incipient wetness, which gave much smaller particle size than if using $PdCl_2 + H_2PtCl_6$.[17] Both catalysts presented high activity in tetralin hydrogenation, but the material prepared using monometallic salts was more selective (higher *trans*- to *cis*-decalin ratio) and more resistant to sulfur poisoning.

The bimetallic paddlewheel complex $[Pd^{II}Co^{II}(\mu\text{-OOCCH}_3)_4] \cdot H_2O \cdot 2CH_3COOH$ (Figure 3.2d), bridged by acetate ligands, was also used as catalyst precursor.[18] It was immobilized on a planar silicon wafer and oxidized to give a $Pd^{IV}Co^{III}/SiO_2$ model catalyst. The immobilization could be carried out by spin-coating or grafting. In the first case, the complex was hydrogen-bonded intact on the surface, while in the second chemical grafting occurred by ligand exchange of a $\mu\text{-OOCCH}_3$ ligand for hydroxyl groups on the surface.

The Pd^{IV}Co^{III}/SiO$_2$ catalyst activated by oxidation presented superior catalytic activity for the aerobic oxidation of octadecanol under solvent-free conditions than a catalyst prepared from a physical mixture of the separated monometallic compounds tripalladium(II) hexaacetate and cobalt(II) diacetate. The observed synergistic effect between Pd^{IV} and Co^{III} in the catalysts was more effective when using the single source bimetallic precursor.

3.2.4 Other Bimetallic Combinations

The group of Mark A. Keane published a series of papers using a family of compounds of similar nature as single-source precursors for bimetallic catalysts: $\{(DMF)_{10}Ln_2[Pd(CN)_4]_3\}\infty$ (where Ln = La, Ce, Sm, Eu, Gd or Yb)[19,20] and $\{(DMF)_xAEMPd(CN)_4\}\infty$ (where AEM = Sr, Ba; x = 4, 3).[21] The first type consists in one-dimensional arrays containing cyanide-bridged lanthanide(III) and transition metals and are prepared by metathesis reactions of 2 : 3 molar ratios of LnCl$_3$ with K$_2$[Pd(CN)$_4$] in DMF at room temperature over several days.[22] The AEM–Pd complex was prepared using a similar procedure by replacing the lanthanide salt by an alkaline earth metal salt. The crystal structures of the Ba–Pd and Sr–Pd compounds obtained are displayed in Figure 3.3, which shows the extended array and cyanide bridge linking Ba–Pd (two-dimensional sheet structure) and Sr–Pd (one-dimensional ladder array). Ln–Pd/SiO$_2$ catalysts prepared from the organometallic precursors were tested in the gas phase hydrodechlorination (HDC) of chlorobenzene (CB), 1,2-dichlorobenzene (1,2-DCB) and 1,3-dichlorobenzene (1,3-DCB). The lanthanide was found to play a promoting role for Pd but had no activity by itself. The role of Ln might be to (i) favor formation of smaller Pd particles, (ii) inhibit Pd hydride formation, (iii) contribute to a greater quantity of surface hydrogen and (iv) participate directly in activation of the C–Cl bond(s) for hydrogenolytic attack. The extent of this promotion depends on the nature of the catalyst precursor and sequence of metal(s) introduction onto the support: a concurrent (using the mixed-metal precursor) or subsequent (using monometallic precursors) addition of Ln gave the best HDC performance. The alkaline earth metal (AEM = Sr and Ba) elements also displayed a promotional effect in HDC, that was explained by a surface Pd/AEM synergy that enhances Pd dispersion and inhibits hydride formation together with an increase in H$_2$ chemisorptive uptake allied to a more effective C–Cl bond activation for hydrogen scission.

3.2.5 Mixed-metal Precursors of Bulk Phases

Although this is not the subject of the present chapter, it is worth noting that heterometallic precursors have also been used in order to form particular phases in bulk form. One peculiar such example involves the use of (NH$_4$)$_6$Mo$_7$O$_{24}$ and melamine to precipitate a white Mo(VI)–melamine hybrid solid as single source precursor of monophasic β-Mo$_2$C.[23] The high purity of the crystalline Mo carbide obtained by optimized heat treatment under H$_2$

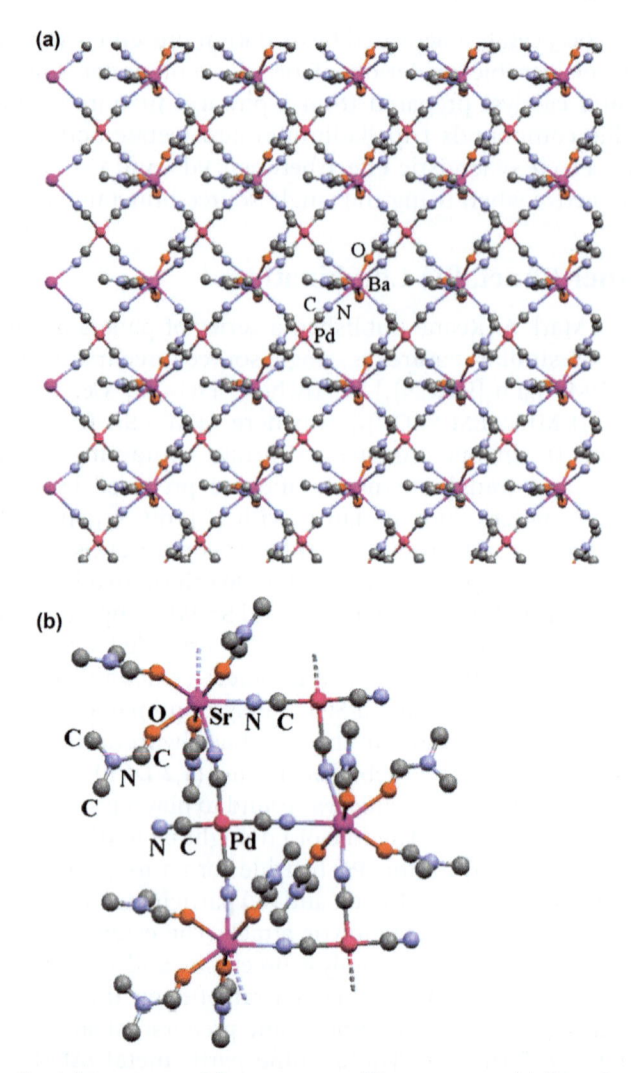

Figure 3.3 Crystal structures of the organometallic precursors: (a) {(DMF)$_3$BaPd(CN)$_4$}∞ and (b) {(DMF)$_4$SrPd(CN)$_4$}∞.
Reproduced with permission from ref. 21.

allowed achieving ultra-high selectivity to tetralin in the hydrogenation of naphthalene. Another interesting example of this type uses a complex where the heteroatom for the final phase, here sulfur, is located within the ligands.[24] The newly synthesized precursor complex [Cd(SOCPh)$_2$Lut$_2$] (where Lut = 3,5-dimethylpyridine (lutidine)) (Figure 3.4) was used for the preparation of CdS nanoparticles (NPs) with different shapes and sizes (rods and spheres). They displayed a much better photocatalytic activity than commercial TiO$_2$ in the degradation of Rose Bengal (RB) dye in presence of white light. This is only one recent example, because many authors have used previously various sulfur-containing Cd(II) complexes as precursors to

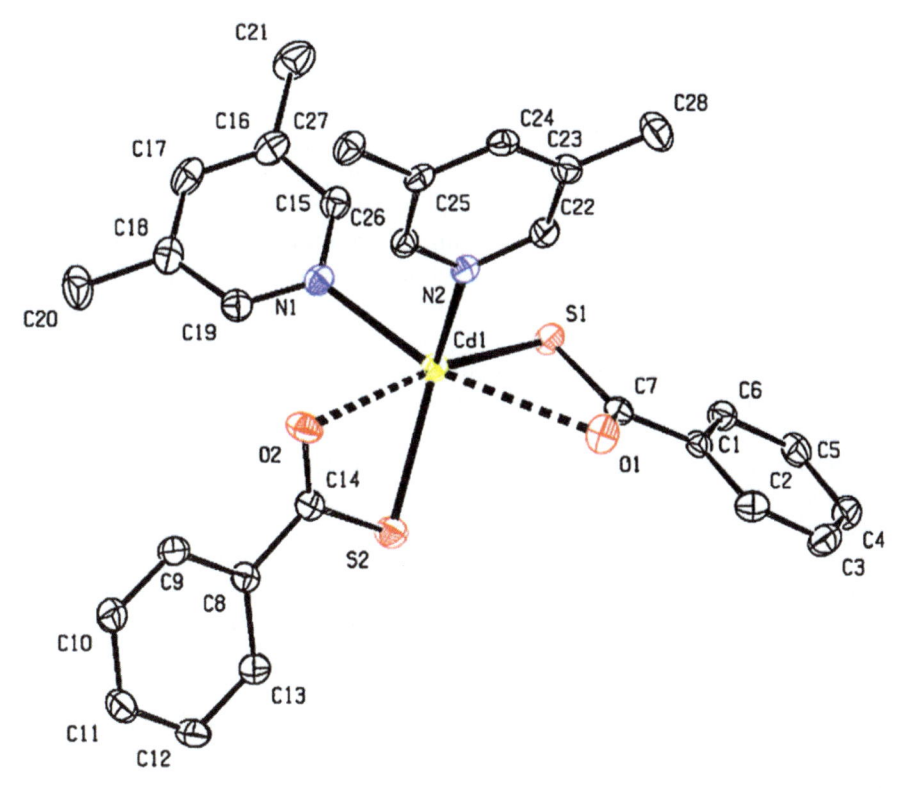

Figure 3.4 Crystal structure of the single-source precursor complex [Cd(SOCPh)$_2$Lut$_2$]. Reproduced with permission from ref. 24.

synthesize variable shapes and sizes of CdS NPs under different conditions. The most widely encountered sulfur-containing ligands in the complexes are the dithiocarbamate, dithiophosphinate, xanthate, thiohydrocarbazide, thiourea, thiocarboxylate, dithiocarbazate and thiosemicarbazide.

The sol–gel method allows obtaining multi-metallic oxides from single source precursors,[25] usually metal alkoxides, in which the control of stoichiometry and high homogeneity of the final products is possible. The most suited single source precursors of spinels are double alkoxides of general formula M[Al(OR)$_4$]$_2$. For example, the hydrolysis of the heterobimetallic alkoxide {Cu[Al(OPr^i)$_4$]$_2$} was studied in detail.[26] The gel after 10 days of hydrolysis yielded the inverse spinel, CuAl$_2$O$_4$, in pure phase, after one time firing at 900 °C. The product so obtained was nanocrystalline and porous in nature without the use of surfactants.

Photocatalysts of the type MTi$_2$O$_5$ (M = Mg, Mn, Fe, Co, Zn, and Sn) were similarly obtained from M[O-Ti(OPr^n)$_3$]$_2$) single source precursors by hydrolysis and supercritical drying.[27] The precursors were synthesized by condensation reaction of Ti(OPr^n)$_4$ with M(OAc)$_2$. The pure phase MgTi$_2$O$_5$ with pseudobrookite structure was obtained only for Mg, while composites MTiO$_3$/TiO$_2$ were obtained for M = Fe, Co and Zn. All presented photocatalytic activity.

3.3 Clusters as Precursors

3.3.1 Definition

The term 'cluster' to define a multinuclear molecular complex presenting M–M bonds was first introduced by Cotton.[28] In this chapter, we will refer to clusters as molecular species presenting at least three metal atoms bonded by a minimum of two M–M bonds and surrounded by a layer of organic ligands. These ligands confer them solubility, hence these compounds are amenable to NMR, infrared and mass spectrometry characterizations. The definite nuclearity and geometry of the metal core as well as three-dimensional arrangement of the ligands can only be precisely determined by crystallography. Clusters can be homometallic (*i.e.*, presenting only metal atoms of one particular nature) or heterometallic. For the purpose of the present chapter, we will mainly focus on the latter.

Clusters are thus intermediate species between coordination compounds ('complexes') and free-standing nanoparticles, which used to be called 'colloids'. Hence confusion arises in the literature because the term 'cluster' is often used in very general terms for a small agglomerate of a few metal atoms, in the gas phase or immobilized on a support. This is not the same definition as presented here, because these agglomerates are devoid of ligands and are usually not monodisperse. They are not molecular species that can be solubilized, purified and crystallized. A definition that has appeared[29] concerning molecular clusters is connected to this, stating that clusters are colloids that can be crystallized and characterized by single-crystal X-ray diffraction. Moreover, spectroscopic characterization by IR, NMR and mass spectrometry showed that the solid-state structure is usually retained in solution. A crystal structure is not obtained for random agglomerates.

However, the border between very large molecular metal clusters and very small ligand-stabilized metal nanoparticles ('colloids') is difficult to draw.[29] Indeed, the number of metal atoms in their cores can be very similar. The largest molecular carbonyl-only metal cluster that has been characterized up to now is $[Ni_{32}Pt_{24}(CO)_{56}]^{6-}$ in which the size is *ca.* 2.1 nm in diameter.[30] These two fields will probably merge, because one of the major current goals of nanochemistry is obtaining atomically monodisperse ultra-small metal nanoparticles (smaller than 5 or even 3 nm).[29,31]

By increasing the number of metal atoms, a transition from an electron-precise molecular species with covalent bonds to a metallic state should occur.[31] From a theoretical point of view, this corresponds to a transition of the electronic state of the cluster from discrete molecular eigenstates (localized electrons) at low nuclearities to highly delocalized electrons subject to quantum size effects at high nuclearities. This means that as the nuclearity increases, the HOMO–LUMO gap should decrease, and eventually a continuum of energy levels should appear as the bulk state is reached. Molecular clusters should therefore undergo insulator-to-metal transitions with increasing nuclearity, passing through a semiconductor regime. However, ligand coordination partially quenches the metallization process.

3.3.2 Synthesis

Many synthetic routes exist to obtain mono- or heterometallic clusters.[32] The most common reactions for the build-up of cluster species involve reduction of metal salts in the presence of potential ligands. As many transition metal clusters are carbonylated species, these reactions are very often simply performed under a carbon monoxide atmosphere. Again, it should be noted that this parallels metal nanoparticle syntheses that also very often involve metal salts and reducing agents. For clusters, and especially mixed-metal compounds, other methods exist that can start from salts, mononuclear complexes or preformed clusters. Redox condensation has been widely used, whereby two fragments containing one different metal each merge together with concomitant exchange of electrons. Other types of condensation reactions, with thermal or high pressure assistance also lead to cluster formation. In most cases, these reactions are not predictable and can yield unexpected products. This is due to the very small energetic differences between similar geometries and/or nuclearities, especially when the nuclearity increases. Isomerization of clusters in solution is a highly facile process that gives rise to fluxionality on the NMR timescale. To construct heterometallic clusters, strategies involving bridging ligands have been more selective, as well as using electrophilic capping groups. However, designing a synthetic procedure for a given cluster structure is not yet feasible. Nevertheless, since this field bloomed in the 1980s, many species, with variable nature of metallic atoms, type of ligands, nuclearities and stoichiometries, have appeared over the years, that can be reliably obtained in a pure form.

3.3.3 Clusters in Catalysis

The first application of clusters in catalysis was called 'the cluster–surface analogy', and consisted in finding parallels between ligands reactivity within clusters and transformations of organic molecules on metallic surfaces.[33] It arises from the observations that the metal packing in clusters often resembles that of the bulk metallic state.[29] In addition, rearrangements of metal cluster cores have been connected to the hopping mechanism of adatom diffusion on metal surfaces. The cluster–surface analogy seems to be valid for structural considerations of ligand bonding. However, the reactivity observed on surfaces and with heterogeneous catalysts is not always the same as on clusters.[29]

Cluster compounds have also been tested quite successfully as organometallic compounds in homogeneous catalysis.[34] The great challenge in this area was to prove that the active species is indeed the intact cluster and not decomposition products, like mononuclear complexes or colloidal nanoparticles formed *in situ*. This can be studied by using various poisoning tests, kinetics and spectroscopic characterizations. Proofs of intact cluster catalysis have been gained from observed synergistic effects with heterometallic clusters or very stable capped triangular clusters. Also, asymmetric induction in a reaction catalyzed by a chiral cluster and experiments using para-hydrogen (p-H_2) NMR methods have given more evidence for cluster species acting as homogeneous catalysts.[29]

Clusters have also been used in catalysis as immobilized complexes. The most famous approach developed in this area is the so-called 'ship-in-a-bottle' technique, where the cluster is built inside the super-cages of zeolites and becomes entrapped due to its large size. Moreover, it has been found that cluster compounds react at the surface of common industrial heterogeneous catalysts supports. Indeed, inorganic oxides, such as silica and alumina, present hydroxyl groups while carbonaceous supports also present oxygen surface groups such as carboxylates, ketones and phenols. The nature of surface species formed after reacting clusters with solid supports has been mainly elucidated spectroscopically, by infra-red and EXAFS (extended X-ray absorption fine structure spectroscopy), but also by solid-state NMR, XPS and Mössbauer.[35] The main surface reactions that have been identified are (i) ion-pairing with surface cationic centers, (ii) weak interaction with surface oxygenated groups, (iii) oxidative addition into an O–H bond giving surface hydrides, (iv) nucleophilic or electrophilic attack on a ligand of the cluster, (v) cleavage of M–C bonds of the cluster by surface groups and formation of Si, Al–O–M bonds, for example, or (vi) abstraction by surface groups of H atoms bonded to the metal. The support enters the coordination sphere of metal atoms in the clusters and can thus be considered as playing a ligand role. The catalytic activity of the grafted molecular cluster will therefore be different to the starting cluster applied in homogeneous phase. The support might play electronic or steric effects and affect the selectivity. Alternatively, common industrial supports have been found to be favoring clusters surface-mediated syntheses. Neutral clusters are obtained on silica, while anionic clusters are obtained on basic supports such as MgO or silica + Na_2CO_3 or K_2CO_3.[36] Usually, a suitable metal salt is impregnated on the support, followed by reductive carbonylation in the solid state. This has allowed the preparation of several immobilized homometallic Pt, Rh and Ir clusters, as well as bimetallic clusters. These supported molecular clusters can be used as heterogenized catalysts.[37] However, by far the most widely-spread application of cluster in catalysis is their use as precursors of well-defined nanoparticles, which will be developed in more details in the next section.

3.3.4 Clusters as Precursors of Supported Bimetallic Nanoparticle Catalysts

3.3.4.1 Introduction

This idea of using clusters as precursors of nanoparticles is the same as developed above for heterometallic complexes: having a preformed precursor that once adsorbed on a chosen support can be treated to yield the active phase in a controlled manner. In this case, the molecular cluster itself is not the catalytically active phase but only a well-defined parent species.

Bimetallic Supported Catalysts from Single-source Precursors

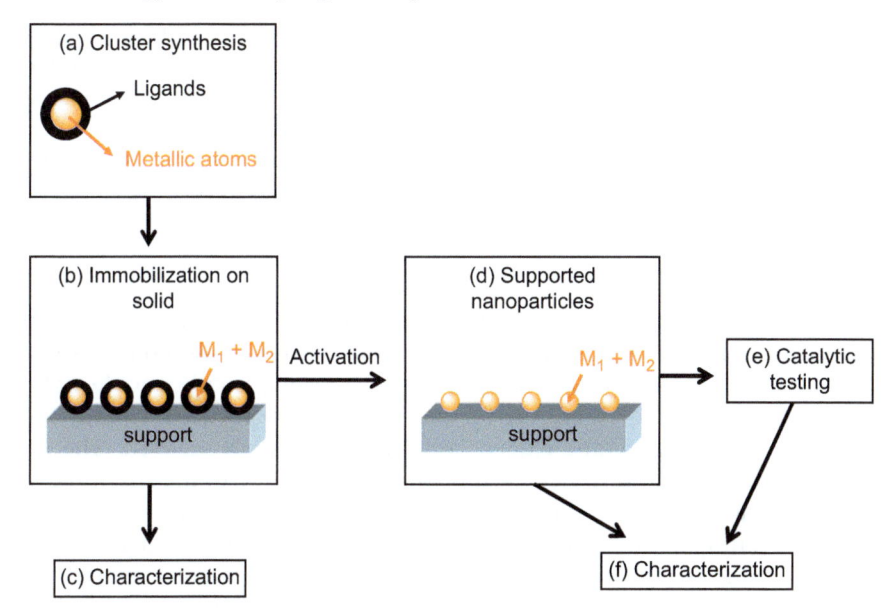

Figure 3.5 General methodology of supported nanoparticles preparation from molecular clusters.

This is now known since a long time, and many book chapters and reviews have appeared on the subject.[31,38–47]

The general experimental methodology involves the following steps (Figure 3.5): (a) synthesis of clusters with appropriate ligands to confer them desired solubility, (b) deposition of clusters on selected support, usually by wet impregnation where the cluster in soluble form is contacted with suspended solid support, (c) characterization of the immobilized species to confirm its intact nature or identify possible surface reaction that might have occurred, (d) transformation of the molecular cluster into a 'naked' activated supported nanoparticle, (e) catalytic evaluation and (f) characterization before and after catalysis. The characterization has relied heavily on infrared spectroscopy and EXAFS, but also XPS, powder XRD, solid-state NMR and TEM microscopy. This latter technique is very useful to determine the size of the nanoparticle obtained. One needs to confirm if indeed the obtained supported active phase is nanostructured with some memory effect of the cluster precursor.

Clusters compared to mononuclear complexes or heterometallic dimers have the advantage of presenting metallic atoms already assembled in a discrete entity (the metal core). Moreover, if using carbonyl clusters, the metal oxidation state is usually low or even zero, hence the activation step needs only to remove the ligand layer without the need for a reducing agent. This ligand removal step can usually be carried out thermally under mild conditions in an inert atmosphere, therefore avoiding metal sintering and rearrangements. This leads to much smaller nanoparticles, ideally of the same size as the starting cluster core. When using mixed-metal clusters, heterometallic supported nanoparticles can be obtained with controlled dimensions and compositions. Bimetallic particles alloyed at the nanoscale can be obtained with metals that do not form alloys in the bulk phase. In the present chapter, we will focus mainly on bimetallic nanoparticles obtained from clusters and applied in heterogeneous catalysis.

3.3.4.2 Key Concepts

One main concept in the area of using mixed-metal clusters as precursors of supported bimetallic nanoparticles has been to use bimetallic combinations where one metal is very active in the catalytic reaction (usually a noble metal) while the other has high affinity for the support. Given that common supports are usually inorganic oxides, oxophilic metals are selected. The key point is that the oxophilic metal will provide stabilization of the ensemble on the surface and metal-support interactions, which will ensure small sizes of the active nanoparticles. The noble metal will be highly dispersed permitting lower loading and giving stable and reusable catalysts. For example, the group of B. C. Gates showed that the clusters $\{Pt[W(CO)_3(C_5H_5)]_2(PhCN)_2\}$ and $\{Pt_2W_2(CO)_6(C_5H_5)_2(PPh_3)_2\}$ could be immobilized intact on MgO or γ-Al_2O_3.[48,49] Infrared spectroscopy and EXAFS mainly were very useful to confirm the intact nature of the clusters compound after immobilization on the support and to follow their fate during post-treatment. The molecular clusters interact weakly by hydrogen bonding of their organic ligands with surface hydroxyl groups of the supports. They could even be recovered by extraction with an organic solvent. After thermal treatment under hydrogen to remove the ligands, it was shown that Pt formed highly dispersed structures, stabilized by interactions with the more oxophilic metal that binds strongly to the support.

Another point, as described above for heteronuclear complexes, is to compare the systems obtained with mixed-metal clusters with more traditional catalysts prepared from separated metal salts or monometallic clusters. This has been exemplified many times in the literature. Coming back to the above-cited example,[48,49] the platinum dispersion and Pt–W interactions were found to be much lower in catalysts prepared from mononuclear analogous complexes. Another very interesting study carried out consecutively by two groups concerns the Pt–Au immiscible system.[50,51] Bimetallic Pt–Au/SiO_2 catalysts were prepared by co-impregnation of hexachloroplatinic (H_2PtCl_6) and tetrachloroauric acids ($HAuCl_4$) or from a $Pt_2Au_4(C\equiv C^tBu)_8$ cluster precursor with similar loadings. These were compared to monometallic equivalents prepared by incipient wetness impregnation of the same metal salts. The cluster-derived sample presented smaller average particle sizes (~ 3 nm) and more narrow size distribution than the two others, as well as higher dispersion. EDXS analysis within the TEM (on a collection of particles) showed the presence of Pt and Au in an approximate ratio of 1:2, while the co-impregnated catalysts presented very large (>10 nm) Au particles and zones with small Pt particles. FTIR studies of adsorbed CO were performed on these three categories of samples, which showed only for the cluster-derived one a completely different behavior with additional peaks and shifts, ascribed to electronic modification of Pt by the presence of Au. The catalyst obtained from $Pt_2Au_4(C=C^tBu)_8$ displayed enhanced selectivity for the production of light hydrocarbons relative to Pt/silica in the hexane conversion reaction, while the co-impregnated sample displayed different selectivity effect.[50] Catalytic results for the selective catalytic reduction of NO by propylene, the oxidation of propylene in the absence of NO, and the $^{16}O/^{18}O$ homo-exchange reaction[51] were in agreement with characterization data: the catalysts prepared by co-impregnation displayed very similar behaviors to their monometallic counterparts, indicating that the two metals remained segregated, while the cluster-derived catalyst behaved completely differently, indicating a bimetallic effect. The catalyst prepared from cluster precursor also presented greatly enhanced resistance to deactivation. It is also worth noting that the selected cluster is devoid of phosphine ligands to avoid contaminations, and because Pt–Au only-carbonyl clusters do not exist, it presents ligands that contain only C and H. A similar study was carried out with the analogous Pt–Cu cluster, $Pt_2Cu_4(C=C^tBu)_8$.[52] Similar effects were observed, *i.e.*, formation of extremely small bimetallic nanoparticles, selectivity shift in hexane conversion reaction for the production of lighter hydrocarbons, but the resistance to deactivation was not obtained in this case. A last example of this philosophy compared catalysts prepared from the organometallic cluster $[Pt_2Mo_2Cp_2(CO)_6(PPh_3)_2]$ with more traditional systems prepared by incipient wetness from the metallic salts precursors $H_2[PtCl_6] \cdot 6H_2O$ and $(NH_4)_6[Mo_7O_{24}] \cdot 4H_2O$ in 1:1 molar ratio.[53] The support was a home-made zirconium-doped mesoporous silica. The catalysts were activated by calcination followed by reduction under H_2. In the catalyst prepared from metal salts, larger particles were observed, giving a medium size of 19.5 nm *vs.* 2.7 nm when starting from cluster. However, this latter catalyst presented a slightly lower activity at all studied temperatures in the hydrogenation of tetralin.

The third key concept in this area is to prove the bimetallic nature of the obtained nanoparticles. Several spectroscopic and imaging techniques are traditionally used in combination to reach this goal. Indeed, even at the cluster immobilization stage, cluster breakdown or agglomeration might occur, leading to local modifications of composition. Obviously, the most deleterious step for homogeneity of composition is activation, where high temperature promotes metal mobility and might lead to segregations or surface enrichment in the metal with lowest surface free energy. Electron microscopy in combination with EDXS spectroscopy has been used extensively to measure particle sizes and confirm homogeneity of composition within a sample. EXAFS allows to highlight the presence of M–M or heterometal M–M' bonds as well as giving structural information on the nanoparticles, because its gives information about the average number of nearest-neighbor atoms and inter-atomic distances. EXAFS data are usually collected at adsorption edges of both metals and fitted with parameters arising from analysis of reference samples. Very often, these fits are confirmed by theoretical calculations. This is much easier to perform at the precursor stage, when the cluster is immobilized but not yet activated. During removal of the ligands, the metal–metal bonds might become shorter, structural changes might occur that are very interesting to follow by EXAFS. In some cases, the replacement of some of the original ligands by surface functions can be proven. If the clusters have been strongly rearranged during activation and/or agglomerated into larger nanoparticles, it might sometimes be difficult to collect high-quality EXAFS data. The structures can be determined more precisely when the final nanoparticles are the most highly dispersed and the sample is uniform. For example, a Pt–Ru/MgO catalyst was prepared by adsorption of $Pt_3Ru_6(CO)_{21}(\mu_3\text{-H})(\mu\text{-H})_3$ followed by ligand removal by treatment at 300 °C.[54] Infrared spectroscopy indicated that it was not intact after adsorption on MgO. EXAFS characterization after activation demonstrated the presence of heterometallic Pt–Ru bonds, which was not the case when preparing the equivalent catalyst from a mixture of $Pt(acac)_2$ and $Ru(acac)_3$. Similarly, the clusters $(NEt_4)_2[Ir_4(CO)_{10}(SnCl_3)_2]$ and $NEt_4[Ir_6(CO)_{15}(SnCl_3)]$ have been used for the preparation of silica-supported Ir–Sn bimetallic catalysts.[55] Characterization relied heavily on EXAFS and XANES, and allowed a structural model for the Ir–Sn species in the activated catalyst to be proposed.

3.3.4.3 Further Examples of Mixed-metal Clusters as Precursors of Nanoparticles

The group of B. C. Gates has been very active in the field.[47] A large number of papers from their group and others such as Ichikawa report 'ship-in-a-bottle' synthesis to form bimetallic clusters inside the super-cages of zeolites (Faujasite, ZSM-5, ALPO-5 *etc.*). This occurs by ion exchange or metal salt/complex impregnation, followed by reductive carbonylation. Once formed, the entrapped clusters can be thermally treated by calcination and reduction to be subsequently transformed into bimetallic very small nanoparticles supported within zeolites. For example, $Rh_{6-x}Ir_x(CO)_{16}$ clusters (with x ranging from 1 to 4) were prepared by reductive carbonylation at 150–227 °C of Rh^{3+} and Ir^{4+} introduced into zeolite NaY by ion exchange.[56] Highly dispersed Rh–Ir nanoparticles were formed by decarbonylation of these clusters in O_2 followed by reduction with H_2.

A seminal study has been carried out by Nashner *et al.* on the preparation of Ru–Pt carbon-supported nanoparticle catalyst from the $PtRu_5C(CO)_{16}$ cluster.[57,58] The molecular cluster precursor was deposited on carbon black (Vulcan XC-72) by incipient wetness from a THF solution (in quantity corresponding to 1–2 wt% metal loading). The sample was dried in air and activated by heating at a rate of 15 °C min^{-1} in flowing H_2 (40 mL min^{-1}) to a temperature of 400 °C and keeping it at that temperature for 1 h (Figure 3.6). The structure of the obtained nanoparticles on an atomic scale was deduced

Bimetallic Supported Catalysts from Single-source Precursors 71

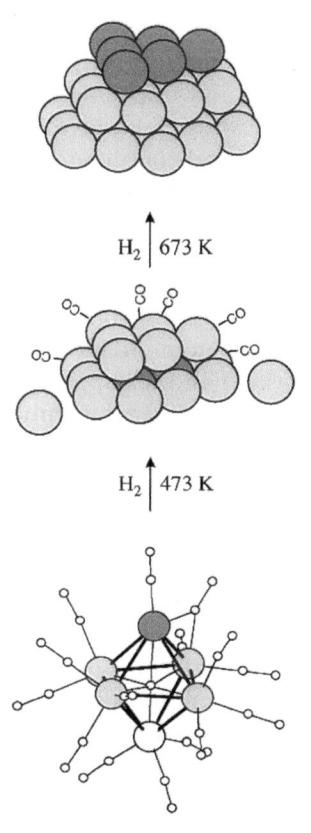

Figure 3.6 Qualitative model of the nucleation and growth processes of nanoparticles formed from PtRu$_5$C(CO)$_{16}$ cluster deposited on carbon: Ru atoms are light gray, whereas Pt is dark gray.
Reprinted with permission from M. S. Nashner et al., J. Am. Chem. Soc., 1998, **120**, 8093. Copyright 1998 American Chemical Society.

by using several independent methods: EXAFS, scanning transmission electron microscopy (STEM), electron microdiffraction and temperature-programmed desorption (TPD). These techniques were implemented *in situ* to follow the transformation of the molecular cluster into nanoparticles during activation in H$_2$. It was shown that an intermediary state (Figure 3.6, middle) exists, corresponding to nucleation of the nanoparticle. TPD indicated that CO ligands are lost between 100 and 400 °C, while XANES reveals apparition of a metallic state with some remaining M$_x$(CO)$_y$ surface complexes at low temperature (<200 °C). At this stage, Pt–Pt bonding is observed by EXAFS, and a model is proposed where Pt atoms have formed a metallic core with incomplete coalescence of Ru atoms on the external surface. STEM allowed visualization of the very small (1.6 nm average diameter) final particles obtained after further heating and their narrow size distribution. A typical particle seems to arise from a condensation of *ca.* six PtRu5 units. Their bimetallic composition was confirmed by energy dispersive X-ray

(EDX) analysis. The measured composition is consistent with Pt–Ru stoichiometry in starting cluster. Microdiffraction indicated intriguingly an fcc structure, while bulk alloys of this composition adopt an hcp structure. Bond distances in the final particles for Pt–Pt, Pt–Ru and Ru–Ru were estimated from EXAFS data, as well as metal coordination numbers. A compositional asymmetry was quantified by comparing the statistically predicted ratio of Ru–Pt (or Pt–Ru) bonds to Ru–Ru (or Pt–Pt) bonds in the first (and higher) shell to the measured EXAFS data. It was found that Pt–Pt and Ru–Ru bonding is preferred over heterometallic bonding in the nanoparticles. The proposed microstructure of the nanoparticle is an fcc hemisphere (Figure 3.6, top) where the Pt atoms (dark gray) segregate to the surface of the nanoparticle and away from the support, maximizing the number of Pt–Pt bonds. Even if segregation occurred within each particle (but not into separated monometallic ones), it should be noted that heterometallic bonding in all shells was strongly evidenced by EXAFS data, providing direct evidence for the formation of bimetallic nanoparticles.

Following this founding study, other works have concerned carbon-supported nanoparticles prepared from clusters, given their importance as electrocatalysts in fuel cells.[59] The oxygen reduction reaction (ORR) in particular is important in fuel cell technology and was studied by many authors with Ru-based systems. Several carbon-supported bimetallic systems were prepared by immobilizing a monometallic carbonyl cluster on a carbon support, followed by reaction with a compound of a second metal before reduction into nano-alloy (or preceding the immobilization by the reaction). Recently, the complexes $[Pd_3(\mu_3\text{-PhPz})_6]$, $[NH_4]_2[CoPd_2(Me_2Ipz)_4Cl_4]$ and $[NH_4]_2[Co_2Pd(Me_2Ipz)_4Cl_4]$ bearing pyrazolate ligands were immobilized on Vulcan carbon and reduced either chemically (with $NaBH_4$) or thermally (with H_2).[60] The reduction process was followed by high resolution XPS: the Pd peak was shown to shift to lower binding energy upon reduction. The thermal treatment gave more reduced metals but larger nanoparticles on the surface than chemical reduction, and with compositional discrepancies with respect to nominal Co–Pd ratios. XRD characterization proved alloy formation in the two bimetallic cases. The obtained Pd/C and PdCo/C catalysts were selective in ORR reaction and presented better methanol tolerance than commercial catalysts. In another study devoted to fuel cells, the clusters $py_2Pt[MoCp(CO)_3]_2$, $(Me)(cod)PtMoCp(CO)_3$ and $\{Pt_3(dppm)_3CO[Mo(Cp)]\}[BPh_4]$ were used for the preparation of $PtMo_2$, PtMo and Pt_3Mo catalysts, respectively, supported on Vulcan carbon.[61] The activation step was carried out thermally by a succession of treatments in air, nitrogen and getter gas (90/10 N_2/H_2 mixture), to give final metal contents of 58 wt%, 43 wt% and 29 wt%. Detailed *in situ* XRD analyses permitted to follow the activation step and to find the best experimental conditions to avoid for example formation of phosphides phases. TEM micrographs revealed small (1–10 nm) nanoparticles whose composition was confirmed by EDXS to be that of the starting cluster precursor. Notably, the second catalyst was air-stable and available in gram-scale quantities. It was tested as hydrogen oxidation electrocatalysts in a working H_2/O_2 proton exchange membrane fuel cell (PEMFC) and was shown to exhibit improved CO tolerance compared to a commercial Pt/C anode catalyst. Similarly, two mixed Pt–Sn clusters were used as single-source precursors for Pt–Sn/Vulcan carbon nanocomposites.[62] The nanoparticles of 5–8 nm diameter were composed of bimetallic Pt–Sn alloy with a loading of 15 wt% but with the concomitant presence of PtP_2 or Pt metal. Their performance as electro-oxidation catalysts in direct methanol fuel cell was comparable to a catalyst prepared by conventional methods.

Finally, the group of B. F. G. Johnson, in collaboration with J. M. Thomas, has also been a major contributor in cluster-derived supported catalysts since the late 1990s. Mixed-metal clusters were mainly immobilized within the mesopores of the siliceous MCM-41 and converted thermally to bimetallic nanoparticles, which presented interesting catalytic activities in a range of reactions. The first examples concerned silver–ruthenium[63,64] and copper–ruthenium[65] clusters. Activation was followed *in situ* by infrared spectroscopy and EXAFS. Infrared spectroscopy is indeed very useful to characterize carbonyl clusters given the very strong bands associated with CO stretching: the activation can be associated with disappearance of these bands. The structural parameters retrieved from the EXAFS data indicated retention of discrete Ag_3Ru_{10} particles in the activated catalyst prepared from the $[Ag_3Ru_{10}C_2(CO)_{28}Cl]^{2-}$ cluster. Also from the EXAFS data, a model of the activated Cu–Ru 1.5 nm particles was proposed (Figure 3.7) where the Cu atoms are in close contact with the surface through Cu–O bonds (there was no direct indication in the EXAFS data for any Ru–O shells). Annular dark-field STEM electron microscopy allowed visualization of the particles very clearly. The idea of using the more oxophilic metal as an anchoring point for the nanoparticles was thus successfully used here, with no indication of sintering or fragmentation. The obtained Cu–Ru/MCM-41 catalyst was found

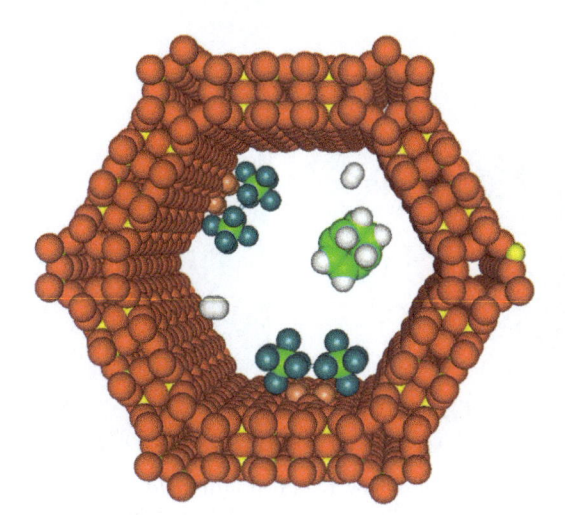

Figure 3.7 Model of the activated Cu–Ru nanoparticles deposited within the mesopores of MCM-41 from the cluster precursor $[Ru_{12}C_2(CO)_{32}Cu_4Cl_2]^{2-}$.

active in the hydrogenation of 1-hexene, diphenylacetylene, phenylacetylene, stilbene, *cis*-cyclooctene and D-limonene. Ruthenium anionic carbonylates clusters were found to align themselves within the mesopores of MCM-41.[66] We immobilized other mixed-metal clusters of interest into MCM-41, to obtain very active and selective catalysts. For example, the [Pd$_6$Ru$_6$(CO)$_{24}$]$^{2-}$ cluster was introduced in MCM-41 and activated into bimetallic Pd–Ru nanoparticles.[67] High-resolution electron microscopy together with EDXS mapping (Figure 3.8) confirmed the bimetallic nature of the obtained nanoparticles.[68] These were very active for polyaromatics hydrogenation, for example naphthalene into decalin. Interestingly, *cis*-decalin was obtained in 86% selectivity. Tri-dimensional TEM (electron tomography) was used to confirm the location of the nanoparticles within the mesopores and not on the external surface of each MCM-41 crystallite. For partial hydrogenation, two different Ru–Sn clusters were synthesized in high yields from [Ru$_6$C(CO)$_{16}$]$^{2-}$ and SnCl$_4$.[69] Their crystal structures were determined in order to confirm the elemental composition of each cluster precursor. After incorporation into MCM-41 (the polar nature of such clusters being a great asset for impregnating a siliceous support), they were activated by heating under vacuum. The state of the supported species before and after activation was studied by *in situ* IR and EXAFS. It was found that the clusters were deposited intact before activation, but that the nanoparticles were anchored

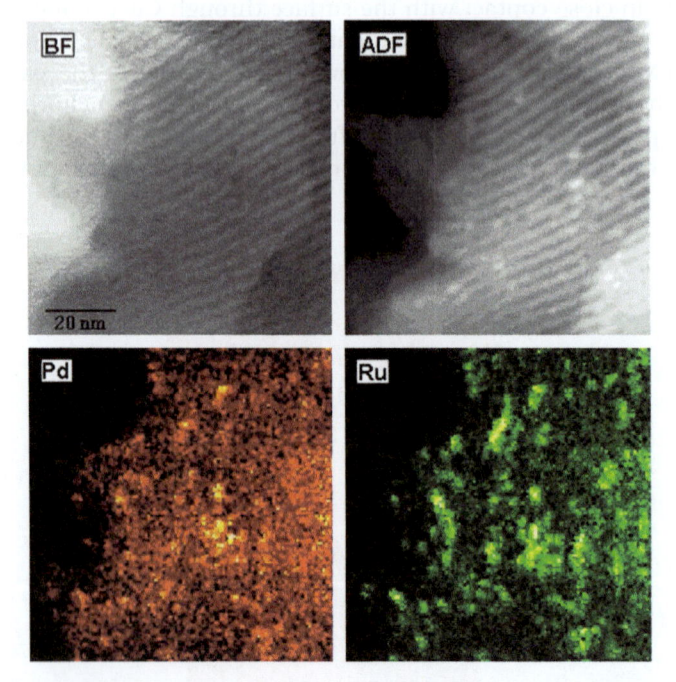

Figure 3.8 Elemental mapping by EDXS: bright field (BF) and annular dark field (ADF) images of the Pd$_6$Ru$_6$/MCM-41 catalyst and corresponding X-ray elemental maps of Pd and Ru.
Reproduced with permission from ref. 68.

Bimetallic Supported Catalysts from Single-source Precursors 75

via the Sn atoms to the surface after activation. The elemental ratio determined by EDXS in the final catalyst was found to be Ru/Sn = 6 : 1 in the case of the [Ru$_6$C(CO)$_{16}$SnCl$_3$]$^-$ cluster, in agreement with the metals ratio in the starting cluster core. The obtained Ru$_6$Sn/MCM-41 catalyst was used successfully for the selective hydrogenation of poly-enes in solvent-free conditions, for example for the transformation of 1,5-cyclooctadiene into cyclooctene, or 2,5-norbornadiene into norbornene.[70] Interestingly, the Pd$_6$Ru$_6$/MCM-41 catalyst gave norbornane selectively in the same reaction conditions. A similar study was conducted with mixed Ru–Pt systems.[71] These were found to be very active for the hydrogenation of cyclohexene to cyclohexane, naphthalene to decalin (with an increased resistance to sulfur poisoning), benzoic acid to cyclohexane carboxylic acid, and dimethyl terephthalate (DMT) to 1,4-cyclohexane-dimethanol (CHDM) in one step (Figure 3.9). Moreover, the catalysts could be recovered and re-used several

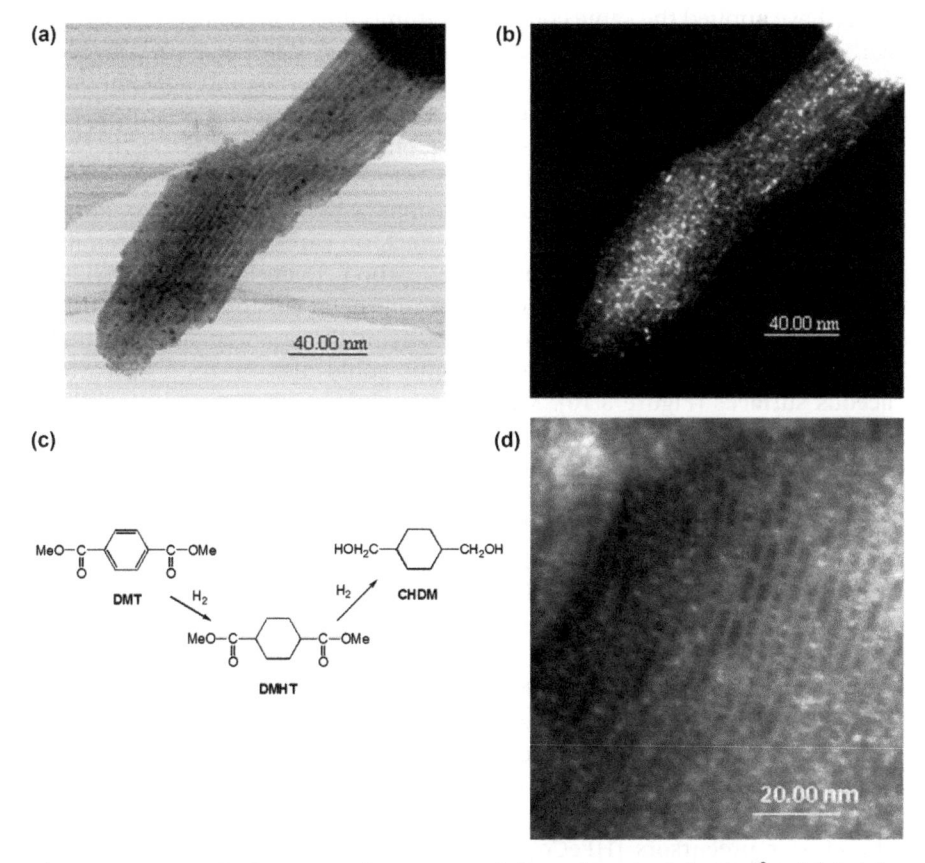

Figure 3.9 Ru–Pt/MCM-41 catalyst prepared from [Ru$_{10}$Pt$_2$C$_2$(CO)$_{28}$]$^{2-}$: (a) bright field (BF), (b) high angle annular dark field (HAADF) electron microscopy images, (c) hydrogenation of dimethyl terephthalate (DMT) to 1,4-cyclohexanedimethanol (CHDM) as catalyzed transformation, and (d) BF electron micrograph after catalysis.
Reproduced with permission from ref. 71.

times without a decrease in selectivity or activity. Alumina-supported catalysts were also prepared from mixed-metal Ru–Pt clusters such as [Ru$_5$PtC(CO)$_{15}$]$^{2-}$, and compared to analogous materials prepared from inorganic salts.[72] Using cluster precursors allowed excellent control of bimetallic composition and smaller particles in the final catalyst, which was more active and selective than commercial or salt-derived counterparts in the production of hydrogen *via* the steam reforming of ethanol. Other connected successful stories concerned the use of Ru–Sn mixed-metal clusters of varying stoichiometries immobilized on Davison 923 mesoporous silica,[73] a trimetallic Ru–Pt–Sn cluster supported on silica,[74] and bimetallic Ir–Bi nanoparticles obtained from clusters deposited onto MCM-41.[75] Gold–iron carbonyl clusters were also used as precursors of bimetallic nanoparticles supported on ceria[76] or the siliceous mesoporous SBA-15[77], and shown to be superior for the total oxidation of methanol and toluene, chosen as representative VOCs.

We have applied the same concepts to immobilize mixed-metal clusters on carbon supports. However, it was found that the less polar carbonaceous surface was not ideal to interact with clusters: sintering and agglomeration could not be avoided on activated carbon used as such.[78] On pristine carbon nanotubes also, agglomerates were found on the external walls together with individual clusters aligned in the tip region.[79] This was ascribed to a stabilizing effect of oxygenated groups known to be present in higher quantities at the tips together with a 'pincer' effect of carbon layers extremities. Therefore, we started functionalizing carbon materials with chelating phosphine groups, in order to anchor mixed-metal clusters on their surface by ligand exchange.[80] In addition, by modifying the last step of the synthesis methodology, charged ammonium groups could also be fixed on carbonaceous surfaces (Figure 3.10). Mixed-metal clusters were thus immobilized on functionalized carbon surfaces before being transformed into supported bimetallic nanoparticles by thermal activation. This led successfully to highly dispersed Ru–Pt nanoparticles derived from mixed-metal clusters supported on activated carbon.[81] However, in the case of Ru–Au clusters, the ligand exchange reaction led to ejection of Au atoms from the bimetallic precursors to lead to very small Ru-rich nanoparticles together with much bigger Au-only sintered particles (Figure 3.11). The same phosphine-functionalized activated carbon support was used to anchor Pd clusters that were activated thermally into Pd nanoparticles active in nitrobenzene hydrogenation.[82] The functionalization methodology could be successfully transposed to carbon nanofibers,[83] ordered mesoporous carbons[84] and carbon nanotubes.[85] In this last case, Fe–Co nanoparticles were obtained on multi-walled carbon nanotubes and carbon nanofibers (Figure 3.12) using the cluster precursors [HFeCo$_3$(CO)$_{12}$] and (NEt$_4$)[FeCo$_3$(CO)$_{12}$], that were shown by magnetic characterization to be blocked super-paramagnetic Fe–Co nanoparticles together with paramagnetic ions. TEM indicated that the nanoparticles were better dispersed and of smaller sizes on functionalized than on pristine carbon supports. Very recently, we have developed a new

Bimetallic Supported Catalysts from Single-source Precursors 77

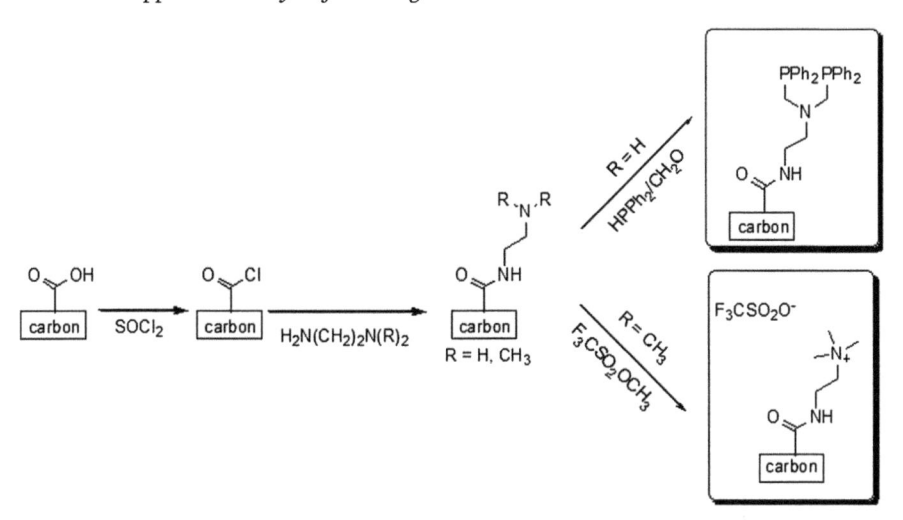

Figure 3.10 Functionalization scheme of carbon materials: the obtained modified surfaces present either chelating phosphines or charged ammonium groups.
Reproduced from ref. 85.

carbon nanotubes functionalization strategy using radicals that could also be used to anchor mixed-metal clusters as precursors of carbon-supported nanoparticles.[86]

As a final illustration, it is worth noting that other authors have also used clusters as precursors of magnetic nanoparticles, supported or unsupported. In the unsupported case, mixed-metal clusters such as $[FeCo_3(CO)_{12}]^-$, $[Fe_3Pt_3(CO)_{15}]^{2-}$, $[FeNi_5(CO)_{13}]^{2-}$ or $[Fe_4Pt(CO)_{16}]^{2-}$ were used for example as precursors of FeCo3, FePt, FeNi4 and Fe4Pt 5–10 nm sized nanoparticles.[87] In a typical synthesis, oleic acid and trioctylphosphine oxide are dissolved in anhydrous 1,2-dichlorobenzene (DCB) and heated to 186 °C, before rapid injection of the cluster dissolved in DCB with vigorous stirring, which are the typical reaction conditions for the preparation of ligand-stabilized nanoparticles. Characterization with a SQUID magnetometer revealed that the obtained nanoparticles are superparamagnetic at room temperature. In the supported case, mesoporous supports such as xerogel or MCM-41 were impregnated with the $[NEt_4][Co_3Ru(CO)_{12}]$ cluster, followed by heating under inert atmosphere to give highly dispersed magnetic nanoparticles.[88] Notably, the experimental conditions were milder than if using metal salts as precursors.

3.3.4.4 Ligand-stabilized Nanoparticles as Precursors of 'Naked' Supported Nanoparticles

Before ending this chapter, it is worth mentioning that several authors came up with the idea of preparing supported catalysts from preformed ligand-stabilized nanoparticles. The experimental methodology involves the

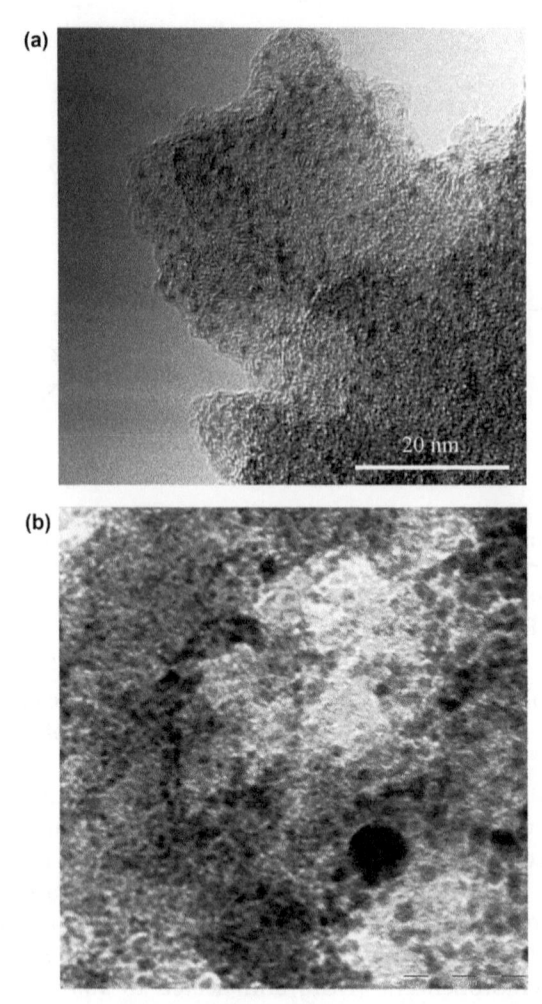

Figure 3.11 TEM images after thermal activation of clusters (a) [Ru$_5$PtC(CO)$_{14}$(COD)] and (b) [Ru$_6$Au$_2$C(CO)$_{16}$(PPh$_3$)$_2$] immobilized on activated carbon *via* chelating phosphine groups.
Reproduced with permission from ref. 81.

same steps than with clusters (see the scheme in Figure 3.5), *i.e.*, (i) synthesizing ligand-stabilized metal-containing species in solution (preferably monodisperse in size), (ii) immobilizing them on a support and (iii) activating them by ligand removal in order to obtain heterogeneous catalysts. This was coined the 'precursor' concept by H. Bonnemann and colleagues.[89] These authors quote that 'an obvious advantage of the precursor concept over the conventional salt-impregnation method is that both the size and the composition of the colloidal metal precursors may be tailored independently of the support'. This is precisely the line of reasoning developed for cluster compounds. As mentioned above, when small colloid metal nanoparticles

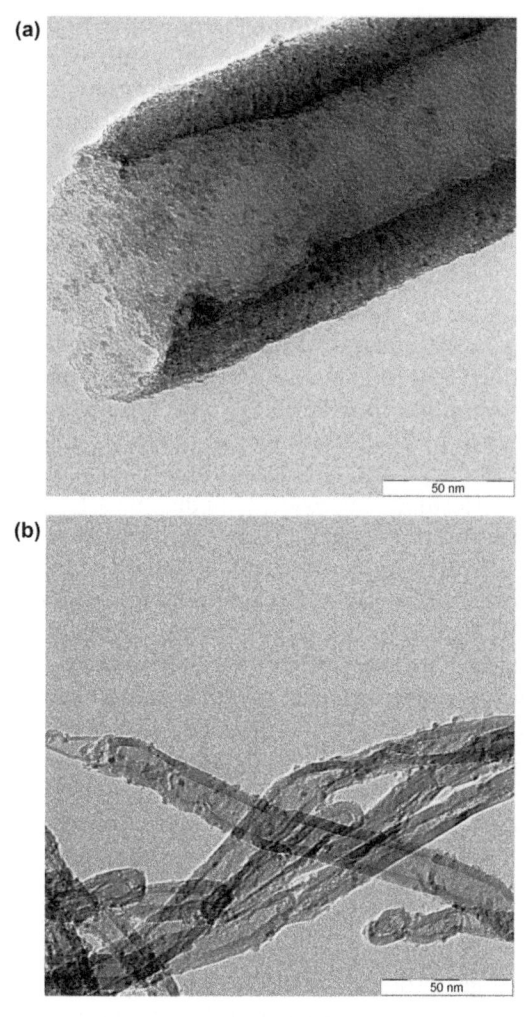

Figure 3.12 TEM images of cluster-derived Fe–Co magnetic nanoparticles supported on (a) carbon nanofibres and (b) multi-walled carbon nanotubes. Reproduced from ref. 85.

are concerned, the boundary between them and giant molecular clusters begin to fade. Hence, a clear distinction between this type of study and what has been described in the preceding sections is not relevant. H. Bonnemann and co-workers used their 'precursor' concept to prepare homogeneous alloys, segregated alloys, layered bimetallics and 'decorated' particles. The solution-stabilized nanoparticles can be characterized in detail by EXAFS, electron microscopy and X-ray diffraction before being immobilized on a support, hence allowing better control on the final nanostructure. In addition, the stabilizing agent might be tailored to provide a lipophilic or hydrophilic protective shell, the nanoparticles can be coated by intermediate layers, such as an oxide, or modified by dopants. Bimetallic colloids are prepared by co-reduction of different metal salts. The immobilization is carried out by dipping the supports into organic or aqueous media containing the suspended precursor colloidal particles at ambient temperature. The surfactant shell is removed by annealing in a tube furnace in a mixture of 10% O_2 in N_2 at 300 °C followed by a reductive step at 300 °C in pure hydrogen. One key advantage of this method is the ease of manipulation of the alloy composition by varying the salts ratio in the co-reduction step.

H. Bonnemann et al. used the 'precursor' concept for manufacturing multimetallic fuel cell catalysts with very small particle sizes (<2 nm) and high metal loading (>20 wt% metal). Bimetallic nanoparticles are expected to provide improvements in catalytic activity, resistance to poisoning and long-term stability in fuel cells. Methods of preparation for monometallic or heterometallic colloidal nanoparticles soluble in organic (termed 'organosols') or aqueous (termed 'hydrosols') media have been reported.[90] Most transition metals are amenable to their synthetic methods. The reducing agent is usually a hydrotriorganoborate. The origin of bimetallic effects was sought at the precursor stage by characterizing in detail bimetallic colloidal precursors by EXAFS: it was found that synergistic effects could be ascribed to a geometric lattice structure effect rather than electronic band structure effects.[91] The catalytic performance of 'sols' immobilized on a support before removing the stabilizing ligand layer has also been investigated. It was found that these systems could already be active for a variety of reactions (depending on the nature of metal(s)) without the activation step. In addition, the resistance to poisoning and deactivation was increased by the protective layer and enantioselective transformations could be carried out. In this latter case, colloidal platinum (particle size 2 nm) was stabilized by a chiral ammonium group derived from dihydrocinchonidine and tested in the enantioselective hydrogenation of ethyl pyruvate: adsorbed on SiO_2 or Al_2O_3 supports, it gave the (R)-enantiomer in 81% and 85% ee respectively. Finally, highly active PEM-FC and DMFC fuel cell anode catalysts with a narrow particle size distribution were obtained for instance by preparing a colloid from $Pt(acac)_2$ and $Ru(acac)_3$ in the presence of $Al(CH_3)_3$ as both reducing agent and stabilizer in toluene.[92] The organic stabilizer was then modified by reacting with polyethylenglycol-dodecylether to obtain more hydrophilic characteristics. The organoaluminium colloids were supported on high-surface-area carbon by dropping it from a dispersion in toluene to a suspension of Vulcan carbon in the same solvent. The dried powder was then heated in an air/Ar mixture followed by pure hydrogen to remove the surfactant stabilizer. XPS characterization after activation indicated typical parameters for zero-valent noble metals, but Al was still present in partially oxidized form. HRTEM demonstrated that adsorption on the Vulcan support leaves the particle sizes and size distributions practically unaffected (10% changes), while the activation procedure causes some temperature-induced sintering of the particles (∼20%) to give final metallic PtRu particles of 1.5–2 nm average size. The obtained $PtRu(AlR_3)$/Vulcan catalysts compares very advantageously for the oxidation of methanol with other state-of-the-art PtRu electrocatalysts prepared *via* traditional routes.

Many other groups have of course exploited this idea. A nice example concerns the preparation of TiO_2-supported PdAu bimetallic catalysts from dendrimer-encapsulated nanoparticles.[93] The nanoparticles were synthesized by co-complexation of the two corresponding metal salts within fourth generation PAMAM dendrimers followed by chemical reduction with $NaBH_4$. The PdAu/TiO_2 composite was prepared by the sol–gel method, from $Ti(O^iPr)_4$, to contain 1 wt% of the metals after thermal treatment in flowing O_2 and then H_2. Thermogravimetric and elemental analyses indicated that the dendrimer was removed during calcination. The dendrimer-encapsulated PdAu alloy nanoparticles presented an average particle size of 1.8 nm, which increased to 3.2 nm during thermal treatment (Figure 3.13).

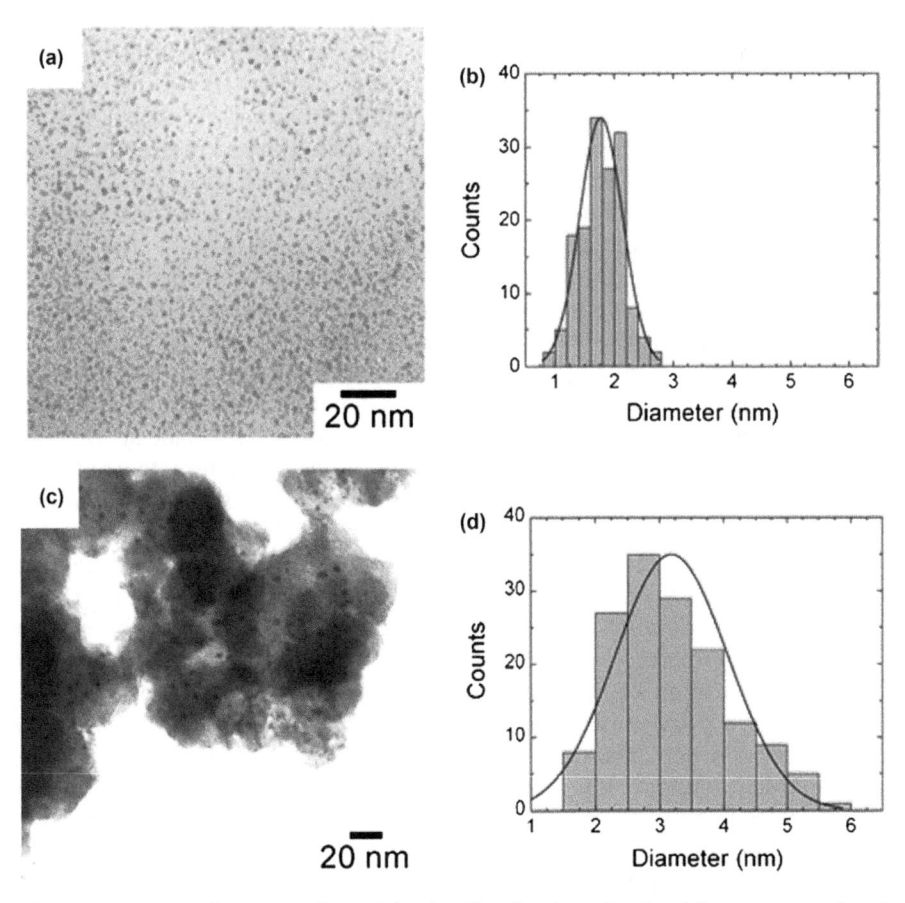

Figure 3.13 TEM images and particle-size distributions for dendrimer-encapsulated nanoparticles prepared by co-complexation of metal salts followed by reduction: (a) and (b) as synthesized, (c) and (d) after incorporation into the titania matrix and activation.
Reprinted with permission from R. W. J. Scott *et al.*, *J. Am. Chem. Soc.*, 2005, **127**, 1380. Copyright 2005 American Chemical Society.

EDX spectroscopy indicated that the composition of the starting nanoparticles (50% Pd–50% Au) was retained in the supported catalyst, with high compositional uniformity. This ratio could easily be modified by changing the amounts of salts complexed originally within the dendrimer. The catalytic activity for CO oxidation was higher for the bimetallic catalyst prepared from dendrimer-encapsulated nanoparticles than for the corresponding Pd-only and Au-only monometallic materials.

3.4 Conclusions and Future Challenges

The general concept developed in this chapter is that of using heteronuclear coordination compounds (complexes) or clusters as precursors of supported bimetallic nanoparticles. The precursor compound can be synthesized, purified, and characterized in detail before being adsorbed on a given support. In some cases, the specific interactions between precursor and solid surfaces have been studied extensively. The supported compounds are then activated into heterogeneous catalysts, usually by thermal treatment, to remove the ligand shell and reduce the metallic atoms if necessary. The advantage of using organometallic (usually carbonyl) clusters over inorganic complexes are that a preformed metal core exists that can lead to control over size and composition of final nanoparticles if sintering can be avoided during the activation step. In addition, the metals are usually already in a zero-oxidation state. Many success stories have appeared in the literature over the years, with sometimes formal proof of bimetallic nature of the obtained nanoparticles, which are usually small and well-dispersed on the support, together with improved catalytic performance when compared to more conventional catalysts prepared from separated metal salts.

The bottleneck in this research area will always be large scale synthesis of molecular well-defined transition metal complexes or clusters. Most of them are air- and moisture-sensitive and can only be produced in small amounts. Moreover, serendipity played an important role in obtaining new heterometallic combinations and stoichiometries. Control and prediction in cluster synthesis is not yet the rule. Bimetallic nanoparticles will play a key role in future developments due to their relative easier synthesis. The drawback of protected but free-standing nanoparticles being their recovery, the immobilization of nanoparticles on a solid support will certainly continue to be a prized line of research, as initiated by Bonnemann with his 'precursor concept'. The remaining questions in this field are the nature of nanoparticles/support interactions, as well as the need to remove the capping agents to boost the catalytic activity, with the risk of losing control over size and composition at the nanoscale.

In the future, using ligand-stabilized colloidal nanoparticles as precursors of tailored supported catalysts will certainly spread. Indeed, their syntheses; although very sensitive to the experimental conditions and sometimes suffering from a lack of reproducibility, are much easier than molecular clusters, especially when upscaling is necessary. In addition, tuning metal composition and even their shapes is now possible. Obtaining supported catalysts with a control on composition, size and shape of bimetallic nanoparticles is therefore envisioned using this method. However, the frontier between molecular clusters and colloidal nanoparticles is shading, as characterization methods pertaining to the molecular world are now being used to characterize more precisely nanoparticles. Therefore, we envision a merging of these two fields of study, with methodologies developed for clusters being transferred to ligand-stabilized colloidal nanoparticles, as well as new applications found for cluster-derived catalysts on various supports.

References

1. J. H. Sinfelt, *Bimetallic Catalysts: Discoveries, Concepts and Applications*, John Wiley and Sons, New York, 1983.
2. W. Yu, M. D. Porosoff and J. G. Chen, *Chem. Rev.*, 2012, **112**, 5780.
3. C. H. Bartholomew and R. J. Farrauto, *Fundamentals of Industrial Catalytic Processes*, Wiley Interscience, Hoboken, NJ, 2006.
4. A. S. Fung, P. A. Tooley, M. R. McDevitt, B. C. Gates, D. C. Koningsberger and M. J. Kelley, *Amercian Chemical Society, Division of Petroleum Chemistry, Preprints*, 1988, **33**, 591.
5. D. M. Somerville and J. R. Shapley, *Catal. Lett.*, 1998, **52**, 123.
6. A. L. Kustov, V. G. Kessler, B. V. Romanovsky, G. A. Seisenbaeva, D. V. Drobot and P. A. Shcheglov, *J. Mol. Catal. A: Chem.*, 2004, **216**, 101.
7. C. Kappenstein, M. Saouabe, M. Guerin, P. Marecot, I. Uszkurat and I. Z. Paal, *Catal. Lett.*, 1995, **31**, 9.
8. T. C. Deivaraj and J. Y. Lee, *J. Electrochem. Soc.*, 2004, **151**, A1832.
9. T. C. Deivaraj and J. Yang Lee, *J. Power Sources*, 2005, **142**, 43.
10. A. R. Salehi Rad, M. B. Khoshgouei and A. R. Rezvani, *J. Mol. Catal. A: Chem.*, 2011, **344**, 11.
11. A. R. Salehi Rad, M. B. Khoshgouei, S. Rezvani and A. R. Rezvani, *Fuel Process. Technol.*, 2012, **96**, 9.
12. D. L. Boxall, G. A. Deluga, E. A. Kenik, W. D. King and C. M. Lukehart, *Chem. Mater.*, 2001, **13**, 891.
13. E. S. Steigerwalt, G. A. Deluga and C. M. Lukehart, *J. Phys. Chem. B*, 2002, **106**, 760.
14. A. D. Anderson, G. A. Deluga, J. T. Moore, M. J. Vergne, D. M. Hercules, E. A. Kenik and C. M. Lukehart, *J. Nanosci. Nanotechnol.*, 2004, **4**, 809.
15. Y. Okawa, T. Masuda, H. Uehara, D. Matsumura, K. Tamura, Y. Nishihata and K. Uosaki, *RSC Adv.*, 2013, **3**, 15094.
16. S.-F. Huang, Y. Chi, C.-S. Liu, A. J. Carty, K. Mast, C. Bock, B. MacDougall, S.-M. Peng and G.-H. Lee, *Chem. Vap. Deposition*, 2003, **9**, 157.
17. M. C. Carrion, B. R. Manzano, F. A. Jalon, D. Eliche-Quesada, P. Maireles-Torres, E. Rodriguez-Castellon and A. Jimenez-Lopez, *Green Chem.*, 2005, 7, 793.

18. E. Erasmus, J. W. Niemantsverdriet and J. C. Swarts, *Langmuir*, 2012, **28**, 16477.
19. S. Jujjuri, E. Ding, E. L. Hommel, S. G. Shore and M. A. Keane, *J. Catal.*, 2006, **239**, 486.
20. S. Jujjuri, E. Ding, S. G. Shore and M. A. Keane, *J. Mol. Catal. A: Chem.*, 2007, **272**, 96.
21. E. Ding, S. Jujjuri, M. Sturgeon, S. G. Shore and M. A. Keane, *J. Mol. Catal. A: Chem.*, 2008, **294**, 51.
22. D. W. Knoeppel, J. Liu, E. A. Meyers and S. G. Shore, *Inorg. Chem.*, 1998, **37**, 4828.
23. M. Pang, X. Wang, W. Xia, M. Muhler and C. Liang, *Ind. Eng. Chem. Res.*, 2013, **52**, 4564.
24. S. K. Maji, A. K. Dutta, S. Dutta, D. N. Srivastava, P. Paul, A. Mondal and B. Adhikary, *Appl. Catal., B*, 2012, **126**, 265.
25. V. G. Kessler, *J. Sol-Gel Sci. Technol.*, 2004, **32**, 11.
26. N. Tomar, E. Ghanti, A. K. Bhagi and R. Nagarajan, *J. Non-Cryst. Solids*, 2009, **355**, 2657.
27. P. N. Kapoor, S. Uma, S. Rodriguez and K. J. Klabunde, *J. Mol. Catal. A: Chem.*, 2005, **229**, 145.
28. F. A. Cotton, *Inorg. Chem.*, 1964, **3**, 1217.
29. P. J. Dyson, *Coord. Chem. Rev.*, 2004, **248**, 2443.
30. C. Femoni, M. C. Iapalucci, G. Longoni and P. H. Svensson, *Chem. Commun.*, 2004, 2274.
31. S. Zacchini, *Eur. J. Inorg. Chem.*, 2011, 4125.
32. P. J. Dyson and J. S. McIndoe, *Transition Metal Carbonyl Cluster Chemistry*, Gordon and Breach Science Publishers, 2000.
33. E. L. Muetterties, T. N. Rhodin, E. Band, C. F. Bruker and W. R. Pretzer, *Chem. Rev.*, 1979, **79**, 91.
34. G. Süss-Fink and G. Meister, *Adv. Organomet. Chem.*, 1993, **35**, 41.
35. J. C. Fierro-Gonzalez, S. Kuba, Y. Hao and B. C. Gates, *J. Phys. Chem. B*, 2006, **110**, 13326.
36. E. Cariati, D. Roberto, R. Ugo and E. Lucenti, *Chem. Rev.*, 2003, **103**, 3707.
37. R. Ugo, C. Dossi and R. Psaro, *J. Mol. Catal. A: Chem.*, 1996, **107**, 13.
38. P. Braunstein and J. Rosé, in *Comprehensive Organometallic Chemistry II*, ed. E. W. Abel, F. G. A. Stone and G. Wilkinson, Elsevier, United Kingdom, 1995, ch. 7, vol. 10, p. 351.
39. P. Braunstein and J. Rosé, in *Metal Clusters in Chemistry*, ed. P. Braunstein, L. A. Oro and P. R. Raithby, Wiley-VCH, Weinheim, Germany, 1999, vol. 2, p. 616.
40. R. Ugo and R. Psaro, *J. Mol. Catal.*, 1983, **20**, 53.
41. L. N. Lewis, *Chem. Rev.*, 1993, **93**, 2693.
42. B. C. Gates, *Chem. Rev.*, 1995, **95**, 511.
43. M. Ichikawa, *Adv. Catal.*, 1992, **38**, 283.
44. M. G. White, *Catal. Today*, 1993, **18**, 1.
45. B. C. Gates, L. Guczi and H. Knözinger, *Stud. Surf. Sci. Catal.*, 1986, **29**, 1.

46. Cristina Femoni, M. Carmela Iapalucci, Francesco Kaswalder, Giuliano Longoni and Stefano Zacchini, *Coord. Chem. Rev.*, 2006, **250**, 1580.
47. O. S. Alexeev and B. C. Gates, *Ind. Eng. Chem. Res.*, 2003, **42**, 1571.
48. O. Alexeev, M. Shelef and B. C. Gates, *J. Catal.*, 1996, **164**, 1.
49. O. S. Alexeev, G. W. Graham, D.-W. Kim, M. Shelef and B. C. Gates, *Phys. Chem. Chem. Phys.*, 1999, **1**, 5725.
50. B. D. Chandler, A. B. Schabel, C. F. Blanford and L. H. Pignolet, *J. Catal.*, 1999, **187**, 367.
51. C. Mihut, C. Descorme, D. Duprez and M. D. Amiridis, *J. Catal.*, 2002, **212**, 125.
52. B. D. Chandler, A. B. Schabel and L. H. Pignolet, *J. Catal.*, 2000, **193**, 186.
53. M. C. Carrion, B. R. Manzanoa, F. A. Jalon, P. Maireles-Torres, E. Rodriguez-Castellon and A. Jimenez-Lopez, *J. Mol. Catal. A: Chem.*, 2006, **252**, 31.
54. S. Chotisuwan, J. Wittayakun and B. C. Gates, *Stud. Surf. Sci. Catal.*, 2006, **159**, 209.
55. A. Gallo, R. Psaro, M. Guidotti, V. Dal Santo, R. Della Pergola, D. Masih and Y. Izumi, *Dalton Trans.*, 2013, **42**, 12714.
56. M. Ichikawa, in *Metal Clusters in Chemistry*, ed. P. Braunstein, L. A. Oro and P. R. Raithby, Wiley-VCH, Weinheim, Germany, 1999, p. 1273.
57. M. S. Nashner, A. I. Frenkel, D. L. Adler, J. R. Shapley and R. G. Nuzzo, *J. Am. Chem. Soc.*, 1997, **119**, 7760.
58. M. S. Nashner, A. I. Frenkel, D. Somerville, C. W. Hills, J. R. Shapley and R. G. Nuzzo, *J. Am. Chem. Soc.*, 1998, **120**, 8093.
59. N. Alonso-Vante, *Fuel Cells*, 2006, **06**, 182.
60. L. Arroyo-Ramírez, R. Montano-Serrano, T. Luna-Pineda, F. R. Román, R. G. Raptis and C. R. Cabrera, *ACS Appl. Mater. Interfaces*, 2013, **5**, 11603.
61. K. C. Kwiatkowski, S. B. Milne, S. Mukerjee and C. M. Lukehart, *J. Cluster Sci.*, 2005, **16**, 251.
62. F. E. Jones, S. B. Milne, B. Gurau, E. S. Smotkin, S. R. Stock and C. M. Lukehart, *J. Nanosci. Nanotechnol.*, 2002, **2**, 81.
63. D. Ozkaya, J. Meurig Thomas, D. S. Shephard, T. Maschmeyer, B. F. G. Johnson, G. Sankar, W. Zhou, R. G. Bell and R. Oldroyd, *Chem. – Eur. J*, 1997, **36**, 2242.
64. B. F. G. Johnson, *Coord. Chem. Rev.*, 1999, **190**, 1269.
65. D. S. Shephard, T. Maschmeyer, G. Sankar, J. Meurig Thomas, D. Ozkaya, B. F. G. Johnson, R. Raja, R. D. Oldroyd and R. G. Bell, *Chem. – Eur. J.*, 1998, **4**, 1214.
66. W. Zhou, J. M. Thomas, D. S. Shephard, B. F. G. Johnson, D. Ozkaya, T. Maschmeyer, R. G. Bell and Q. Ge, *Science*, 1998, **280**, 705.
67. R. Raja, G. Sankar, S. Hermans, D. S. Shephard, S. Bromley, J. M. Thomas, B. F. G. Johnson and T. Maschmeyer, *Chem. Commun.*, 1999, 1571.
68. D. Ozkaya, W. Zhou, J. M. Thomas, P. Midgley, V. J. Keast and S. Hermans, *Catal. Lett.*, 1999, **60**, 113.
69. S. Hermans and B. F. G. Johnson, *Chem. Commun.*, 2000, 1955.

70. S. Hermans, R. Raja, J. M. Thomas, B. F. G. Johnson, G. Sankar and D. Gleeson, *Angew. Chem., Int. Ed.*, 2001, **40**, 1211.
71. R. Raja, T. Khimyak, J. M. Thomas, S. Hermans and B. F. G. Johnson, *Angew. Chem., Int. Ed.*, 2001, **40**, 4638.
72. A. C. W. Koh, L. Chen, W. Kee Leong, T. Peng Ang, B. F. G. Johnson, T. Khimyak and J. Lin, *Int. J. Hydrogen Energy*, 2009, **34**, 5691.
73. R. D. Adams, E. M. Boswell, B. Captain, A. B. Hungria, P. A. Midgley, R. Raja and J. M. Thomas, *Angew. Chem., Int. Ed.*, 2007, **46**, 8182.
74. M. Manzoli, V. N. Shetti, J. A. L. Blaine, L. Zhu, D. Isrow, V. Yempally, B. Captain, S. Coluccia, R. Raja and E. Gianotti, *Dalton Trans.*, 2012, **41**, 982.
75. R. D. Adams, M. Chen, G. Elpitiya, M. E. Potter and R. Raja, *ACS Catal.*, 2013, **3**, 3106.
76. R. Bonelli, S. Albonetti, V. Morandi, L. Ortolani, P. M. Riccobene, S. Scire and S. Zacchini, *Appl. Catal., A*, 2011, **395**, 10.
77. R. Bonelli, C. Lucarelli, T. Pasini, L. F. Liotta, S. Zacchini and S. Albonetti, *Appl. Catal., A*, 2011, **400**, 54.
78. C. Willocq, A. Delcorte, S. Hermans, P. Bertrand and M. Devillers, *J. Phys. Chem. B*, 2005, **109**, 9482.
79. S. Hermans, J. Sloan, D. S. Shephard, B. F. G. Johnson and M. L. H. Green, *J. Chem. Soc., Chem. Commun.*, 2002, 276.
80. C. Willocq, S. Hermans and M. Devillers, *J. Phys. Chem. C*, 2008, **112**, 5533.
81. C. Willocq, D. Vidick, B. Tinant, A. Delcorte, P. Bertrand, M. Devillers and S. Hermans, *Eur. J. Inorg. Chem.*, 2011, 4721.
82. C. Willocq, V. Dubois, Y. Z. Khimyak, M. Devillers and S. Hermans, *J. Mol. Catal. A: Chem.*, 2012, **365**, 172.
83. D. Vidick, S. Hermans and M. Devillers, *Stud. Surf. Sci. Catal*, 2010, **175**, 827.
84. D. Vidick, A. F. Leonard, C. Poleunis, A. Delcorte, M. Devillers and S. Hermans, *Catal. Today*, 2014, 112.
85. D. Vidick, M. Herlitschke, C. Poleunis, A. Delcorte, R. P. Hermann, M. Devillers and S. Hermans, *J. Mater. Chem. A*, 2013, **1**, 2050.
86. B. Vanhorenbeke, C. Vriamont, F. Pennetreau, M. Devillers, O. Riant and S. Hermans, *Chem. – Eur. J.*, 2013, **19**, 852.
87. I. Robinson, S. Zacchini, L. D. Tung, S. Maenosono and N. T. K. Thanh, *Chem. Mater.*, 2009, **21**, 3021.
88. F. Schweyer, P. Braunstein, C. Estournes, J. Guille, H. Kessler, J.-L. Paillaud and J. Rose, *Chem. Commun.*, 2000, 1271.
89. R. Richards, R. Mortel and H. Bonnemann, *Fuel Cell Bull.*, 2001, **4**(37), 7.
90. H. Bonnemann, G. Braun, W. Brijoux, R. Brinkmann, A. Schulze Tilling, K. Seevogel and K. Siepen, *J. Organomet. Chem.*, 1996, **520**, 143.
91. J. Rothe, G. Kohl, J. Hormes, H. Bonnemann, W. Brijoux and K. Siepen, *J. Phys. IV*, 1997, **7**, C2–959.
92. U. A. Paulus, U. Endruschat, G. J. Feldmeyer, T. J. Schmidt, H. Bonnemann and R. J. Behm, *J. Catal.*, 2000, **195**, 383.
93. R. W. J. Scott, C. Sivadinarayana, O. M. Wilson, Z. Yan, D. W. Goodman and R. M. Crooks, *J. Am. Chem. Soc.*, 2005, **127**, 1380.

CHAPTER 4

Atomically Precise Gold Catalysis

KATLA SAI KRISHNA,[a,b] JING LIU,[a,b] PILARISETTY TARAKESHWAR,[c] VLADIMIRO MUJICA,[c] JAMES J. SPIVEY[b] AND CHALLA S. S. R. KUMAR*[a,b]

[a] Center for Advanced Microstructures and Devices, Louisiana State University, Baton Rouge, LA 70806, USA; [b] Center for Atomic-Level Catalyst Design, # 324, Cain Department of Chemical Engineering, Louisiana State University, 110 Chemical Engineering, South Stadium Road, Baton Rouge, LA 70803, USA; [c] Department of Chemistry and Biochemistry, Arizona State University, Tempe, Arizona 85287-1604, USA
*Email: ckumar1@lsu.edu

4.1 Introduction

Gold nanoparticle-based catalysts are widely known in literature since early reports by Haruta and co-workers.[1–5] The majority of investigations to date focused on size- and shape-controlled gold nanoparticle catalysis. Although advances have been made in preparing monodisperse nanoparticles, there is still a growing need to synthesize nanoparticles with atomic precision in order to understand catalytic significance of specific sites on the gold cluster. Such precision is also essential to correlate first-principles computational research with observable catalytic mechanisms. Compared to traditional nanoparticle catalysts,[6a,b] tasks such as determining size-dependent effects, identifying active species, and understanding the active site mechanism becomes more precise with atomically precise catalysts.

RSC Catalysis Series No. 22
Atomically-Precise Methods for Synthesis of Solid Catalysts
Edited by Sophie Hermans and Thierry Visart de Bocarmé
© The Royal Society of Chemistry 2015
Published by the Royal Society of Chemistry, www.rsc.org

Therefore accurate correlation of catalytic properties with the exact atomic structure can only be achieved by developing nanoparticle catalysts with atomic precision.

Recently, remarkable advances have been made by researchers in solution-phase synthesis of atomically precise gold catalysts from thiol-stabilized gold nanoclusters. These gold nanoclusters are composed of a precise number of metal atoms (n) and of ligands (m), denoted as $Au_n(SR)_m$, where n can have values up to a few hundred atoms (~ 2 nm). The thiol-protected nanoclusters are atomically well-defined in terms of their molecular purity and are stable under ambient conditions. Several reviews have been published in recent times on atomically precise gold clusters.[6-8] Most of these reviews are focused either on the synthesis and characterization aspects of the clusters or discussed their electronic properties and crystal structure information. There are only few that discussed catalytic applications of the clusters.[9,10] However, there is no review/book chapter published until now focusing on the correlation of electronic/magnetic properties on catalytic activity of the atomically precise gold clusters. With the growing number of research publications arising from this field, there is now a need for a review on the topic of atomically precise gold catalysis. This chapter provides an in-depth analysis and reviews synthesis, electronic, magnetic properties of atomically precise gold cluster catalysts and the correlation of these properties with their catalytic activity.

The chapter is divided into six major sections starting with an introduction followed by an overview of synthesis and catalytic applications of the atomically precise gold clusters. In the second section, detailed analysis of synthetic approaches and selected examples of catalysis involving the clusters are discussed. The third section comprises of an overview of electronic and magnetic structure of the clusters where the electronic structure information is derived from X-ray absorption spectroscopy and photoemission spectroscopy studies and magnetic structure information is derived from theoretical calculations and experimental superconducting quantum interference device (SQUID) measurements of the clusters. The fourth section describes correlation of electronic structure of the clusters with their catalytic activity. Similarly, the fifth section describes an opportunity for correlation of magnetic structure of the clusters with their catalytic activity. The final section provides our conclusions and our analysis of future perspectives for the field.

4.2 Overview of Synthesis and Catalysis

4.2.1 Synthesis of Atomically Precise Gold Nanoclusters

The number of gold atoms that constitute a gold nanocluster provides significant information about its unique structural and electronic properties. The gold nanoclusters, generally, possess a non-fcc atomic packing structure unlike nanoparticles which have an fcc structure. Synthesis of atomically precise gold nanoclusters enables precise correlation of accurate crystal

Atomically Precise Gold Catalysis

structures of these clusters with their catalytic and other properties. In recent times, atomically precise gold clusters have been widely investigated due to their unique properties, which are very different from traditional nanoparticles.

Initial works on the synthesis of narrowly distributed atomically precise gold clusters was carried out by Whetten and co-workers through solvent fractionation process.[11,12] Their method yielded nanoclusters (named as 'Nanocrystal Au Molecules') in the size range of 1–3.5 nm with masses between 5 and 93 kDa characterized by laser desorption ionization (LDI) technique. Murray and co-workers have reported on phenylethylthiol-capped 25-atom nanoclusters[13] which were previously mischaracterized as 38-atom gold nanoclusters.[14] Tsukuda and co-workers reported on the synthesis of a series of glutathione capped atomically precise gold clusters, which were separated and purified using polyacrylamide gel electrophoresis (PAGE) process and characterized using electrospray ionization mass spectroscopy (ESI-MS).[15] Recently, thiolate-stabilized gold nanoclusters of different sizes were reported by Jin and co-workers.[16–18] Their group had isolated and thoroughly investigated the crystal structures, electronic, magnetic and catalytic properties of thiol-capped Au_{25}, Au_{38}, Au_{102} and Au_{144}. Their group was also the first one to report on 'size-focusing' method to achieve atomically precise gold clusters.[19] Figure 4.1 gives a schematic of the synthesis approaches used to form atomically precise gold cluster catalysts.

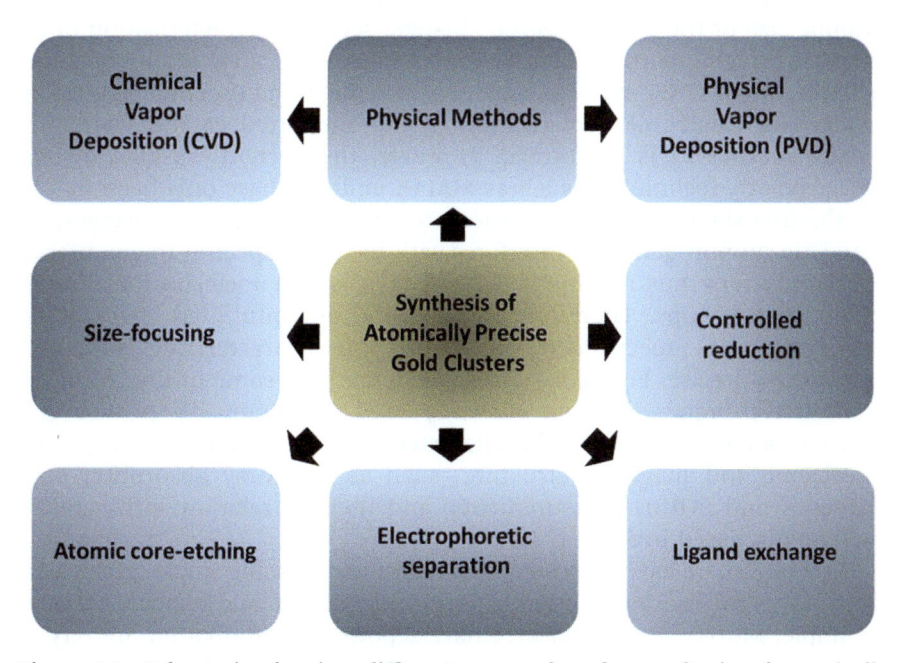

Figure 4.1 Schematic showing different approaches for synthesis of atomically precise gold clusters catalysts.

Traditionally nanoparticle characterization for size determination has been predominantly carried out using high-resolution transmission electron microscopy (TEM). The advantage of using electron microscopy over other techniques is that the particle size of the nanoparticles along with their size distribution can be readily identified through a visual approach unlike spectroscopy tools like UV-visible or FT-IR. The electron density differences present in different elements also makes it convenient to distinguish accurate sizes of organic-ligand capped nanoparticles, core–shell and bimetallic/multi-metallic nanoparticles. However, due to the high-energy electron beam source, the nanoparticles tend to interact with the electron beam causing their aggregation and thus leading to size aberration. This problem is often noticed during characterization of atomically precise gold clusters.[20] Therefore, these clusters need a more precise and nondestructive characterization tool that can accurately determine their size based on the number of atoms present in them. Mass spectrometry appears to be the most reliable alternative technique available to date for this purpose. Different mass spectrometry techniques such as laser desorption ionization (LDI), matrix-assisted laser desorption ionization (MALDI) and electrospray ionization (ESI) have been utilized. Of the three, the LDI and MALDI are relatively harsher techniques that tend to break the Au–S and S–C bonds of the clusters during analysis. On the other hand, ESI is a much software and more preferred technique.

In general, characterization of atomically precise gold cluster catalysts involves three distinct stages: first, to determine the accuracy of their atomic size of the as-synthesized clusters; second, characterization of the clusters after supporting them on high-surface area catalytic supports such as TiO_2, SiO_2, CeO_2 etc. and calcining to remove the organic capping ligands; and finally, characterization of the supported catalysts after the catalysis reaction is performed. However, the currently available mass spectrometry techniques (such as MALDI and ESI) have only been helpful in the first stage of characterization. For the second and third stages, the catalytic support generally interferes with the mass spectral fragmentation of the clusters thus making it difficult for their characterization through mass spectroscopy. Therefore, there is still an unfulfilled need for developing mass spectrometry technique for characterizing supported atomically-precise. Figure 4.2 gives a schematic of some of the characterization techniques that are currently used for atomically precise gold cluster catalysts. More on the characterization techniques for exploring electronic and magnetic structure (such as the X-ray absorption spectroscopy and SQUID measurements) and their correlation with catalysis will be discussed in Section 4.3.

Recently, there have been significant advances in using single-crystal X-ray techniques for characterisation of atomically-precise gold nanoclusters. For example, Murray and co-workers have reported the formation and crystal structure (Figure 4.3) of atomically precise anionic Au_{25} clusters capped by

Atomically Precise Gold Catalysis 91

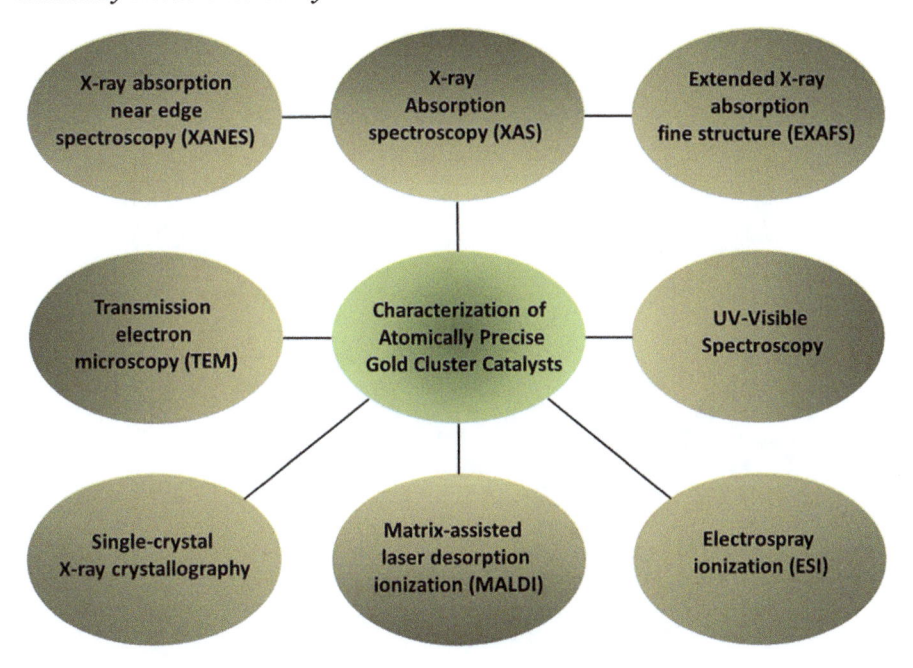

Figure 4.2 Schematic showing different characterization techniques for atomically precise gold cluster catalysts.

phenylethylthiol ligands and stabilized by tetraoctylammonium bromide cation (TOA$^+$).[21]

4.2.2 Catalysis

Most catalysts are based on metal nanoparticles with widely varying cluster sizes (~1–100 nm) supported on various supports. Attributing the observed catalytic properties to a specific surface site, particularly the mechanism or nature of active site, is virtually impossible due to polydispersity with respect to atomic precision. It is well established that sites of different coordination or proximity to the support are catalytically different. Therefore, it is critical to have access to catalytically active metals such as gold containing a specific (and small) number of atoms in order to investigate the fundamental steps occurring at specific sites. Atomically precise gold nanoclusters, containing from a few to a few hundred atoms, are promising for nanocatalysis. These clusters exhibit excellent catalytic activity in oxidation and hydrogenation reactions. These clusters significantly promote molecule activation by enhancing adsorption energy of reactant molecules on the catalyst surface. The knowledge of the atomic-level structure of these clusters allows a precise correlation of structure with catalytic properties and also permits the identification of catalytically

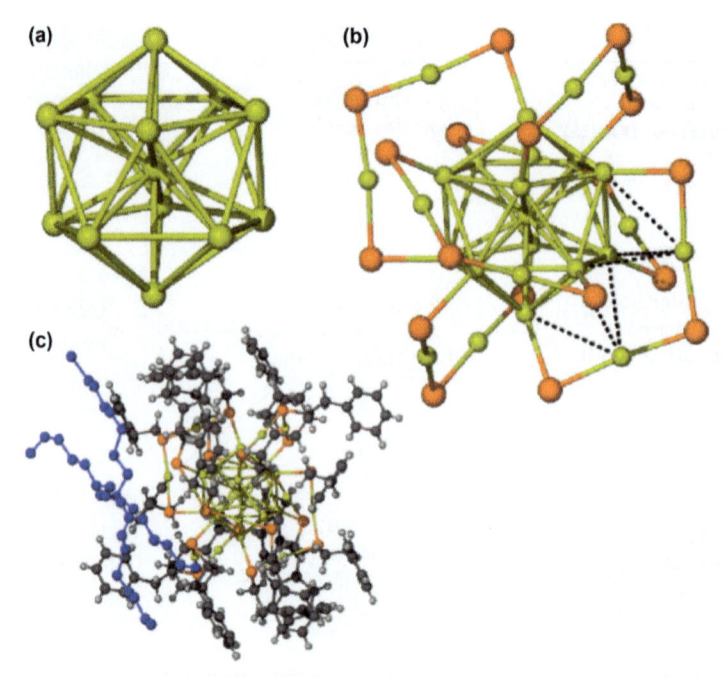

Figure 4.3 Breakdown of the X-ray crystal structure of $[TOA^+][Au_{25}(SCH_2CH_2Ph)_{18}^-]$ as seen from [001]. (a) Arrangement of the Au_{13} core with 12 atoms on the vertices of an icosahedron and one in the center. (b) Depiction of gold and sulfur atoms, showing six orthogonal $Au_2(SCH_2CH_2Ph)_3$ 'staples' surrounding the Au_{13} core (two examples of possible aurophilic bonding shown as dashed lines). (c) $[TOA^+][Au_{25}(SCH_2CH_2Ph)_{18}^-]$ structure with the ligands and TOA^+ cation (depicted in blue) (Legend: gold, yellow; sulfur, orange; carbon, gray; hydrogen, off-white; the TOA^+ counter-ion is over two positions with one removed for clarity).
(Reproduced with permission from ref. 21, Copyright American Chemical Society, 2008).

active sites on the gold particle at an atomic level. These fundamental properties can also pave way for designing newer types of novel, highly active, and selective gold cluster catalysts.

Generally, gold cluster catalysts are obtained using two different synthesis approaches *i.e.*, physical approach and the chemical approach. The difference between both approaches is that the clusters formed using physical approach does not possess any stabilizing ligands whereas the clusters formed through chemical approach have organic ligands stabilizing them. Physical approaches normally include gas-phase deposition techniques such as chemical vapor deposition (CVD) and physical vapor deposition (PVD) and chemical approaches include solution-based reduction of metal salts in the presence of organic ligands. Due to the absence of ligands on their surface, the bare clusters formed using physical methods have large surface energy and are highly reactive for catalytic applications. By controlling the size of clusters, tuning of catalytic activity and selectivity is also feasible. For instance, Goodman and co-workers have observed size-dependent catalytic activity of gold clusters supported over TiO_2 for CO oxidation reaction in a gas phase reaction.[22] In several of those studies, complex formation between gold clusters and molecular oxygen has been reported due to the interaction of dioxygen with the gold clusters.[23,24]

The number of investigations on the use of atomically precise catalysts for catalyzing a variety of chemical reactions continue to increase. The chemical conversion reactions that are currently being explored using atomically precise gold cluster catalysts were previously carried out using traditional gold nanoparticle catalysts. A major set of catalytic reactions that are widely investigated using atomically precise catalysts include (i) the CO oxidation reaction, (ii) selective oxidation of styrene, and (iii) selective hydrogenation of α,β-unsaturated ketones/aldehydes; which are discussed in detail in the following section. In addition, catalytic reactions such as the selective oxidation of sulfide to sulfoxide,[25] Ullmann-type homocoupling reaction of aryl iodides[26] and electrocatalytic reduction of CO_2[27] are also being explored by researchers.

4.2.2.1 The CO Oxidation Reaction

Haruta et al. have pioneered the work on carbon monoxide oxidation catalyzed by Au nanoparticles supported on oxides.[1,2] Since then, gold nanocatalysts have been widely investigated, particularly due to their unusually high activity for CO oxidation even at temperatures as low as −70 °C. This reaction has practical applications as well, e.g., eliminating CO impurities from H_2-rich syngas for fuel cells.[28] Despite significant progress in gold catalysis, the fundamentally important mechanistic aspects of CO oxidation by nanoparticles are still unclear.[29] Although conventional gold catalysts prepared using wet impregnation methods produce active catalysts for CO oxidation, these catalysts are not useful in understanding the mechanism clearly. Some of the disadvantages include:

- A broad size distribution of the gold nanoparticles (with respect to atomic-precision), posing a major challenge in determining the active sites (such as metallic nanoparticles or gold ions) that catalyzes the CO oxidation.
- Lack of accurate information on the active sites (such as gold surface or gold/support interface) and the nature of support.
- Inability to utilize computational tools for modeling "atomically-imprecise" traditional nanoparticle catalysts.

Among several supports for gold catalysts that have been tested for CO oxidation, TiO_2 appears to be the best in terms of catalytic activity. Detailed investigation into CO oxidation using supported gold nanoclusters also revealed that the reaction is sensitive to the size of the clusters and significant differences in turnover frequencies was observed based on the size of the gold nanocluster.[7] Recently, Hakkinen and co-workers[30] have also shown that ligands can alter the electronic structure of gold nanoclusters in a CO oxidation reaction using DFT computations. Their results showed that in ligand-stabilized gold nanoclusters ($d = 1.2$–2.4 nm), the differences in the energy gap between highest occupied molecular orbital (HOMO) and lowest occupied molecular orbital (LUMO) significantly affects the binding energy between molecular O_2 and the gold nanoclusters. It is, therefore, clear that in order to understand the intricate mechanistic details of CO oxidation using gold catalysts, preparation of well-defined atomically precise gold catalysts is a must.

Two other important tasks for atomically precise catalysis are obtaining uniformly distributed nanoclusters on the support material and removal of the organic capping ligands to expose the gold surface to reactants. High temperature calcination is commonly preferred for this purpose due to its convenience. However, this process affects the size and uniformity of the nanoclusters. Recently, Jin and co-workers demonstrated high catalytic activity of supported $Au_{25}(SR)_{18}$ nanoclusters for the CO oxidation reaction.[31] In their work, they directly deposited solution phase nanoclusters onto various oxide supports (such as TiO_2, CeO_2, and Fe_2O_3). The catalysts were then evaluated for the CO oxidation reaction in a fixed bed reactor. The catalysts were pretreated with O_2 for 1.5 h at 150 °C before passing CO into the reactor. The supports exhibited a strong effect, and the $Au_{25}(SR)_{18}/CeO_2$ catalyst showed higher catalytic activity than others. Interestingly, they found that the O_2 pretreatment significantly enhanced catalytic activity. Since this pretreatment temperature is well below the thiolate desorption temperature (~ 200 °C), the thiolate ligands were expected to remain on the Au_{25} cluster surface, indicating that the CO oxidation reaction is catalyzed by the intact $Au_{25}(SR)_{18}$ on supports. They also found that increasing the O_2 pretreatment temperature to 250 °C (above the thiolate desorption temperature) did not lead to any improvement in activity at all reaction temperatures from room temperature to 100 °C. Their results showed $\sim 94\%$ CO conversion at 80 °C and $\sim 100\%$ conversion at 100 °C for the $Au_{25}(SR)_{18}/CeO_2$ catalyst that is O_2-pretreated at 150 °C. The effect of water vapor on the catalytic performance was also investigated. These results are in striking contrast with the common understanding that surface ligands must be removed to obtain higher catalytic activity. Their results also suggests that the perimeter sites of the interface of $Au_{25}(SR)_{18}/CeO_2$ are the active centers (Figure 4.4). It is clear that the intact structure of the $Au_{25}(SR)_{18}$ catalyst during the process of CO oxidation allows researchers to gain mechanistic insights into the catalytic reaction.

Gaur *et al.* have recently reported CO oxidation reaction using TiO_2 supported Au_{38} nanoclusters. They prepared nearly monodisperse $Au_{38}(SC_{12}H_{25})_{24}$ clusters (1.7 ± 0.2 nm) using a modified Brust process and utilized thiol-etching process for the ligand exchange.[32] This nanocluster

Atomically Precise Gold Catalysis 95

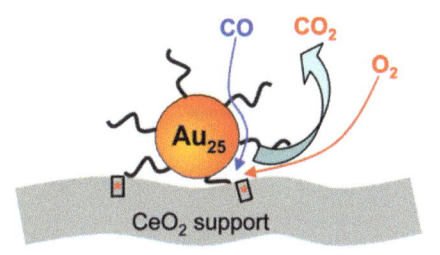

Figure 4.4 Proposed model for CO oxidation at the perimeter sites of Au$_{25}$(SR)$_{18}$/CeO$_2$ catalyst.
(Reproduced with permission from ref. 31, Copyright American Chemical Society, 2012).

solution was used to impregnate microporous TiO$_2$ support to obtain 0.7 wt% Au$_{38}$/TiO$_2$ catalyst. The catalyst was subsequently dried in air and treated with H$_2$/He at 400 °C to remove most of the sulfur ligands. The process, however, increased the gold cluster size to 3.9 ± 0.96 nm. The X-ray photoelectron spectroscopy (XPS) and extended X-ray absorption and fine structure (EXAFS) analysis of the supported catalyst showed trace levels of residual sulfides, apparently located at the Au–TiO$_2$ interface. The CO oxidation experiments on these supported clusters also showed an activation energy and range of TOFs comparable to those reported by other researchers.

In a different report, Gaur *et al.* have also performed CO oxidation reaction in an *in situ* FTIR (DRIFTS) cell to monitor the species that were formed during the reaction, and the results were compared to commercially available Au/TiO$_2$ catalyst.[33] The Au/TiO$_2$ catalyst, prepared from Au$_{38}$ clusters, was found to be less active than a commercial catalyst for CO oxidation. This was due to the presence of bidentate carbonate species. DRIFTS studies on the commercial catalyst showed strong peaks at 1718 and 1690 cm^{-1} that were not present in their catalyst. They proposed that these bands were due to bridging CO$_2$ between Au–Ti and Au–Au centers formed *via* direct oxygen atom transfer from the titania surface to Au–CO (Figure 4.5). On the surface of the catalyst, however, a different reaction mechanism involving sulfur-mediated oxygen transfer at the Au–TiO$_2$ interface was also proposed (Figure 4.6).

4.2.2.2 Selective Oxidation of Styrene

Styrene epoxidation is a commercially important catalytic process for chemical industry. Gold catalysts supported on various oxide materials have shown to be active catalysts for styrene epoxidation reactions.[34–37] However, the gold catalysts previously reported were structurally polydisperse and hence understanding structure–property relationships were difficult. Some fundamental questions regarding the nature of gold catalysts and the

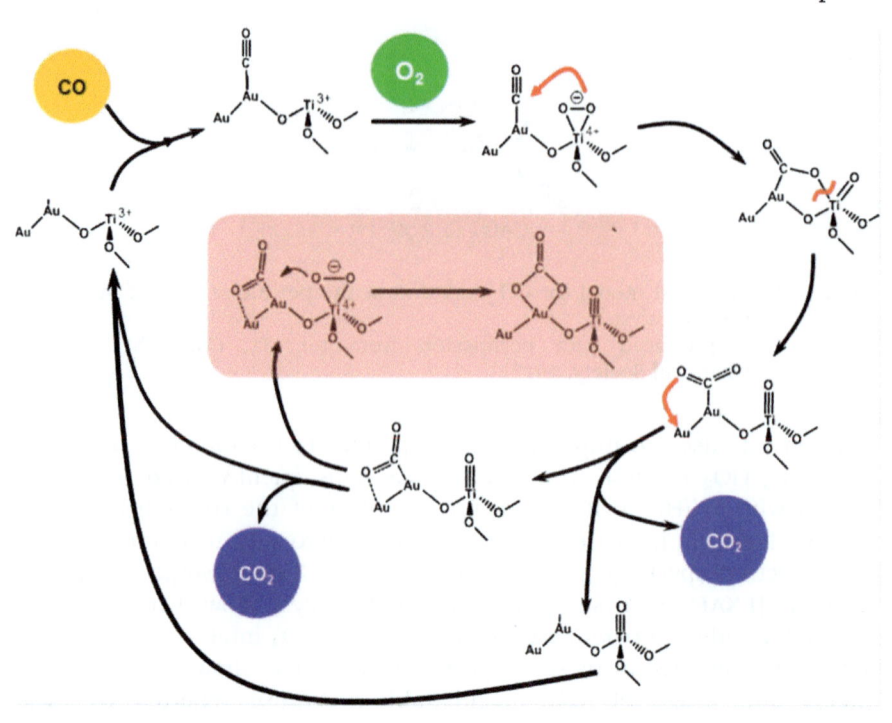

Figure 4.5 Proposed reaction mechanism for CO oxidation on a commercial Au/TiO$_2$ catalyst.
(Reproduced with permission from ref. 33, Copyright Elsevier, 2013).

mechanism of epoxidation still need to be addressed. The development of atomically precise gold catalysts now offers a unique opportunity to study the fundamental aspects of gold catalysis, in particular regarding how structure and the electronic properties of nanoclusters influence their catalytic performance.

Au$_{25}$(SR)$_{18}$ nanoclusters immobilized on hydroxyapatite support was reported by Tsukuda and co-workers[35] for the selective oxidation of styrene in toluene. They achieved ∼100% conversion of styrene with 92% selectivity towards epoxide product. Recently, Jin and co-workers[37] have reported selective oxidation of styrene using three different sized nanoclusters (*i.e.*, Au$_{25}$, Au$_{38}$ and Au$_{144}$). They found that the catalytic activity of the catalysts exhibited a strong dependence on the cluster size. In general, smaller nanoclusters exhibited higher conversions. With a catalyst loading of ∼1% for all the three cluster sizes, the Au$_{25}$ nanoclusters showed the highest conversion of styrene, followed by Au$_{38}$ and Au$_{144}$ respectively. They also compared the catalysis results with larger crystalline ∼3 nm sized gold nanocrystals which were not atomically monodisperse and were capped by octylthiolates. The 3-nm gold nanocrystals showed a much lower conversion. They investigated the effect of the nature of thiolate ligands, aromatic *versus* long-chain alkanethiolates, and found that the chemical

Atomically Precise Gold Catalysis 97

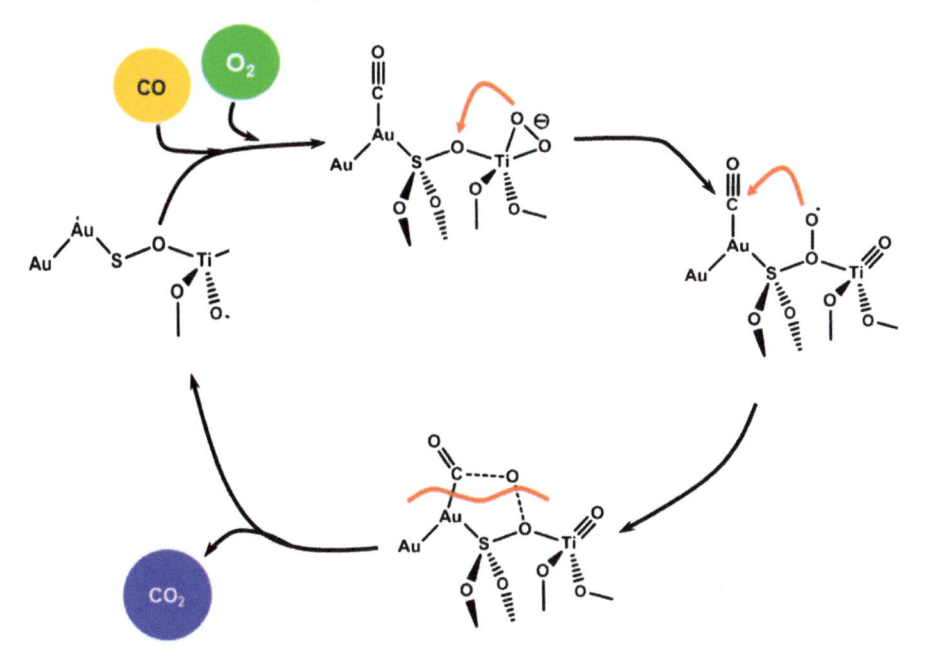

Figure 4.6 Oxygen transfer *via* SO$_3^-$ linkages in thiolate-derived Au/TiO$_2$ catalyst. (Reproduced with permission from ref. 33, Copyright Elsevier, 2013).

nature of the ligands do not affect the catalytic activity and selectivity.[38] Therefore, it appears from these investigations that the catalytic properties of nanoclusters were primarily determined by the metal core rather than by the ligand shell. It was not clear if the ligands were lost during the catalysis. However, recent X-ray absorption spectroscopy studies demonstrate that the thiolate ligands are likely to remain on the clusters during the oxidation process.[32,39] These results are not surprising since the reaction temperature is well below the thiolate desorption temperature.[40] The authors also found that the observed size dependency is not a consequence of the higher surface area of the small-sized nanoclusters. By considering the correction factor for the number of surface Au atoms, they observed that the activities of Au$_{25}$(SR)$_{18}$ and Au$_{38}$(SR)$_{24}$ are much higher than their expected activities, using Au$_{144}$(SR)$_{60}$ as a basis for comparison and that the enhanced activity was not due to the smaller size of the clusters but likely due to intrinsic electronic and structural properties. This was demonstrated by taking Au$_{25}$ clusters as an example. The Au$_{25}$ clusters are usually obtained either as anionic or neutral clusters depending on the synthesis process. These clusters show a core–shell type crystal structure with Au$_{13}$ forming the core and Au$_{12}$ forming the shell, each with different charge distribution. The delocalized electrons from the 6s valence state for the Au$_{13}$ core can contain either eight (for anionic) or seven (for neutral) electrons. Therefore the activation of O$_2$ for an oxidation process occurs at the electron-rich Au$_{13}$ core. The anionic and neutral forms of Au$_{25}$ can be

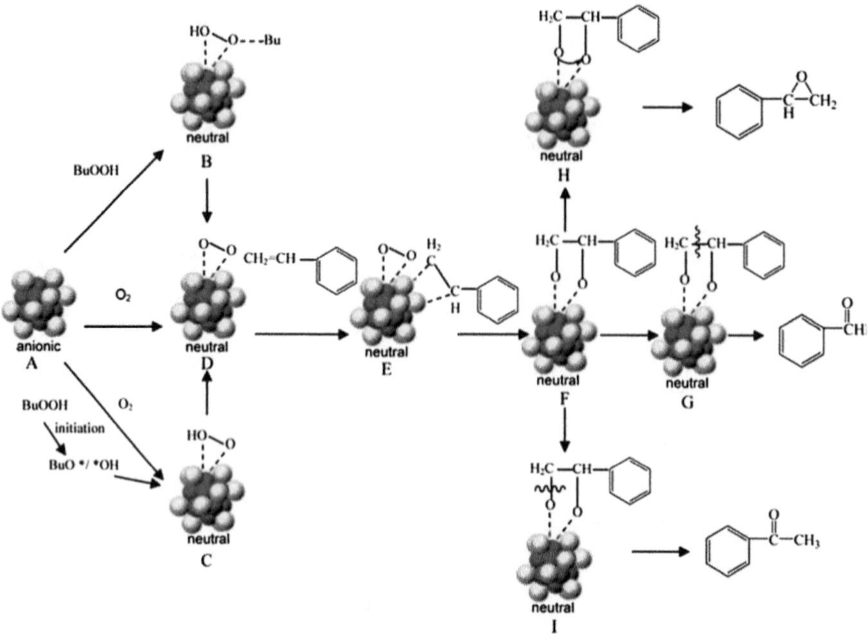

Figure 4.7 Proposed mechanism of selective oxidation of styrene catalyzed by a Au$_{25}$(SR)$_{18}$ cluster. For clarity, the thiolate ligands are not shown. (Dark gray: Au atoms of the core, light gray: Au atoms of the shell). (Reproduced with permission from ref. 37, Copyright Wiley-VCH Verlag GmbH & Co. KGaA, 2010).

interconverted by using O$_2$/peroxide as an oxidant. This process could be responsible for the O$_2$ or *tert*-butyl hydroperoxide (TBHP) activation prior to styrene oxidation during the catalytic process. Jin and co-workers have also proposed a mechanism for the selective oxidation of styrene catalyzed with Au$_{25}$(SR)$_{18}$ nanoclusters (Figure 4.7).[37]

Recently Liu *et al.*[39] have investigated styrene oxidation reaction using two different Au$_{25}$ clusters capped with different ligands. They made an attempt to correlate fundamental electronic structure with styrene oxidation catalysis by atomically precise mixed-ligand (thiol and phosphine) biicosahedral structure [Au$_{25}$(PPh$_3$)$_{10}$(SC$_{12}$H$_{25}$)$_5$Cl$_2$]$^{2+}$ and thiol-stabilized icosahedral core–shell structure [Au$_{25}$(SCH$_2$CH$_2$Ph)$_{18}$]$^-$ clusters by using a combination of X-ray absorption spectroscopy (XAS) and ultraviolet photoemission spectroscopy (UPS).

4.2.2.3 Selective Hydrogenation of α,β-Unsaturated Ketones/ Aldehydes

Selective hydrogenation of α,β-unsaturated ketones/aldehydes to the corresponding alcohols is a well investigated chemical reaction catalyzed by a

Atomically Precise Gold Catalysis

variety of conventional gold catalysts (both bulk and nano) to predominantly form α,β-unsaturated alcohols.[41–43] The side products of the reaction are saturated ketones from the olefin hydrogenation, and saturated alcohols from extended hydrogenation. The conventional nanoparticle-based gold catalysts have achieved high conversion and selectivity for hydrogenation reaction of α,β-unsaturated ketones to form unsaturated alcohols. However, complete selectivity for unsaturated alcohols is yet to be achieved. Additionally, polydispersity of the sizes of nanoparticles hindered fundamental investigation into the effects of electronic properties on the catalytic performance. To overcome this, several research groups have studied the reaction using atomically precise gold nanoclusters.

Jin and co-workers have investigated hydrogenation of α,β-unsaturated ketones (or aldehydes) using $Au_{25}(SR)_{18}$ catalysts. They have obtained 99 + % selectivity for α,β-unsaturated alcohols with chemoselective hydrogenation of the C=O bond at room temperature.[44] The Au_{25} nanoclusters were found to be intact during the reaction as confirmed by mass spectrometry analysis after the reaction. Therefore, a correlation between electronic structure and catalytic performance of the $Au_{25}(SR)_{18}$ nanoclusters could be established. The mechanism of the hydrogenation process was explained as follows: adsorption of the C=O groups occur on eight uncapped Au_3 faces of the icosahedron by the interaction of the active site with the O atom of the C=O group in α,β-unsaturated ketones. The electron transfer from Au_{13} to the O atom of C=O activates the C=O bond and then the weakly nucleophilic hydrogen attacks the activated C=O group, leading to the unsaturated alcohol product (Figure 4.8). In general, the presence of electron-rich Au_{13} core is believed to favor the selective hydrogenation of the C=O bond over the

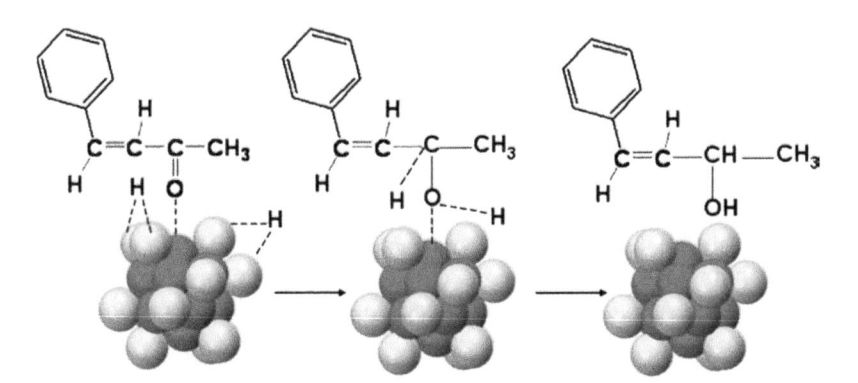

Figure 4.8 The proposed mechanism of the chemoselective hydrogenation of an α,β-unsaturated ketone to an unsaturated alcohol catalyzed by $Au_{25}(SR)_{18}$ nanoparticles. For clarity, the thiolate ligands are not shown. Dark gray: Au atoms of the core; light gray: Au atoms of the shell.
(Reproduced with permission from ref. 44, Copyright Wiley-VCH Verlag GmbH & Co. KGaA, 2010).

C=C bond in α,β-unsaturated ketones at mild temperatures. The Au atoms on the surface have a low-coordination ($N = 3$) behavior, which enables the adsorption of H_2 and later for its dissociation.

In a different line of investigation, Jin et al. have also studied the $Au_{25}(SR)_{18}$ catalysts for the stereoselective formation a single cyclic alcohol isomer in the oxidation of bicyclic ketones.[45] The mechanism of its formation was explained as follows: Similar to the case of hydrogenation of α,β-unsaturated ketones, C=O bond becomes activated by the Au_{13} core and the H_2 adsorption and dissociation happens on the surface sites of Au_{12} shell. The C=O bond is then stereoselectively attacked by a H atom to yield the desired product. The axial attack of the H atom yields *exo*-alcohol and equatorial attack results in *endo*-alcohol. The spatial environment around $Au_{25}(SR)_{18}$ catalyst appears to strongly influence the direction of H atom attacks and the products formed showed preference for axial attack over equatorial attack.

4.3 Overview of Electronic and Magnetic Structure of Catalysts

Generally, gold nanoparticles (>2 nm particles) possess a metal core with their electron conduction band containing the quasi-continuous energy levels. This leads to the situation where the average spacing between the energy levels (δ) is far less than $k_B T$ (where $k_B T$ is the thermal energy at the given temperature T and k_B is the Boltzmann constant). Hence, the electrons in the conduction band roam freely within the metal core and when light shines upon them, they are collectively excited to show plasmon resonance.[46] Contrastingly, gold nanoclusters (<2 nm particles) do not show any plasmon resonance due to quantization of energy levels within their conduction band. Therefore, they exhibit discrete electronic structure with molecule-like properties with single electron HOMO–LUMO transitions instead of collective plasmon excitations of electrons in nanoparticles (Figure 4.9).[47] One can observe multiple optical absorption peaks in the optical spectrum of the gold nanoclusters whereas gold nanoparticles (spherical) produce a single surface plasmon resonance peak.

The distinctive features of the nanoclusters emerge predominantly because of their quantum size, which can be estimated using the free electron theory.[48] The average spacing between the electronic energy levels (δ) is expressed as:

$$\delta \approx E_F/N \tag{1}$$

where, E_F is the Fermi energy and N is the number of gold atoms. Since N relates to the particle size (d) as d^3, the spacing (δ) is inversely related to the cube of the particle size. Thus, with decreasing size, the spacing between the energy levels increases. Considering room temperature thermal energy ($k_B T$)

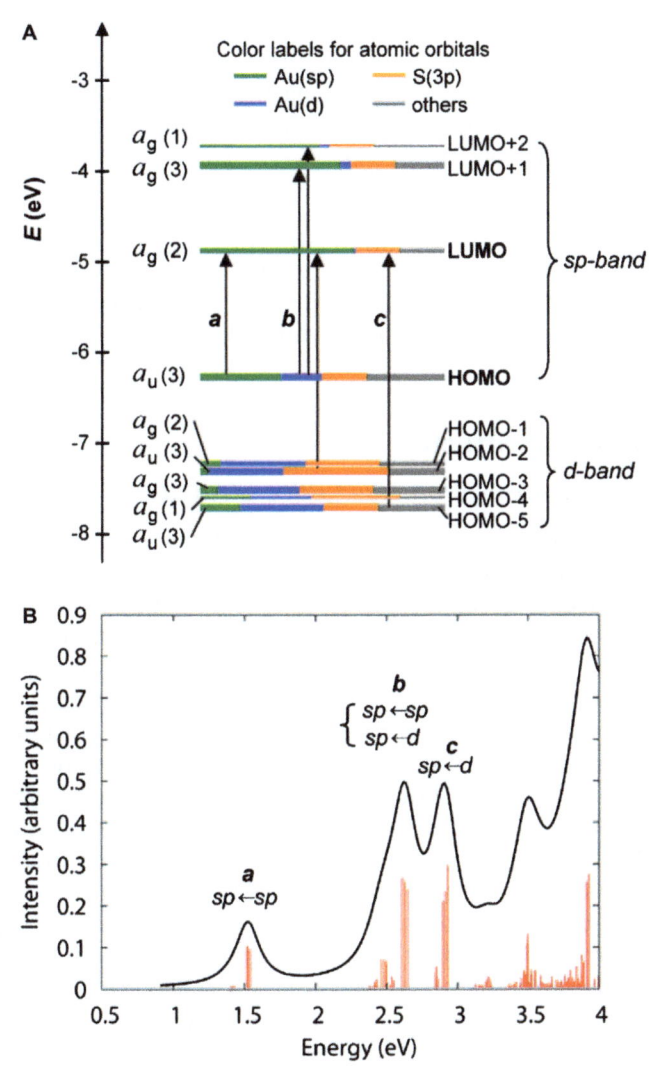

Figure 4.9 (A) Kohn–Sham orbital energy level diagram for a model compound $Au_{25}(SH)_{18}^-$. The energies are in units of eV. Each KS orbital is drawn to indicate the relative contributions (line length with color labels) of the atomic orbitals of Au (6sp) in green, Au (5d) in blue, S (3p) in yellow, and others in gray (those unspecified atomic orbitals, each with a <1% contribution). The left column of the KS orbitals shows the orbital symmetry (g, u) and degeneracy (in parenthesis); the right column shows the HOMO and LUMO sets. (B) The theoretical absorption spectrum of $Au_{25}(SH)_{18}^-$. Peak assignments: peak 'a' corresponds to 1.8 eV observed, peak 'b' corresponds to 2.75 eV (observed), and peak 'c' corresponds to 3.1 eV (observed).
(Reproduced with permission from ref. 47, Copyright American Chemical Society, 2008).

of the nanoclusters (*i.e.*, at 298 K), the average spacing (δ) between the energy levels become much larger than the thermal energy resulting in energy quantization becoming distinct. By replacing the value of Fermi energy for gold as 5.5 eV into the equation, the critical number of gold atoms can be obtained as ~200–300 atoms (*i.e.*, ~1.8–2.1 nm) from the following equation:

$$N = (59 \text{ atoms nm}^{-3}) \, V \, (\text{nm}^3) \qquad (2)$$

Therefore, ~2 nm is the crucial size at which quantization of electronic energy becomes significant and below which size, the collective plasmons do not exist.

4.3.1 Electronic Structure

4.3.1.1 X-Ray Absorption Near Edge Structure

X-Ray absorption fine structure (including X-ray absorption near edge structure (XANES) and extended X-ray absorption fine structure (EXAFS)) is an element-specific spectroscopic technique to explore the chemical environment, electronic and local atomic structure of different kinds of gold nanomaterials.[32,39,49–59] XANES usually refers to the region up to 40–50 eV above the absorption edge in X-ray absorption fine structure. The factors affecting the spectra shape and edge position of XANES include electronic structure, oxidation state, local symmetry and ligand bonding environment of the interesting species. Thus, XANES has recently been selected as one of the important techniques to investigate atomically precise Au nanoclusters.[32,39,50–55,57] More specifically, the intensity of the first resonance (so-called white line) in Au $L_{3,2}$-edge XANES is usually used to evaluate the unoccupied d-band states of Au clusters,[39,49–51,54] as the white line intensity of Au $L_{3,2}$-edge is associated with $2p_{3/2}$ and $2p_{1/2}$ to 5d transition. In addition, any shift of edge position can be associated with change of average oxidation state of Au atoms. Moreover, the XANES of the elements involved in Au–ligand bonding (*e.g.*, S K-edge in thiolate ligand) was also utilized to provide alternative perspectives for studying the Au–ligand interaction in Au nanoclusters.[39,50,51,56]

To best of our knowledge, the thiolate protected Au nanoclusters ($Au_n(SR)_m$) are the most extensively studied ligand protected Au nanoclusters by XANES. The common structure of $Au_n(SR)_m$ nanoclusters consists of a Au core and $[RS(AuSR)_x]$ oligomers (staple). Zhang and Sham[50] applied the Au L_3-edge XANES to study the size-dependent charge redistribution of a series of thiol-capped Au nanoparticles (size: 1.6, 2.4 and 4.0 nm). The most important observation of Au L_3-edge XANES study was that the white line intensity increases as the particle size decreases and is always higher than that for Au metal. They also estimated an increase of 11.2%, 9.0%, and 7.2% in the d-hole for the 1.6, 2.4 and 4.0 nm nanoclusters respectively. Such a trend in the d band depletion was attributed to the charge transfer from

Atomically Precise Gold Catalysis

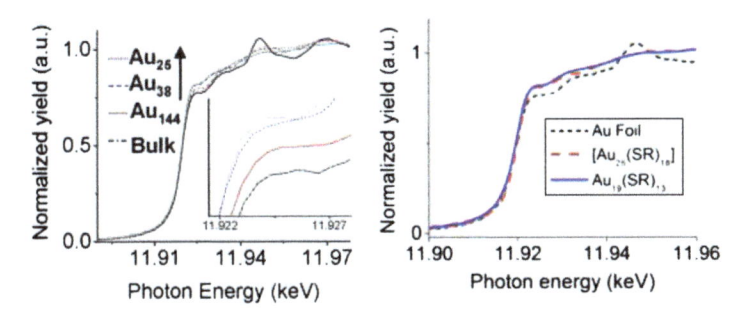

Figure 4.10 Au L$_3$-edge XANES of the thiolate ligand protected Au$_{19}$, Au$_{25}$, Au$_{38}$ and Au$_{144}$ nanoclusters compared with the bulk Au.
(Reproduced with permission from ref. 51 and 60, Copyright American Chemical Society, 2010, and 2012).

nanoparticles to thiol. They also examined the Au–S interaction using S K- and L$_{3,2}$-edge XANES which confirmed the Au–S bonding. The Au L$_3$-edge XANES of Au$_{19}$(SR)$_{13}$,[60] Au$_{25}$(SR)$_{18}$,[39,51] Au$_{38}$(SR)$_{24}$[32,51,61] and Au$_{144}$(SR)$_{60}$[51] have been reported by different groups. As seen in Figure 4.10,[51,60] the spectral features of all thiolate Au nanoclusters are broader in comparison with the bulk and the edge is slightly shifted to higher energy. These are related to the small particle size (or low Au coordination number) and slightly oxidized Au. There is a very clear trend of increasing white line intensity relative to the bulk Au as the cluster size reduces; which is consistent with the work of Zhang and Sham.[50] The estimated gain of d-hole counts relative to the bulk increased by 0.028 e$^-$ for Au$_{144}$, 0.082 e$^-$ for Au$_{38}$, and 0.089 e$^-$ for Au$_{25}$. It is known that the white line of bulk Au is contributed by s–p–d hybridization.[62,63] For bare Au nanoparticles,[49,57] the d–d interaction is very strong and s–d hybridization is relatively weak, so the corresponding white line intensity is lower than bulk Au. However, in the case of thiolate ligand protected Au nanoclusters,[51] the strong interaction between Au atoms and thiolate ligands causes the d band electrons of Au atoms to transfer to the thiolate ligands and thus result in increase in white line intensity. For smaller Au nanoclusters, higher the fraction of 'staple' Au atoms, higher is the observed d-band depletion. In addition to the Au L$_3$-XANES, the S K-edge XANES spectra of the nanoclusters (Figure 4.11) showed the lower energy shift of Au–S features at around 2470–2471 eV accompanied by an increase of Au–S relative peak in the order of Au$_{25}$, Au$_{38}$ and Au$_{144}$.[51]

The Au L$_3$-edge XANES of phosphine ligand protected Au$_{55}$ nanoclusters (Au$_{55}$(PPh$_3$)$_{12}$Cl$_6$) was reported by Benfield's group.[64] It was observed that the white line intensity of Au$_{55}$ was much smaller relative to KAu^{III}Cl$_4$ and (PPh$_3$)Au^{I}Cl and closely resembles that of bulk Au. It is consistent with the lower Au oxidization state in Au$_{55}$ (+6/55). The estimation of densities of unoccupied 5d states by the L$_3$ − kL$_2$ calculations pointed out the relative lower value for Au$_{55}$ (2.7) compared to bulk Au (4.1). The smaller size of Au$_{55}$ instead of the ligand effect was considered as dominant reason for it.

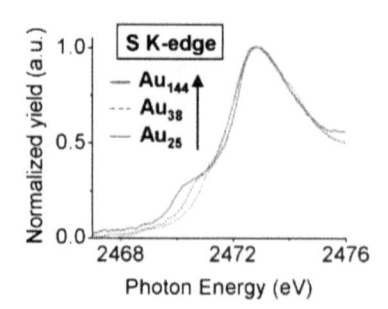

Figure 4.11 S K-edge XANES of the thiolate ligand protected Au$_{25}$, Au$_{38}$ and Au$_{144}$ nanoclusters.
(Reproduced with permission from, ref. 51 Copyright American Chemical Society, 2010).

However, an *ab initio* theoretical calculation of the partial density of state (PDOS) of an edge Au atom in the triangle Au$_6$ for Au$_{55}$(PPh$_3$)$_{12}$Cl$_6$ showed that the Fermi energy of Au is shifted to a higher energy compared to the bare Au$_{55}$. It indicated that the Au$_{55}$ core is negatively charged due to the electron back donation from the –P(Ph$_3$)$_3$ groups.[65]

Recently, phosphine-stabilized gold clusters [Au$_6$(Ph$_2$P-O-tolyl)$_6$](NO$_3$)$_2$ were investigated by time-resolved *in situ* XANES during the low-temperature peroxide-assisted removal of the phosphines.[66] The pattern of removal of the phosphines was evaluated by the change of spectra features. It was noticed that the white line intensity of spectra decreases as a function of time and finally resembles that of metallic gold. Further, the linear combination analysis of Au L$_3$-edge XANES showed that the fraction of the starting cluster compound decreases with the increasing fraction of metallic Au present in the system as a function of time.

As the phosphine and thiolate ligands favor different ways of bonding, the mixed ligand protected Au nanoclusters exhibit different structures without staple.[53,56] The Au$_{13}$[PPh$_3$]$_4$[S(CH$_2$)$_{11}$CH$_3$]$_2$Cl$_2$ and Au$_{13}$[PPh$_3$]$_4$[S(CH$_2$)$_{11}$CH$_3$]$_4$ nanoclusters were characterized by Au L$_3$-edge XANES.[53] The results suggested that the mixed ligands play the role as an electron acceptor in both of the Au$_{13}$ clusters resulting in increase in the whiteline intensity. The study of mixed-ligand biicosahedral structure [Au$_{25}$(PPh$_3$)$_{10}$(SC$_{12}$H$_{25}$)$_5$Cl$_2$]$^{2+}$ by Au L$_3$-XANES[35] revealed higher d-band vacancies in this mixed-ligand nanoclusters compared to Au$_{25}$(SR)$_{18}$ and bulk gold, indicating a significant ligand effect in the electron structure. The S and P K-edge XANES showed the formation of Au–S(P) bondings and features different compared to the thiolate ligand–Au bonding.

Besides the nanoclusters mentioned above, the protein stabilized gold nanoparticles have also attracted scientific and technical interests due to their potential application in biomedical imaging and analytical detection.[67,68] Some researchers have utilized Au L$_3$-edge XANES to characterize these novel biomaterials.[54,69] For example, the BSA-stabilized Au

Atomically Precise Gold Catalysis 105

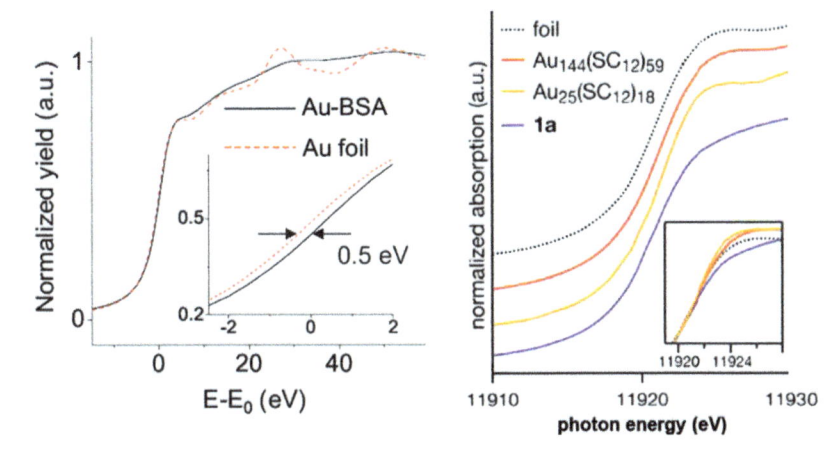

Figure 4.12 Au L$_3$-edge XANES of the BSA (left) and PVP (right) stabilized Au nanoclusters compared with the bulk Au.
(Reproduced with permission from, ref. 54 and 58, Copyright AIP Publishing LLC, 2009 and American Chemical Society, 2009).

nanoclusters[54] (shown in Figure 4.12) was found to possess lower d-electron density than that of the bulk by 0.047 e⁻ and thus reflects higher d-band depletion. This was consistent with the theoretical calculation of projected local density of states (L-DOS). The Au–S bonding effect (charge transfer) associated with the 'staple' motif structure analogous to Au$_{25}$(SR)$_{18}$ was well correlated with the increasing white line intensity in XANES and positive edge shift in XANES and Au 4f XPS.[54]

Interestingly, the characteristics of white line intensity in Au L$_3$-edge XANES (Figure 4.12) observed for 2-nm AuNPs capped with dendrimer was very different from the thiolate and mixed ligands stabilized Au clusters. The white line intensity is less intense compared to the bulk, indicating that the Au atoms gain 5d electron relative to the bulk when capped with weakly interacting dendrimers.[70] Similar phenomenon was reported in the case of PVP polymer stabilized Au nanoparticles.[58] It is also worth mentioning that the Au L$_3$-XANES of the solvated Au$_{38}$ has shown noticeable lower white line intensity in comparison with the solid-state Au$_{38}$.[61] It indicates a relatively higher 5d electron intensity in solvated Au$_{38}$. It is likely due to the charge transfer being more efficient in solid-state Au$_{38}$ having a shorter Au–S bond.

4.3.1.2 Photoemission Spectroscopy for Valence Band Study

The ultraviolet and valence band X-ray photoemission spectroscopy can be used to directly study the valence band electronic structure including the 6s and 5d bands density of states of Au. A series of ligand protected Au nanoclusters have been investigated by the valence band XPS, as shown in Figure 4.13.[50,51,60] The X-ray source was used in these experiments to more

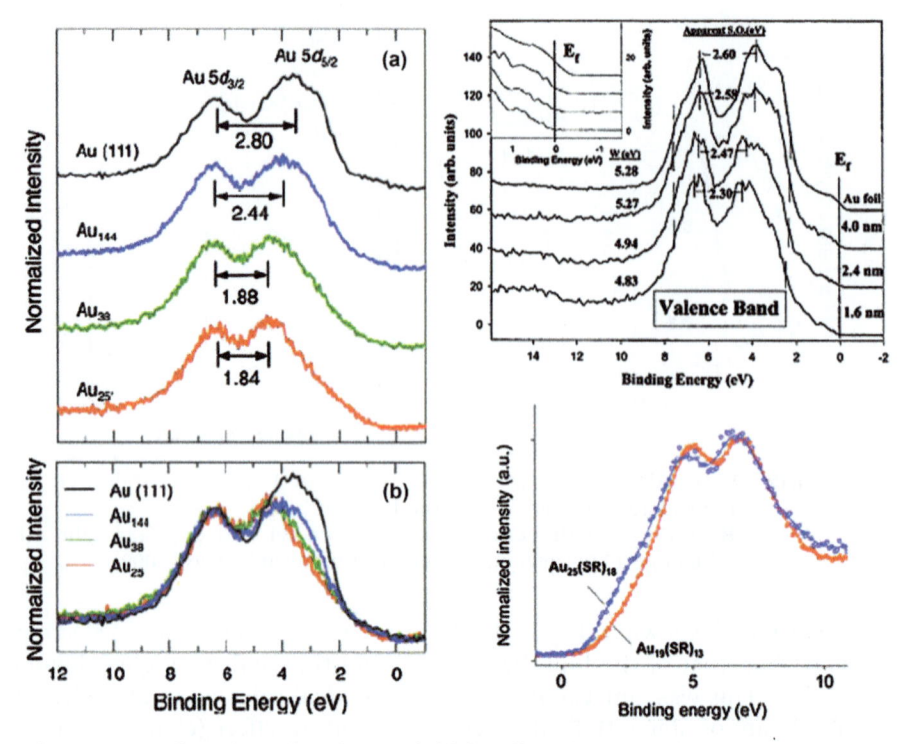

Figure 4.13 The valence band XPS of thiolate ligand protected Au nanoclusters. (Reproduced with permission from, ref. 50, 51 and 60, Copyright American Physical Society 2003, American Chemical Society 2010 and 2012).

efficiently collect the photoelectrons from the buried surface in the measurements.[60] The $5d_{5/2}$ and $5d_{3/2}$ orbitals were identified in all spectra. The earlier work by Zhang and Sham[50] revealed that the d band width and apparent spin–orbit splitting became narrower and the centroid of d band shifted to higher energy relative to the bulk when particle size decreases. This is in agreement with the results obtained for bare Au nanoparticles. The valence band XPS were carried out to study the valence band of Au_{25}, Au_{38} and Au_{144} with size of 1.0, 1.3 and 1.6 nm.[51] They pointed out that the 5d band separation does not depend on size but rather on average nearest neighbor by comparing the Au_{25} and Au_{38} nanoclusters. Furthermore, it was observed that the intensity of Au $5d_{5/2}$ in valence band XPS decreases as the cluster size decreases. This is because the 5d band electron depletion became more significant when the cluster size became smaller. And the charge transfer from Au to S is preferred from Au $5d_{5/2}$ rather than the more stable $5d_{3/2}$ level. The comparison between the valence band XPS of Au_{19} and Au_{25} nanoclusters[60] demonstrated that the composition of the clusters affect the feature at 2–3 eV range which is present in Au_{25} while absent in Au_{19}. The valence band electronic structure of phosphine stabilized Au_{55}

clusters was characterized by the valence band XPS as well. The shifts of peaks and Fermi level to higher binding energy were interpreted by final state effect.[71]

More recently, the ultraviolet photoemission spectroscopy (UPS) was applied in the study of Au_{38}, Au_{25} and mixed ligand protected Au_{25}. By using VUV as photon source, this spectroscopic tool is particularly sensitive in the surface region (~ 1 nm), providing information about the valence band structure near the Fermi level, which has been successfully employed to investigate bare Au nanoparticles.[72,73] The UPS is capable to yield fine structure of valence band of ligand protected Au nanoclusters. For instance, the UPS study of thiolate-stabilized Au_{38} nanoclusters supported on Si(111) wafer showed that the valence-band spectra of Au_{38} barely resembled the gold valence band features before sputtering.[74] The intensity of 5d band features at about 3 and 6 eV starts to increase after first gentle Ar sputtering cycle and began to be more pronounced at about 3.4 eV for $5d_{j/2}$ orbital, shown in Figure 4.14. A very important finding is that the existence of 4f threshold photoemission resonances strong at $4f_{7/2}$ and weaker at $4f_{5/2}$ threshold in the partially undressed $Au_{38}(SR)_{24}$ clusters deposited on the native SiO_2 oxide layer. It demonstrated that noble metal clusters, even of ultra-small size, will exhibit s–d hybridization.

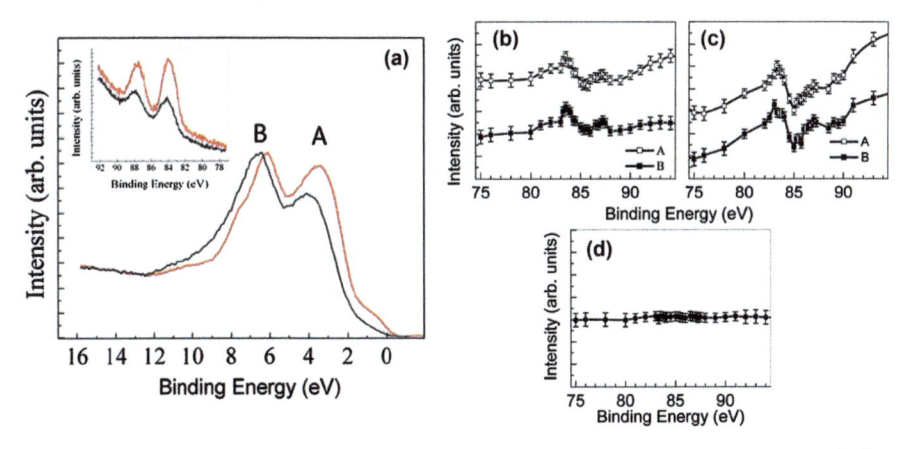

Figure 4.14 (a) Valence band photoemission spectrum of thiol-stabilized $Au_{38}(SR)_{24}$ clusters deposited from solution onto the SiO_2 native oxide surface prepared on Si(111), after sputtering to remove the alkyl thiol ligands (black). The spectrum is compared to a micrometer thick (111) textured gold film (red). The photon energy is 73 eV. The inset shows the 4f core level region for photon energy of 125 eV. Photoemission intensities at binding energies of ~ 3 eV (A) and 6.5 eV (B) (see (a)) vs. photon energies for (b) Au_{38} nanoclusters after sputtering, (c) thick (111) textured gold film. (c) Photoemission intensities at binding energy of ~ 6.5 eV (B) vs. photon energies for ligand protected $Au_{38}(SR)_{24}$ clusters. (Reproduced with permission from, ref. 74. Copyright American Chemical Society, 2010).

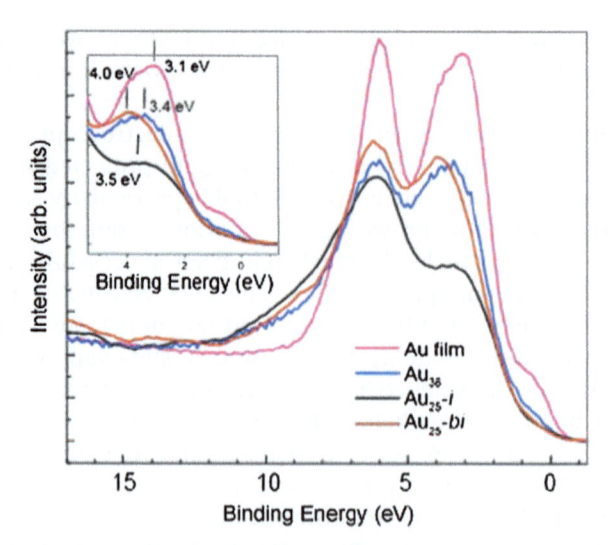

Figure 4.15 Valence-band photoemission spectra of Au$_{25}$-bi (red), Au$_{25}$-i (black; without sputtering), and Au$_{38}$ clusters (after sputtering to remove the thiolate ligands (blue)) deposited from solution on SiO$_2$ native oxide surface prepared on Si(111), compared to a micron thick (111) textured gold film (magenta).
(Reproduced with permission from, ref. 39, Copyright Wiley-VCH Verlag GmbH & Co. KGaA, 2010).

A detailed comparison (Figure 4.15) between the thiolate ligand protected Au$_{25}$ (Au$_{25}$-i), Au$_{38}$ and mixed ligand biicosahedral Au$_{25}$ (Au$_{25}$-bi)[39] revealed that a trend of narrower d-band width and d-band apparent spin–orbit splitting and higher binding energy of d-band center position in the order of for Au$_{25}$-bi and Au$_{25}$-i. The difference in electronic structure between the Au$_{25}$-i and Au$_{25}$-bi was possibly due to the differences in the nature of ligands and the local atomic structure though they have similar cluster size and same number of Au atoms. A close look at the spectra indicated higher 6s-band density of states at the Fermi level region for Au$_{38}$ cluster than Au$_{25}$-bi and Au$_{25}$-i nanoclusters. It suggested that Au$_{38}$ could be considered as more metallic than Au$_{25}$-bi and Au$_{25}$-i in accordance with observed trend for d-band binding energies.

4.3.2 Magnetic Structure

Most of the magnetic properties of gold nanoparticles or for that matter gold nanofilms can be attributed to either the surface atoms or surface-bound molecules.[75–77] Though there have been numerous computational studies of the structure and dynamics of small gold clusters, the origin of magnetism in gold nanoparticles is still not well-understood. The electronic configuration of gold atom indicates that it has a single unpaired electron in the 6s orbital. However, the relativistic band structure of bulk gold reveals an equal number of spin-UP (alpha) and DOWN (beta) electrons. While this predicted diamagnetism is in excellent agreement with the experimentally observed negative susceptibility and diamagnetism of bulk gold,[78,79] it is interesting to note that a recent experimental X-ray magnetic circular dichroism (XMCD) spectroscopic investigation reveals a Pauli paramagnetism in bulk Au, which is masked by a large diamagnetic response.[80] This Pauli paramagnetism predominantly emerges from the surface atoms, whose spin polarization is stabilized by a large ($\sim$30%) orbital contribution from Au 5d electrons. This orbital contribution is sustained in smaller gold nanoparticles and plays an important role in stabilizing spontaneous spin polarization. The role of surface atoms is further magnified in smaller gold nanoparticles because a decrease in the size of the pure gold nanoparticles from 2.6 nm (Au_{561}) to 0.75 nm (Au_{13}), leads to a two-fold increase (45% to 92%) in the ratio of surface atoms to the total number of atoms.

To the best of our knowledge, there has been only a single investigation of magnetism in pure uncapped icosahedral gold nanoparticles.[81] The spin arrangements therein indicated that the core and surface moments pointing in opposite directions. While the net magnetic moment 16 µB per particle indicated ferrimagnetism, it is interesting to note that the net contribution of the surface atoms was found to be much larger than that of the core atoms. Nearly all the other investigations of magnetism in small gold nanoparticles involved either polymer-capped or ligand-covered systems (Table 4.1). This raises a very important question on whether the observed magnetism is an intrinsic property of the gold atoms or the nature of the ligands in the ligand-covered gold nanoparticle.

The presence of ligands has been shown to increase the number of holes in the 5d band of gold nanoparticles.[70] The strong affinity between the Au surface atoms and the ligand atoms induces a charge transfer from the Au surface atoms to the ligand atoms where the participation of 5d electrons can also be implied leading to generation of unoccupied densities of d states on Au atoms resulting in magnetism. Atomically precise gold clusters, unlike gold nanoparticles, are made up of a specific number of atoms and ligands, whose electronic configuration can be theoretically calculated, which in turn could be used to predict their magnetic behavior.[82] Our group[83] has experimentally investigated the chemically induced magnetism in ligand-capped atomically precise gold clusters, viz. Au_{25}, Au_{38} and Au_{55} using SQUID. We found ferromagnetic behavior in ligand-capped Au_{55} clusters. Figure 4.16 shows their M vs. H curves obtained for four different gold clusters at 5 K. The mixed ligand-stabilized Au_{25} clusters (black) show typical diamagnetic behavior. The thiol-stabilized Au_{25} clusters (red) show paramagnetic behavior with an experimentally observed saturation magnetic moment of $\mu_B = 0.0516$/cluster. The thiol-stabilized Au_{38} clusters (green) show diamagnetic behavior and the phosphine-stabilized Au_{55} clusters (blue) show a ferromagnetic behavior with a saturation magnetic moment of $\mu_B = 0.0584$/cluster.

Table 4.1 Recent investigations on magnetism in gold clusters/nanoparticles.

Gold clusters/nanoparticles	Ligand	Magnetic property	Ref.
Au$_2$ and Au$_3$	Polyvinylpyrrolidone (PVP)	Paramagnetic	113
Au$_{11}$ (~1.4 nm)	Triphenylphosphine	Diamagnetic	114
Au$_{18}$	Glutathione	Paramagnetic	115
Au$_{25}$ (neutral)	Phenylethylthiol	Paramagnetic	83
Au$_{25}$ (biicosahedral)	Triphenylphosphine	Diamagnetic	83
Au$_{38}$	Phenylethylthiol	Diamagnetic	83
Au$_{55}$	Triphenylphosphine	Ferromagnetic	83
Au$_{25}$ (anionic)	Phenylethylthiol	Non-magnetic	99
Au$_{25}$ (neutral)	Phenylethylthiol	Paramagnetic	99
12 nm particles	Dodecanethiol	Diamagnetic	116
5 nm	Dodecanethiol	Super-paramagnetic	116
2 nm particles	Glutathione	Paramagnetic	117
4 nm particles	Glutathione	Paramagnetic	117
1.9 nm particles	Dodecanethiol	Ferromagnetic	109
2 nm organosol	Dodecanethiol	Ferromagnetic	118
6.7 nm	Oleic acid and oleylamine	Ferromagnetic	119
1.4 nm	Dodecanethiol	Ferromagnetic	92
1.5 nm and 5 nm	Tetraoctyl ammonium bromide	Diamagnetic	92
2 nm and 5 nm	Tiopronin	Paramagnetic	120
2 nm	Dodecanethiol	Ferromagnetic	120
2.1 nm	Triphenylphosphine	Ferromagnetic	114
1.9 nm	PAAHC (polyallyl amine hydrochloride)	Ferromagnetic	121
1.7 nm	Azobenzene-thiol	Ferromagnetic	122,123
5 nm	Azobenzene-thiol	Diamagnetic	122
2.0 nm	Dodecanethiol	Ferromagnetic	124
2.3 nm	Octanethiol	Ferromagnetic	124
2.5 nm	Tiopronin	Paramagnetic	124
1.9 nm	Octanethiol/11-thioundecanoic acid	Ferromagnetic	124
3.5 nm	(Without ligand)	Ferromagnetic	81
14.8 nm	(Without ligand)	Ferromagnetic	125
2.1	Dodecanethiol	Paramagnetic	77
2.2 nm	Dodecanethiol	Diamagnetic	77
3.5 nm	Dodecanethiol	Diamagnetic	77
2.5 nm	PVP	Ferromagnetic	126
2 nm	Dodecanethiol	Ferromagnetic	127
2.5 nm	Tiopronin	Paramagnetic	127
3–4 nm	Dodecanethiol/polyethylene	Ferromagnetic	128
2.1 nm	Dodecanethiol	Ferromagnetic	129
1.5 and 5 nm	Tetraoctyl ammonium bromide	Diamagnetic	127

Density functional theory (DFT) calculations on the size-dependent magnetic properties of bare gold nanoparticles found that sizes with an even number of electrons have magnetic ground states of variable multiplicity.[84,85] These states are symmetry-breaking variants of the closed-shell singlet responsible for the diamagnetic behavior of bulk gold. In systems having an odd number of electrons, the (doublet) ground state is magnetic

Atomically Precise Gold Catalysis 111

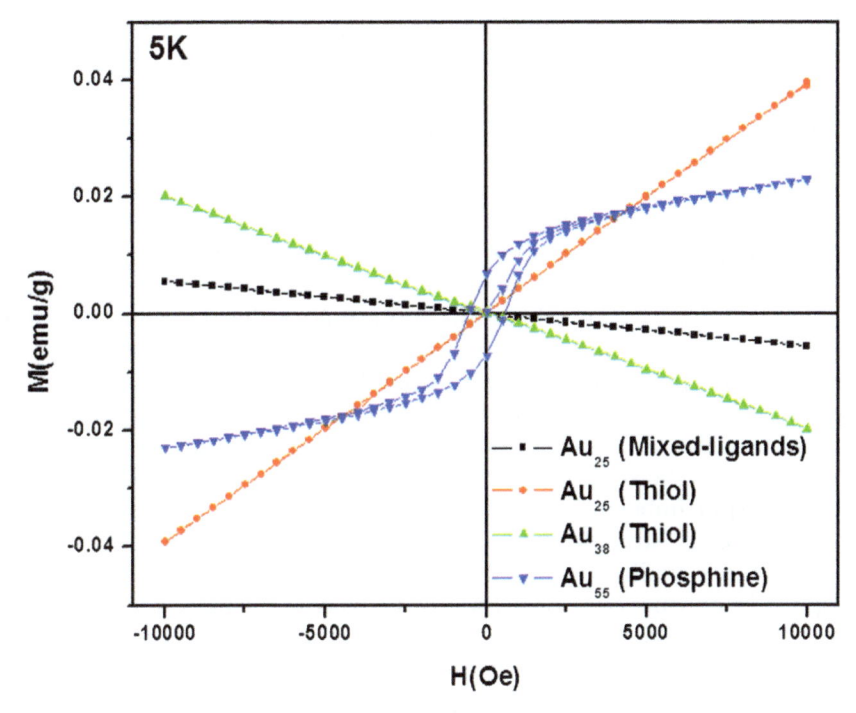

Figure 4.16 M *vs.* H curves for ligand-capped Au$_{25}$, Au$_{38}$ and Au$_{55}$ clusters at 5 K. (Reproduced with permission from ref. 83).

in all cases, and in some cases could also lead to enhanced magnetic moment. One of the notable observations made in the course of these calculations is that magnetism in small gold nanoparticles is essentially a property of the surface atoms which is associated to small changes in the geometry of the surface. This translates into spin symmetry breaking at the Fermi energy, which in turn causes the appearance of a non-zero magnetic moment for the nanoparticle. Subsequent calculations on a system wherein a phenylthiol (C$_6$H$_5$S) group bound to a 13 gold atom nanoparticle indicates the breaking of the molecular spin symmetry due to the formation of Au–S bond.[86] This leads to the appearance of a weak magnetic moment of 2.5×10^{-3} µB. Interestingly, the replacement of the thiol group by an amino group leads to the disappearance of the magnetic moment.[86] Thus, the presence of ligand could strongly modify the properties of the ground state through the partial transfer of electrons or holes, which in turn induce a change of the spin population at the Fermi level of the cluster.[86,87]

Based on DFT calculations of bare gold nanoparticles, it was proposed that a degenerate and partially filled highest occupied molecular orbital (HOMO) could explain the observed magnetism.[88] The rationale for the magnetism is that these molecular orbitals are filled by s-electrons with spins aligned according to Hund's rule. Interestingly, the role of surface atoms becomes important because the spin-aligned s-electrons are predominantly localized on them.[88] A more recent DFT calculation of thiolated gold nanoparticles comprised of three to six gold atoms indicates the absence of polarization on the sulfur atom of the thiol group.[89] Basically, a sulfur electron (possessing sp character) is transferred to the gold nanoparticle which then orbits below the surface atoms and is therefore responsible for the observed magnetism. The only shortcoming with this interesting result is that the calculations have been carried out on very small gold nanoparticles possessing only six atoms.

The magnetism observed in ligand-coated gold nanoparticles can also be explained using the 'Fermi hole effect'.[90] The formation of a gold–ligand covalent bond leaves an exchange or Fermi hole in the electronic d shell of the gold atom.[90,91] Contrary to the conventional notion that Au forms a covalent bond purely through its 6s orbital, the above work invokes 6s–5d hybridization. The idea is that the neighboring electrons surrounding the hole bear antiparallel spins, leading to a net total spin of zero. When the hole is close to the surface, the hole is only half-covered by electrons, which creates a spin imbalance and a net spin polarization. The degree of s–d hybridization is strongly modulated by the ligand and thus gold nanoparticles functionalized with dodecanethiol are less magnetic than particles wrapped in polymers like polyacrylonitrile or polyvinylpyrolidone. The concept of localized 5d hole formation in the vicinity of the Au–S bonds leading to the emergence of a magnetic moment was invoked by Crespo *et al.* to explain the magnetism in thiol-capped gold nanoparticles.[92]

Another theory which explains the orbital magnetism is based on the strong spin–orbit coupling of gold.[93,94] However, its validity is limited to ordered arrays of Au–S bonds on gold surfaces. A recent extension of this theory to magnetism in gold nanoparticles indicates that the grafting of ligands partially depletes the Fermi level, and this unfilled Fermi level develops magnetic moments.[95] Thus, the observed magnetism is of an orbital nature, which can reach very large values depending on the number of electrons in the Fermi level. The occupation at the Fermi level depends on the size of the nanoparticle and other parameters like the nature and number of ligands. Thus, nanoparticles could be either diamagnetic or paramagnetic depending on the number of electrons occupying the Fermi level.

A similar conclusion was reached in a recent work on bare gold nanoparticles of varying sizes (Au_{38}, Au_{55}, Au_{79}).[96] The calculations showed that for Au_{38}, there is a spin symmetry breaking at the Fermi energy and that a majority spin population is observed with respect to the closed-shell singlet. It is useful to recall that the appearance of magnetic moment is because of the spin imbalance resulting from splitting of the electron bands due to spin-symmetry breaking.[97]

One of the remarkable results that has been observed experimentally is that the onset of magnetism in gold nanoparticles is frequently accompanied by the quenching of the absorption peak (~520 nm) in the visible spectrum associated with the plasmon resonance.[92,98] While the appearance of this peak is determined by a condition associated with the behavior of the imaginary component of the dielectric function of the metal, it was found

Atomically Precise Gold Catalysis

that the oscillator strengths depend drastically on the spin multiplicity. Thus, the oscillator strengths for the relevant transitions are substantially larger for a regular singlet than for either the polarized singlet or the higher multiplicity states.[92,96]

Against the backdrop of this discussion on magnetism in gold nanoparticles being largely due to surface atoms, it is interesting to note that Zhu *et al.* found in an investigation of Au_{25} decorated with phenylethylthiols that the unpaired electron is delocalized in a p-orbital localized at the center of the cluster.[99]

The salient conclusions of this section are that bare gold nanoparticles can occasionally acquire a magnetic polarization which can be either preserved or reinforced by ligands. There is growing consensus that the magnetism is mainly due to surface atoms and is of orbital nature. Depending on whether the Fermi level is filled or unfilled, a recent theory can predict on whether a gold nanoparticle is paramagnetic or diamagnetic.[95] However, there is still no cogent explanation on the origin of the ferromagnetism in these gold nanoparticles. In most cases, ferromagnetism is linked to the presence of magnetic walls or domains. However, these nanoparticles are too small and the insulating nature of most of the commonly used organic ligands is not conducive to inter-particle electronic communication. Therefore, it is likely that dipolar interaction is the likely mechanism responsible for the stabilization of bulk magnetization.[100,101]

4.4 Correlation of Electronic Structure–Catalysis Relationship

Correlating the electronic and magnetic structure with catalysis has only been investigated by very few research groups. Nevertheless, there are definitive conclusions suggesting the involvement of the electronic structure with catalysis. For instance, Falicov and Somorjai[102] have proposed a correlation between catalytic activity and low-energy local electronic fluctuations in transition metals. They presented a theory and calculations which indicate that maximum electronic fluctuations took place at high-coordination metal sites. They also have investigated three different reactions, *viz.* synthesis of ammonia from N_2 and H_2 over iron and rhenium surfaces, $H_2/2H_2$ exchange over stepped platinum crystal surfaces at low pressures, and the hydrogenolysis (C–C bond breaking) of isobutane at kinked platinum crystal surfaces to support their theory. Based on the observations, they have concluded that (i) atomically rough surfaces exposes reactant molecules with large numbers of surface atoms, and (ii) stepped and kinked surfaces are the most active in carrying out structure-sensitive catalytic reactions.

The formation of bimetallic surfaces leads to a change in the electronic properties of the parent metal surface, which results in changes in chemical reactivity. Goda *et al.*[103] used DFT calculations to study correlation of electronic properties of bimetallic Pt–Ni surfaces with the catalytic reaction pathways of selective reduction of acetic acid to acetaldehyde. DFT calculations on modes and energies of adsorption of the involved intermediates along with the overall reaction energies were used for this study. In a different report, Malheiro et al.[104] have investigated the effect of oxygen reduction reaction (ORR) in fuel cells containing Pt–M/C catalysts (M = 3d transition metals) by characterizing the electronic properties of Pt–Fe/C catalysts using in situ dispersive X-ray absorption spectroscopy (DXAS) to assess their dependence on Fe content and to analyze its correlation with ORR activity.

Recently, Liu et al.[39] explored the catalytic oxidation of styrene using two different gold cluster catalysts which have same atomic core size of 25 atoms, but different geometric and electronic structure. Using a combination of synchrotron radiation-based X-ray absorption fine structure spectroscopy (XAFS) and ultraviolet photoemission spectroscopy (UPS), the authors have found a direct correlation between the catalytic selectivity of the styrene oxidation reaction with the electronic structure of the nanoclusters. They have concluded that more electropositive nature of the cluster would yield more oxidized products in the reaction and *vice versa*.

Due to the advances being made in the synthesis of atomically precise catalysts, it has now become possible to investigate more thoroughly the correlation between their electronic structure and catalytic properties. These studies in near future are anticipated to pave way for a better understanding of the atomically-precise information related to molecular activation and reaction mechanisms leading to superior design of catalysts with required performance.

4.5 Correlation of Magnetic Structure–Catalysis Relationship

The correlation between magnetism and catalysis has been extensively investigated since past century and was reviewed as early as 1946 by P. W. Selwood[105] from Northwestern University, USA. There are several examples that have been elucidated in that review to highlight the effect of magnetism on catalytic activity of different magnetic metals and metal oxides. For instance, a linear relation between magnetization and catalytic activity was observed for hydrogenation of benzene on Ni–Cu catalyst at 175 °C.[106] The results showed that an active catalyst is quite strongly ferromagnetic, and that heating the catalyst produced a loss of ferromagnetism. This loss of ferromagnetism occurred in a linear relationship to the loss of catalytic activity (Figure 4.17).

In a different example, the catalytic activity of pure Ni metal for nitrous oxide decomposition near its Curie point was found to increase exponentially[107] due to sharp increase in the temperature coefficient of reaction velocity at the Curie temperature. The catalytic activity was also shown to be affected by magnetism through the change in binding energies of the

Atomically Precise Gold Catalysis 115

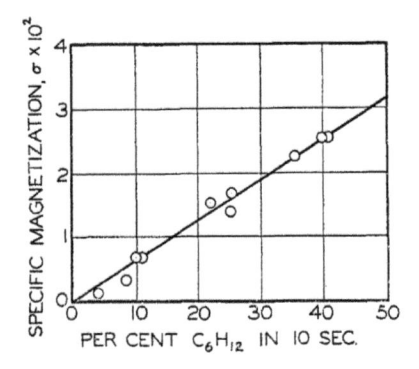

Figure 4.17 Linear relationship between magnetization and catalytic activity. (Reproduced with permission from, ref. 106, Copyright American Chemical Society, 1943).

catalyst. K. N. Saroj *et al.* studied the relationship between magnetism, topology, and reactivity of Rh clusters for chemisorption of H_2 using theoretical calculations.[108] They found that the binding energy of H_2 to the non-magnetic Rh_4 is almost a factor of 2 larger than its magnetic counterpart which shows that the reactivity of H_2 may depend on the underlying magnetic structure of the cluster.

Of late, atomically-precise gold clusters (nanoparticles) of size ~2 nm, capped with dodecanethiol were reported to show ferromagnetic behavior.[109] Though theoretical reports suggested the possible ferromagnetism in such atomically precise gold clusters,[110] there was a lack of supporting experimental studies. Recently, our group has experimentally investigated the chemically induced magnetism in ligand-capped atomically precise gold clusters[83] and explored the correlation between their magnetic structure and catalytic activity.

Here we provide a snap shot of results from our preliminary investigation on the influence of magnetism on the catalysis of atomically precise Au_{25} clusters. The atomically precise Au_{25} clusters, which showed paramagnetic behavior, were chosen for investigating the effect of magnetism on their catalytic activity for styrene oxidation reaction. The Au_{25} clusters were supported on fumed silica (~14 nm particles) and the styrene oxidation reaction was performed both in the presence and absence of magnetic field (~13 200 gauss). The results of the catalysis reaction are given in Table 4.2. While the experimental data is preliminary and needs further revaluation, one can notice a clear indication of correlation between magnetism of the clusters and their catalytic activity. The effect of magnetism primarily observed on the selectivity of the products formed. The presence of magnetic field has suppressed the formation of benzaldehyde from 70% to 56% and increased the yields of other products such as styrene oxide (from 22% to 29%), benzene acetaldehyde (from 3% to 4%) and acetophenone (from 5% to 11%). However, in either case the over-all conversion remained almost same (*i.e.*, 76% and 77%).

Table 4.2 Catalysis results of Au$_{25}$ clusters for styrene oxidation reaction in the presence/absence of magnetic field.

Catalysis conditions	Reaction conditions	Selectivity for products formed (%)				Total conversion (%)
		Benzal-dehyde	Styrene oxide	Benzene acetaldehyde	Aceto-phenone	
Magnetic	Styrene, TBHP, Acetonitrile, 75 °C, Au$_{25}$(SR)$_{18}$@SiO$_2$	56	29	4	11	76
Non-magnetic	Styrene, TBHP, Acetonitrile, 75 °C, Au$_{25}$(SR)$_{18}$@SiO$_2$	70	22	3	5	77

4.6 Conclusions and Future Perspective

Atomically precise nanoclusters are a new class of nanocatalysts that hold great promise for investigating the fundamental aspects of nanocatalysis unlike the traditional nanoparticle catalysts.[111] Based on our discussion in this chapter, it is clear that there is a great opportunity to correlate both electronic and magnetic structure of the clusters with their catalytic activity. In addition, we believe that the field of atomically precise catalysis is moving in the right direction and its impact will be felt in the following ways:

1. With sizes ranging from about a ten to a few hundred atoms, atomically precise nanoclusters are rapidly gaining attraction among researchers. Well-defined atomically precise clusters with desired sizes can be achieved through careful control over synthetic conditions. Also, atomically precise clusters supported on oxide materials serve as a new model system for nanocatalysis to probe the interaction of the catalysts and supports.
2. With extensive literature now available on the electronic and geometric structures of the atomically precise clusters, a correlation of its structure–property relationships could be more thoroughly and precisely investigated. The correlation of the crystal structure of these nanoclusters with their catalytic properties will provide significant guidelines for the future design of catalysts, with atomic precision, for any specific chemical reaction. A computational analysis in combination with this correlation will be critical for a better understanding of the design elements.
3. The ability to follow the catalytic processes with atomic precision is an unfulfilled need in the field of nanoscience. The use of XAS techniques (EXAFS and XANES) for obtaining fundamental electronic information

on catalysts is widely known. With application of techniques such as *in situ* XAS especially if those coupled with catalysis in microfluidics,[112] catalysis reactions could be followed with atomic precision in future.

4. New insights into the magnetic properties of these atomically precise catalysts provide a means to chemically turn-on and tune-in their magnetism and thereby providing an opportunity to not only tailor-make atomically precise nanomagnetic clusters but also to investigate magnetic structure dependent catalysis.

Finally, we envisage tremendous opportunities for the atomically precise catalysts in the future where catalysis can be controlled by manipulating systematically the fundamental electronic and magnetic structure of the catalysts.

Acknowledgements

This research work is supported as part of the Center for Atomic Level Catalyst Design, an Energy Frontier Research Center funded by the U.S. Department of Energy, Office of Science, Office of Basic Energy Sciences under award number DE-SC0001058 and also supported by Board of Regents under grants award number LEQSF (2009-14)-EFRC-MATCH and LEDSF-EPS(2012)-OPT-IN-15.

References

1. M. Haruta, T. Kobayashi, H. Sano and N. Yamada, *Chem. Lett.*, 1987, **16**, 405–408.
2. M. Haruta, N. Yamada, T. Kobayashi and S. Iijima, *J. Catal.*, 1989, **115**, 301–309.
3. A. Corma and H. Garcia, *Chem. Soc. Rev.*, 2008, **37**, 2096–2126.
4. C. Della Pina, E. Falletta, L. Prati and M. Rossi, *Chem. Soc. Rev.*, 2008, **37**, 2077–2095.
5. M. D. Hughes, Y. J. Xu, P. Jenkins, P. McMorn, P. Landon, D. I. Enache, A. F. Carley, G. A. Attard, G. J. Hutchings, F. King, E. H. Stitt, P. Johnston, K. Griffin and C. J. Kiely, *Nature*, 2005, **437**, 1132–1135.
6. (a) Special issue on catalysis, *Nanotechnol. Rev.*, 2013, **2**(5), 485–614; (b) R. Jin, *Nanotechnol. Rev.*, 2012, **1**(1), 31–56; (c) R. Jin, *Nanoscale*, 2010, **2**, 343–362.
7. R. Jin, Y. Zhu and H. Qian, *Chem. – Eur. J.*, 2011, **17**, 6584–6593.
8. R. Jin, H. Qian, Y. Zhu and A. Das, *J. Nanosci. Lett.*, 2011, **1**, 72–86.
9. Y. Zhu, H. Qian and R. Jin, *J. Mater. Chem.*, 2011, **21**, 6793–6799.
10. G. Li and R. Jin, *Acc. Chem. Res.*, 2013, **46**, 1749–1758.
11. T. G. Schaaff, M. N. Shafigullin, J. T. Khoury, I. Vezmar, R. L. Whetten, W. G. Cullen, P. N. First, C. Gutierrez-Wing, J. Ascensio and M. J. Jose-Yacaman, *J. Phys. Chem. B*, 1997, **101**, 7885–7891.

12. R. L. Whetten, J. T. Khoury, M. M. Alvarez, S. Murthy, I. Vezmar, Z. L. Wang, P. W. Stephens, C. L. Cleveland, W. D. Luedtke and U. Landman, *Adv. Mater.*, 1996, **8**, 428–433.
13. J. F. Parker, C. A. Fields-Zinna and R. W. Murray, *Acc. Chem. Res.*, 2010, **43**, 1289–1296.
14. R. L. Donkers, D. Lee and R. W. Murray, *Langmuir*, 2004, **20**, 1945–1952.
15. Y. Negishi, K. Nobusada and T. Tsukuda, *J. Am. Chem. Soc.*, 2005, **127**, 5261–5270.
16. M. Zhu, E. Lanni, N. Garg, M. E. Bier and R. Jin, *J. Am. Chem. Soc.*, 2008, **130**, 1138.
17. H. Qian, M. Zhu, U. N. Andersen and R. Jin, *J. Phys. Chem. A*, 2009, **113**, 4281–4284.
18. H. Qian and R. Jin, *Nano Lett.*, 2009, **9**, 4083–4087.
19. R. Jin, H. Qian, Z. Wu, Y. Zhu, M. Zhu, A. Mohanty and N. Garg, *J. Phys. Chem. Lett.*, 2010, **1**, 2903–2910.
20. D. Stellwagen, A. Weber, L. S. Bovenkamp, R. Jin, H. Bitter and C. S. S. R. Kumar, *RSC Adv.*, 2012, **2**, 2276–2283.
21. M. W. Heaven, A. Dass, P. S. White, K. M. Holt and R. W. Murray, *J. Am. Chem. Soc.*, 2008, **130**, 3754–3755.
22. M. Valden, X. Lai and D. W. Goodman, *Science*, 1998, **281**, 1647–1650.
23. R. Meyer, C. Lemire, S. K. Shaikhutdinov and H. J. Freund, *Gold Bull.*, 2004, **37**, 72.
24. D. Stolcic, M. Fischer, G. Gautefor, Y. Kim, Q. Sun and P. Jena, *J. Am. Chem. Soc.*, 2003, **125**, 2848.
25. G. Li, H. Qian and R. Jin, *Nanoscale*, 2012, **4**, 6714–6717.
26. G. Li, C. Liu, Y. Lei and R. Jin, *Chem. Commun.*, 2012, **48**, 12005–12007.
27. D. R. Kauffman, D. Alfonso, C. Matranga, H. Qian and R. Jin, *J. Am. Chem. Soc.*, 2012, **134**, 10237–10243.
28. F. Gao, T. Wood and D. Goodman, *Catal. Lett.*, 2010, **134**, 9–12.
29. M. C. Kung, R. J. Davis and H. H. Kung, *J. Phys. Chem. C*, 2007, **111**, 11767–11775.
30. O. Lopez-Acevedo, K. A. Kacprzak, J. Akola and H. Hakkinen, *Nat. Chem.*, 2010, **2**, 329–334.
31. X. Nie, H. Qian, Q. Ge, H. Xu and R. Jin, *ACS Nano*, 2012, **6**, 6014–6022.
32. S. Gaur, J. T. Miller, D. Stellwagen, A. Sanampudi, C. S. S. R. Kumar and J. J. Spivey, *Phys. Chem. Chem. Phys.*, 2012, **14**, 1627–1634.
33. S. Gaur, H. Wu, G. G. Stanley, K. More, C. S. S. R. Kumar and J. J. Spivey, *Catal. Today*, 2013, **208**, 72–81.
34. M. Turner, V. B. Golovko, O. P. H. Vaughan, P. Abdulkin, A. B. Murcia, M. S. Tikhov, B. F. G. Johnson and R. M. Lambert, *Nature*, 2008, **454**, 981.
35. Y. Liu, H. Tsunoyama, T. Akita and T. Tsukuda, *Chem. Commun.*, 2010, **46**, 550–552.
36. J. Liu, F. Wang, T. Xu and Z. Gu, *Catal. Lett.*, 2010, **134**, 51.
37. Y. Zhu, H. Qian and R. Jin, *Chem. - Eur. J.*, 2010, **16**, 11455–11462.
38. Y. Zhu, H. Qian, M. Zhu and R. Jin, *Adv. Mater.*, 2010, **22**, 1915.

39. J. Liu, K. S. Krishna, Y. B. Losovyj, S. Chattopadhyay, N. Lozova, J. T. Miller, J. J. Spivey and C. S. S. R. Kumar, *Chem. - Eur. J.*, 2013, **19**, 10201–10208.
40. Z. Wu and R. Jin, *ACS Nano*, 2009, **3**, 2036.
41. P. Lignier, F. Morfin, L. Piccolo, J. L. Rousset and V. Caps, *Catal. Today*, 2007, **122**, 284.
42. C. Aprile, M. Boronat, B. Ferrer, A. Corma and H. Garcia, *J. Am. Chem. Soc.*, 2006, **128**, 8388.
43. P. Lignier, F. Morfin, S. Mangematin, L. Massin, J. Rousset and V. Caps, *Chem. Commun.*, 2007, 186.
44. Y. Zhu, H. Qian, B. A. Drake and R. Jin, *Angew. Chem., Int. Ed.*, 2010, **49**, 1295.
45. Y. Zhu, Z. Wu, C. Gayathri, H. Qian, R. R. Gil and R. Jin, *J. Catal.*, 2010, **271**, 155.
46. S. Eustis and M. A. El-Sayed, *Chem. Soc. Rev.*, 2006, **35**, 209–217.
47. M. Zhu, C. M. Aikens, F. J. Hollander, G. C. Schatz and R. Jin, *J. Am. Chem. Soc.*, 2008, **130**, 5883–5885.
48. H. Qian, M. Zhu, Z. Wu and R. Jin, *Acc. Chem. Res.*, 2012, **45**, 1470–1479.
49. J. T. Miller, A. J. Kropf, Y. Zha, J. R. Regalbuto, L. Delannoy, C. Louis, E. Bus and J. A. van Bokhoven, *J. Catal.*, 2006, **240**, 222–234.
50. P. Zhang and T. K. Sham, *Phys. Rev. Lett.*, 2003, **90**, 245502-1–245502-4.
51. M. A. MacDonald, P. Zhang, H. Qian and R. Jin, *J. Phys. Chem. Lett.*, 2010, **1**, 1821–1825.
52. C. López-Cartes, T. C. Rojas, R. Litrán, D. Martínez-Martínez, J. M. de la Fuente, S. Penadés and A. Fernández, *J. Phys. Chem. B*, 2005, **109**, 8761–8766.
53. L. D. Menard, H. Xu, S. Gao, R. D. Twesten, A. S. Harper, Y. Song, G. Wang, A. D. Douglas, J. C. Yang, A. I. Frenkel, R. W. Murray and R. G. Nuzzo, *J. Phys. Chem. B*, 2006, **110**, 14564–14573.
54. G. Simms, J. D. Padmos and P. Zhang, *J. Chem. Phys.*, 2009, **131**, 214703-1–214703-9.
55. P. D. Cluskey, R. J. Newport, R. E. Benfield, S. J. Gurman and G. Schmid, *Z. Phys. D: At., Mol. Clusters*, 1993, **26**, 8–11.
56. A. I. Frenkel, L. D. Menard, P. Northrup, J. A. Rodriguez, F. Zypman, D. Glasner, S. P. Gao, H. Xu, J. C. Yang and R. G. Nuzzo, *AIP Conf. Proc.*, 2007, **882**, 749–751.
57. J. A. Van Bokhoven and J. T. Miller, *J. Phys. Chem. C*, 2007, **111**, 9245–9249.
58. H. Tsuoyama, N. Ichikuni, H. Sakurai and T. Tsukuda, *J. Am. Chem. Soc.*, 2009, **131**, 7086–7093.
59. A. Visikovskiy, H. Matsumoto, K. Mitsuhara, T. Nakada, T. Akita and Y. Kido, *Phys. Rev. B*, 2011, **83**, 165428-1–165428-9.
60. D. M. Chevrier, M. A. MacDonald, A. Chatt and P. Zhang, *J. Phys. Chem. C*, 2012, **116**, 25137–25142.
61. M. A. MacDonald, P. Zhang, N. Chen, H. Qian and R. Jin, *J. Phys. Chem. C*, 2011, **115**, 65–69.

62. L. F. Mattheiss and R. E. Dietz, *Phys. Rev. B: Condens. Matter Mater. Phys.*, 1980, **22**, 1663–1676.
63. M. G. Mason, *Phys. Rev. B: Condens. Matter Mater. Phys.*, 1983, **27**, 748–762.
64. R. E. Benfield, D. Grandjean, M. Kroll, R. Pugin, T. Sawitowski and G. Schmid, *J. Phys. Chem. B*, 2001, **105**, 1961–1970.
65. Y. Gao, N. Shao, Y. Pei, Z. Chen and X. C. Zeng, *ACS Nano*, 2011, **5**, 7818–7829.
66. J. Kilmartin, R. Sarip, R. Grau-Crespo, D. Di Tommaso, G. Hogarth, C. Prestipino and G. Sankar, *ACS Catal.*, 2012, **2**, 957–963.
67. D. M. Chevrier, A. Chatt and P. Zhang, *J. Nanophotonics*, 2012, **6**, 064504-1.
68. P. L. Xavier, K. Chaudhari, A. Baksi and T. Pradeep, *Nano Rev.*, 2012, **3**, 14767.
69. C. Sun, H. Yang, Y. Yuan, X. Tian, L. Wang, Y. Guo, L. Xu, J. Lei, N. Gao, G. J. Anderson, X. Liang, C. Chen, Y. Zhao and G. Nie, *J. Am. Chem. Soc.*, 2011, **133**, 8617–8624.
70. P. Zhang and T. K. Sham, *Appl. Phys. Lett.*, 2002, **81**, 736.
71. M. Quinten, I. Sander, P. Steiner, U. Kreibig, K. Fauth and G. Schmid, *Z. Phys. D: At., Mol. Clusters*, 1991, **20**, 377–379.
72. L. Guczi, G. Peto, A. Beck, K. Frey, O. Geszti, G. Molnar and C. Daroczi, *J. Am. Chem. Soc.*, 2003, **125**, 4332.
73. A. Visikovskiy, H. Matsumoto, K. Mitsuhara, T. Nakada, T. Akita and Y. Kido, *Phys. Rev. B: Condens. Matter Mater. Phys.*, 2011, **83**, 165428.
74. Y. B. Losovyj, S.-C. Li, N. Lozova, K. Katsiev, D. Stellwagen, U. Diebold, L. Kong and C. S. S. R. Kumar, *J. Phys. Chem. C*, 2012, **116**, 5857–5861.
75. S. Trudel, *Gold Bull.*, 2011, **44**, 3.
76. Y. Yamamoto and H. Hori, *Rev. Adv. Mater. Sci.*, 2006, **12**, 23.
77. G. L. Nealon, B. Donnio, R. Greget, J.-P. Kappler, E. Terazzi and J.-L. Gallani, *Nanoscale*, 2012, **4**, 5244.
78. C. B. Sommers and H. Amar, *Phys. Rev.*, 1969, **188**, 1117.
79. *CRC Handbook of Chemistry and Physics*, ed. W. M. Haynes, CRC Press, Boca Raton, 93rd edn, 2011.
80. M. Suzuki, N. Kawamura, H. Miyagawa, J. S. Garitaonandia, Y. Yamamoto and H. Hori, *Phys. Rev. Lett.*, 2012, **108**, 047201.
81. C.-M. Wu, C.-Y. Li, Y.-T. Kuo, C.-W. Wang, S.-Y. Wu and W.-H. Li, *J. Nanopart. Res.*, 2010, **12**, 177.
82. S. S. Pundlik, K. Kalyanaraman and U. V. Waghmare, *J. Phys. Chem. C*, 2011, **115**, 3809.
83. K. S. Krishna, P. Tarakeshwar, V. Mujica and C. S. S. R. Kumar, *Small*, 2014, **10**, 907.
84. R. J. Magyar, V. Mujica, M. Marquez and C. Gonzalez, *Phys. Rev. B: Condens. Matter Mater. Phys.*, 2007, **75**, 144421.
85. F. Michael, C. Gonzalez, V. Mujica, M. Marquez and M. A. Ratner, *Phys. Rev. B: Condens. Matter Mater. Phys.*, 2007, **76**, 224409.
86. C. Gonzalez, Y. Simón-Manso, M. Marquez and V. Mujica, *J. Phys. Chem. B*, 2006, **110**, 687.

87. L. Puerta, H. J. Franco, J. Murgich, C. Gonzalez, Y. Simón-Manso and V. Mujica, *J. Phys. Chem. A*, 2008, **112**, 9771.
88. W. Luo, S. J. Pennycook and S. T. Pantelides, *Nano Lett.*, 2007, 7, 3134.
89. A. Ayuela, P. Crespo, M. A. García, A. Hernando and P. M. Echenique, *New J. Phys.*, 2012, **14**, 013064.
90. H. Hori, Y. Yamamoto, T. Iwamoto, T. Miura, T. Teranishi and M. Miyake, *Phys. Rev. B: Condens. Matter Mater. Phys.*, 2004, **69**, 174411.
91. M. K. Harbola and V. Sahni, *Phys. Rev. B: Condens. Matter Mater. Phys.*, 1988, **37**, 745.
92. P. Crespo, R. Litrán, T. C. Rojas, M. Multigner, J. M. de la Fuente, J. C. Sánchez-López, M. A. García, A. Hernando, S. Penadés and A. Fernández, *Phys. Rev. Lett.*, 2004, **93**, 087204.
93. Z. Vager and R. Naaman, *Phys. Rev. Lett.*, 2004, **92**, 087205.
94. A. Hernando, P. Crespo and M. A. García, *Phys. Rev. Lett.*, 2006, **96**, 057206.
95. A. Hernando, P. Crespo, M. A. García, M. Coey, A. Ayuela and P. M. Echenique, *Phys. Status Solidi B*, 2011, **248**, 2352.
96. A. Roldan, F. Illas, P. Tarakeshwar and V. Mujica, *J. Phys. Chem. Lett.*, 2011, **2**, 2996.
97. E. C. Stoner, *Proc. R. Soc. London, Ser. A*, 1939, **A169**, 339.
98. M. A. Garcia, J. de la Venta, P. Crespo, J. LLopis, S. Penadés, A. Fernández and A. Hernando, *Phys. Rev. B: Condens. Matter Mater. Phys.*, 2005, **72**, 241403(R).
99. M. Zhu, C. M. Aikens, M. P. Hendrich, R. Gupta, H. Qian, G. C. Schatz and R. Jin, *J. Am. Chem. Soc.*, 2009, **131**, 2490–2492.
100. J. P. Bouchaud and P. G. Zérah, *Phys. Rev. B: Condens. Matter Mater. Phys.*, 1993, **47**, 9095.
101. P. Panissod and M. Drillon, in *Magnetism: Molecules to Materials*, ed. J. S. Miller and M. Drillon, Wiley VCH, 2003, vol. 4, pp. 233–270.
102. L. M. Falicov and G. A. Somorjai, *Proc. Natl. Acad. Sci. U. S. A.*, 1985, **82**, 2207–2211.
103. A. Goda, Ph.D., University of Delaware, 2008, 184 pages; UMI 337429.
104. A. R. Malheiro, R. Gentil, F. I. Pires, J. Perez, and H. M. Villullas, Brazilian Synchrotron Light Laboratory, LNLS 2009 Activity Report.
105. P. W. Selwood, *Chem. Rev.*, 1946, **38**, 41–82.
106. H. Morris and P. W. Selwood, *J. Am. Chem. Soc.*, 1943, **65**, 2245.
107. G. Cohn and J. A. Hedvall, *J. Phys. Chem.*, 1942, **46**, 841.
108. N. K. Saroj, S. E. Weber, P. Jena, K. Wildberger, R. Zeller, P. H. Dederichs, V. S. Stepanyuk and W. Hergert, *Phys. Rev. B: Condens. Matter Mater. Phys.*, 1997, **56**, 8849–8854.
109. J. S. Garitaonandia, M. Insausti, E. Goikolea, M. Suzuki, J. D. Cashion, N. Kawamura, H. Ohsawa, I. G. de Muro, K. Suzuki, F. Plazaola and T. Rojo, *Nano Lett.*, 2008, **8**, 661–667.
110. S. S. Pundlik, K. Kalyanaraman and U. V. Waghmare, *J. Phys. Chem. C*, 2011, **115**, 3809–3820.

111. (a) H. Li, L. Li and Y. Li, *Nanotechnol. Rev.*, 2013, **2**(5), 515–528; (b) P. Zhai, G. Sun, Q. Zhu and D. Ma, *Nanotechnol. Rev.*, 2013, **2**(5), 547–576.
112. (a) H. Alex, N. Steinfeldt, K. Jähnisch, M. Bauer and S. Hübner, *Nanotechnol. Rev.*, 2013, **3**(1), 99–110; (b) K. S. Krishna, C. V. Navin, S. Biswas, V. Singh, K. Ham, G. L. Bovenkamp, C. S. Theegala, J. T. Miller, J. J. Spivey and C. S. S. R. Kumar, *J. Am. Chem. Soc.*, 2013, **135**(14), 5450–5456.
113. B. S. González, M. J. Rodríguez, C. Blanco, J. Rivas, M. A. López-Quintela and J. M. G. Martinho, *Nano lett.*, 2010, **10**, 4217–4221.
114. M. A. Munoz-Marquez, E. Guerrero, A. Fernandez, P. Crespo, A. Hernando, R. Lucena and J. C. Conesa, *J. Nanopart. Res.*, 2010, **12**, 1307–1318.
115. Y. Negishi, H. Tsunoyama, M. Suzuki, N. Kawamura, M. M. Matsushita, K. Maruyama, T. Sugawara, T. Yokoyama and T. Tsukuda, *J. Am. Chem. Soc.*, 2006, **128**, 12034.
116. P. Dutta, S. Pal, M. S. Seehra, M. Anand and C. B. Roberts, *Appl. Phys. Lett.*, 2007, **90**, 213102.
117. Z. Wu, J. Chen and R. Jin, *Adv. Funct. Mater.*, 2011, **21**, 177–183.
118. U. Maitra, B. Das, N. Kumar, A. Sundaresan and C. N. R. Rao, *ChemPhysChem*, 2011, **12**, 2322–2327.
119. P. de la Presa, M. Multigner, J. de la Venta, M. A. García and M. L. Ruiz-González, *J. Appl. Phys.*, 2006, **100**, 123915.
120. P. Crespo, E. Guerrero, M. A. Muoz-Marquez, A. Hernado and A. Fernandez, *IEEE Trans. Magn.*, 2008, **44**, 2768.
121. Y. Yamamoto, T. Miura, M. Suzuki, N. Kawamura, H. Miyagawa, T. Nakamura, K. Kobayashi, T. Teranishi and H. Hori, *Phys. Rev. Lett.*, 2004, **93**, 116801–116804.
122. M. Suda, N. Kameyama, A. Ikegami, M. Suzuki, N. Kawamura and Y. Einaga, *Polyhedron*, 2009, **28**, 1868.
123. M. Suda, N. Kameyama, M. Suzuki, N. Kawamura and Y. Einaga, *Angew. Chem., Int. Ed.*, 2008, **47**, 160.
124. R. B. Rakhi, A. L. M. Reddy, M. M. Shaijumon, K. Sethupathi and S. Ramaprabhu, *J. Nanopart. Res.*, 2008, **10**, 179.
125. K. Kowlgi, L. Zhang, S. Picken and G. Koper, *Colloids Surf., A*, 2012, **413**, 248–251.
126. H. Hori, T. Teranishi, Y. Nakae, Y. Seino, M. Miyake and S. Yamada, *Phys. Lett. A*, 1999, **263**, 406.
127. E. Guerrero, M. A. Muñoz-Márquez, M. A. García, P. Crespo, E. Fernández-Pinel, A. Hernando and A. Fernández, *Nanotechnology*, 2008, **19**, 175701.
128. J. de la Venta, A. Pucci, E. F. Pinel, M. A. Garcia, C. de Julian Fernandez, P. Crespo, P. Mazzoldi, G. Ruggeri and A. Hernando, *Adv. Mater.*, 2007, **19**, 875.
129. B. Donnio, P. Garcia-Vazquez, J. L. Gallani, D. Guillon and E. Terazzi, *Adv. Mater.*, 2007, **19**, 3534.

CHAPTER 5

Atomically Precise Gold Nanoclusters: Synthesis and Catalytic Application

GAO LI AND RONGCHAO JIN*

Department of Chemistry, Carnegie Mellon University, 4400 Fifth Ave, Pittsburgh, PA 15213, USA
*Email: rongchao@andrew.cmu.edu

5.1 Introduction

Gold nanoparticles have attracted significant interest in catalysis research.[1–5] A major aspect of nanocatalysts lies in the size distribution of the particles. For fundamental studies, the polydispersity of conventional nanocatalysts poses some tough issues. First, the size-dependent catalytic activities of nanoparticles are averaged out in the case of polydisperse catalysts. Second, the surface of the nanocatalyst is difficult to identify clearly, albeit crystal facets such as (111) and (100) can be assigned. Thus, it is difficult to relate the observed catalytic performance to the structure and intrinsic properties of nanocatalysts. To overcome these obstacles, well-defined gold nanocatalysts are highly desirable, especially in fundamental studies of catalysis.

In recent years, remarkable progress has been achieved in solution-phase synthesis of atomically precise gold nanoparticles. These unique nanoparticles resemble molecular compounds and are often called nanoclusters, notably the thiolate-protected gold nanoclusters.[6–8] Specific-size Au nanoclusters are composed of precise numbers of gold atoms (n) and of

RSC Catalysis Series No. 22
Atomically-Precise Methods for Synthesis of Solid Catalysts
Edited by Sophie Hermans and Thierry Visart de Bocarmé
© The Royal Society of Chemistry 2015
Published by the Royal Society of Chemistry, www.rsc.org

thiolate (–SR) ligands (m), denoted as Au$_n$(SR)$_m$, with n ranging up to a few hundred atoms (e.g., <200, equivalent size up to 2 nm). These thiolate-protected nanoclusters are well-defind to the atomic level (i.e., of molecular purity), rather than the usual nanometer precision in conventional nanoparticles. The Au$_n$(SR)$_m$ nanoclusters are particularly robust[8] (discussed below).

In this chapter, we first briefly discuss the synthesis of Au$_n$(SR)$_m$ nanoclusters, and then focus our discussion on the catalytic reactions of such nanoclusters. We note that research on the catalytic application of Au$_n$(SR)$_m$ nanoclusters is still in its infancy, but the reported work has shown great promise of such nanoclusters. We use Au$_{25}$(SR)$_{18}$ nanoclusters as a typical example to illustrate its promising catalytic properties, as Au$_{25}$(SR)$_{18}$ has been widely investigated among all the gold nanoclusters.

Compared to conventional metallic Au nanoparticle catalysts, Au$_n$(SR)$_m$ nanoclusters possess several distinct features that are of particular interest to catalysis. First of all, metallic gold nanoparticles (2 to 100 nm) adopt a face-centered cubic (fcc) structure, but Au$_n$(SR)$_m$ nanoclusters (1 to 2 nm) often adopt different atom-packing structures; for example, an icosahedral structure was found in Au$_{25}$(SR)$_{18}$ nanoclusters (~1.27 nm metal core, see detailed discussion in Section 5.2).[9] Second, the ultra-small size induces strong electron-energy quantization in nanoclusters, as opposed to the continuous conduction band in metallic gold nanoparticles.[9,10] Thus, Au$_n$(SR)$_m$ nanoclusters become semiconductors and possess a sizable bandgap (e.g., ~1.3 eV for Au$_{25}$(SR)$_{18}$ and ~0.9 eV for Au$_{38}$(SR)$_{24}$).

The atomically precise Au$_n$(SR)$_m$ nanoclusters are expected to become a promising class of model catalysts, although much work remains to be carried out in future. These well-defined nanoclusters will provide new opportunities for achieving fundamental understanding of metal nanocatalysis, such as the insight into the size dependence and deep understanding of the molecular activation, active centers, and catalytic mechanism by correlation with the structures of nanoclusters. Future research on atomically precise nanocluster catalysts will contribute to the fundamental catalysis and the new design of highly selective catalysts for specific chemical processes.

5.2 Synthesis of Atomically Precise Gold Nanoclusters: Size-focusing Method

We briefly introduce the 'size focusing' methodology for synthesizing atomically precise Au$_n$(SR)$_m$ nanoclusters.[6] In this method, a proper distribution of size-mixed nanoclusters is first made by kinetically controlling the reduction reaction of gold precursor (typically Au(I)) with NaBH$_4$, which is the key step for the final product. Then the size-mixed nanoclusters are subjected to size-focusing under harsh conditions (e.g., at 80 °C and in the presence of excess thiol), under which the unstable nanoclusters decompose or convert to the stable one, and eventually only nanoclusters of the most stable size in the distribution survive the 'focusing' process. By adjusting the initial size range through kinetic control in the first step, the subsequent size-focusing step can give rise to a series of size-discrete, robust $Au_n(SR)_m$ nanoclusters, such as $Au_{25}(SR)_{18}$,[9,11] $Au_{28}(SR)_{20}$,[12] $Au_{36}(SR)_{24}$,[13] $Au_{38}(SR)_{24}$,[14] $Au_{144}(SR)_{60}$,[15] $Au_{333}(SR)_{79}$,[16] etc. Herein we chose $Au_{25}(SR)_{18}$ and $Au_{38}(SR)_{24}$ nanoclusters as examples for a more detailed discussion.

5.2.1 The Case of $Au_{25}(SR)_{18}$ Nanoclusters

Size-focusing was first observed in the synthesis of $Au_{25}(SR)_{18}$ nanoclusters in our early work using the two-phase[17] and one-phase[18] synthetic methods. In both systems, the first step consists of the reduction of Au(III) to Au(I) with excess thiol to form a Au(I)/SR polymeric intermediate at a low temperature (e.g., 0 °C or room temperature). The subsequent reduction of Au(I)/SR to Au(0) by $NaBH_4$ gives rise to thiolate-protected nanoclusters. After prolonged aging of the reaction product, we observed several prominent peaks in the UV-visible spectrum of the crude product. A detailed investigation into this aging process revealed a gradual growth of monodisperse $Au_{25}(SR)_{18}$ nanoclusters from the early stage size-mixed product, Figure 5.1a. Three absorption peaks (at 670, 450 and 400 nm) were observed in the spectrum,[9] which are characteristic of $Au_{25}(SR)_{18}$ nanoclusters.

This size-focusing growth process was found to be quite common (e.g., not limited by the type of thiol). Many types of thiols (HS-R) have been investigated in the synthesis of $Au_{25}(SR)_{18}$ (where R = $-C_2H_4Ph$, $-C_6H_{13}$, $-C_{12}H_{25}$, $-G$ (glutathione), $-C_{10}H_{22}COOH$, etc.).[18] The size-focusing growth process of $Au_{25}(SR)_{18}$ in one-phase synthesis was later investigated further by Dass and co-workers using matrix-assisted laser desorption ionization (MALDI) mass spectrometry.[19] It was observed that at earlier reaction times, a mixture

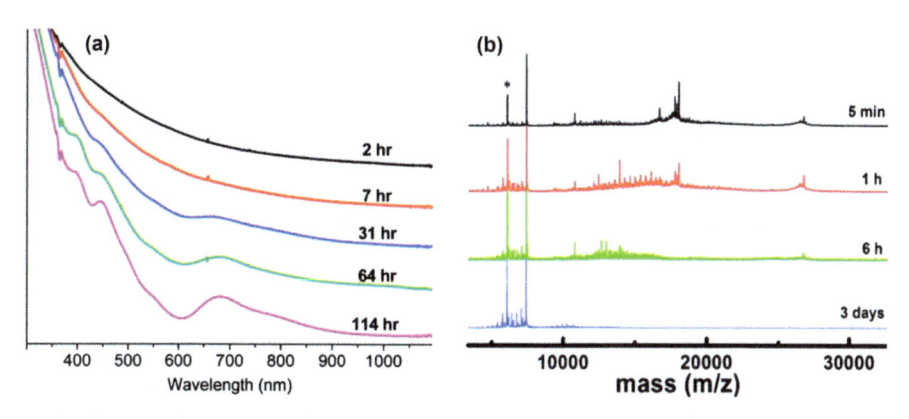

Figure 5.1 UV-visible spectra (a) and mass spectrometric evidence (b) of the 'size focusing' process of $Au_{25}(SR)_{18}$ nanoclusters with aging time. Peaks marked by asterisks are fragments caused by MALDI.
Adapted from ref. 18 and 19.

ranging from Au_{25} to Au_{102} was formed, and subsequently, size evolution led to monodisperse $Au_{25}(SR)_{18}$ nanoclusters (Figure 5.1b).

The case of $Au_{25}(SR)_{18}$ synthesis has illustrated the basic principle of the size-focusing growth process, which is based upon the stability property of different sized nanoclusters. In order to attain atomic monodispersity, synthetic control is needed so that only one specific size of nanoclusters survives the size-focusing process. An important issue is how to render just one size to survive the focusing process. In our work, we found that it is critical to control the size distribution of the starting $Au_n(SR)_m$ mixture. Below, we shall elaborate on this with the case of $Au_{38}(SR)_{24}$ nanocluster synthesis as a typical example.

5.2.2 The Case of $Au_{38}(SR)_{24}$ Nanocluster

The $Au_{38}(SR)_{24}$ formula was first unambiguously determined by Chaki et al.,[20] but the synthetic yield was quite low. Qian et al.[21] improved the $Au_{38}(SR)_{24}$ synthetic method and increased the yield of $Au_{38}(SC_{12}H_{25})_{24}$ to ~10% (Au atom basis) via a two-step approach, and the $Au_{38}(SC_{12}H_{25})_{24}$ composition was verified by electrospray ionization (ESI) MS and other characterizations. In this method, a crude mixture of glutathionate-protected $Au_n(SG)_m$ nanoclusters was first made; the mixed nanoclusters were then subjected to a thermal thiol etching process (e.g., 80 °C) in a two-phase (water/toluene) system. After the ligand exchange (from –SG to –$SC_{12}H_{25}$) on the nanoclusters was completed, the subsequent etching process (in neat dodecanethiol) caused gold core etching, and eventually the starting polydisperse nanoclusters were converted to monodisperse $Au_{38}(SC_{12}H_{25})_{24}$ with high purity.

The two-step method was further extended to the synthesis of $Au_{38}(SC_2H_4Ph)_{24}$ with a higher yield (~25% based on Au) after optimization of some reaction parameters.[22] A detailed mechanistic investigation on the conversion process clearly showed a size-focusing process, evidenced by both optical spectroscopy and MALDI-MS analyses (Figure 5.2). The optical spectrum of the starting $Au_n(SC_2H_4Ph)_m$ showed a decaying curve, indicating a mixture; with reaction going on, several distinct peaks started to emerge in the spectrum, and the final product showed a distinctive optical spectrum characteristic of $Au_{38}(SC_2H_4Ph)_{24}$ nanoclusters (Figure 5.2a).[21,22] MALDI-MS analysis reveals that, with the increase of reaction time, the relatively large Au nanoclusters ($n > 38$) seem to decompose and convert to $Au_{38}(SC_2H_4Ph)_{24}$ (Figure 5.2b). After ~40 h, very pure $Au_{38}(SC_2H_4Ph)_{24}$ nanoclusters (MW, 10 780 Da) were obtained (Figure 5.2b); note that the fragment at 9342 Da was caused by the MALDI method.

As discussed above, a key condition to obtain single-sized nanoclusters is to control the size distribution of the $Au_n(SG)_m$ mixture prior to the size-focusing step. We realized that the solvent might play an important role in controlling the size range of the $Au_n(SG)_m$ starting mixture.[22] In previous work, the $Au_n(SG)_m$ nanoclusters were typically prepared in methanol.[23]

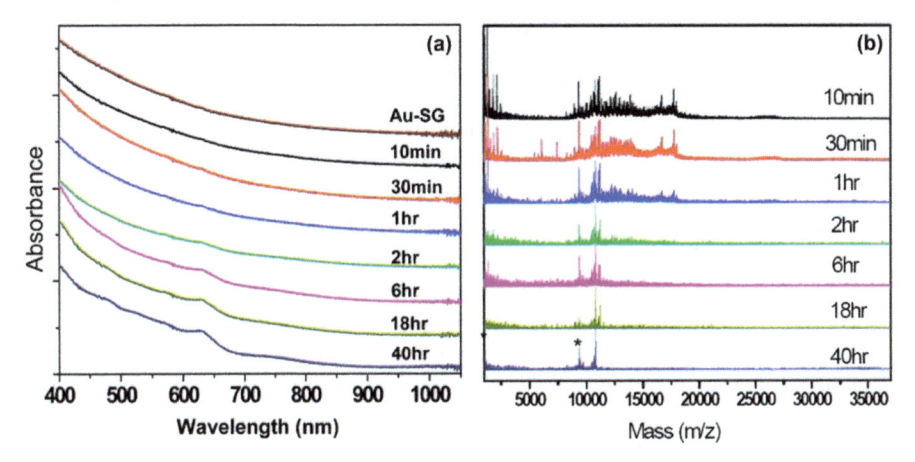

Atomically Precise Gold Nanoclusters: Synthesis and Catalytic Application 127

Figure 5.2 UV-visible spectra (a) and MALDI mass spectra (b) of Au nanoclusters during the size-focusing process. Au–SG stands for the starting $Au_n(SG)_m$ mixture in aqueous solution. 10 min to 40 h indicates the time that $Au_n(SG)_m$ reacts with excess PhC_2H_4SH. The asterisk (*) shows a fragment of $Au_{38}(SC_2H_4Ph)_{24}$ which was caused by the MALDI method. Adapted from ref. 22.

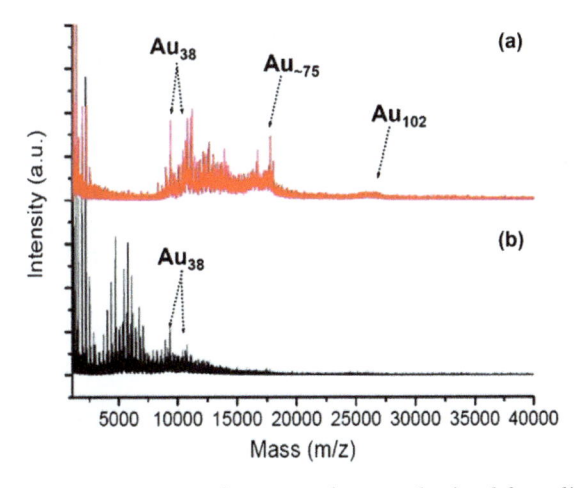

Figure 5.3 MALDI mass spectra of Au nanoclusters obtained from ligand exchange of $Au_n(SG)_m$ with PhC_2H_4SH (reaction for 10 min). The starting $Au_n(SG)_m$ nanoclusters were made (a) in methanol and (b) in acetone. Adapted from ref. 22.

Guided by the size-focusing principle, other solvents were investigated, and we found that acetone was indeed a good solvent for high-yielding synthesis of $Au_{38}(SR)_{24}$ as acetone solvent gave rise to a proper size distribution of $Au_n(SG)_m$ (Figure 5.3a) as the starting material for the size-focusing step. The acetone-mediated synthesis of $Au_n(SG)_m$ produced a dominant size range from 8 to 18 kDa (charge $z=1$), while the methanol system produced a dominant size range below 8 kDa (Figure 5.3b). Regarding the specific role(s) of the solvent, we believe that the solvent influences the [Au(I)–SR]$_x$ aggregate size and/or structure, which subsequently affects the cluster sizes, but details remain to be elucidated. We attribute the high yield of Au$_{38}$(SC$_2$H$_4$Ph)$_{24}$ to the down-conversion of those higher-mass Au$_n$(SG)$_m$ nanoclusters ($38 < n < 102$). For the starting nanoclusters made in methanol, the final yield of Au$_{38}$ was low because the majority of the starting Au$_n$(SG)$_m$ were smaller than 38 atoms (Figure 5.3b). Thus, it is important to prepare somewhat larger Au$_n$ nanoparticles with $n > 38$ but not too large, so that these appropriate nanoclusters can be readily converted to Au$_{38}$.[22]

From the above examples of the nanocluster syntheses, a key is to control the size distribution of the starting size-mixed nanoclusters in the first step (*i.e.*, prior to the size-focusing step). It is worth noting that the other reaction parameters (*e.g.*, reaction temperature, growth kinetics, ligand bulkiness, etching time, thiol/gold ratio, *etc.*) also play important roles in the size-focusing step; for example, the Au$_{38}$(SCH$_2$CH$_2$Ph)$_{24}$ nanoclusters could be transformed to Au$_{36}$(SPh-^{t}Bu)$_{24}$ nanoclusters during the ligand-exchange process.[13]

5.3 Crystal Structures of Gold Nanoclusters

We first briefly discuss the structure, electronic and optical properties of the Au$_{25}$(SR)$_{18}$ nanoclusters.[9] X-Ray crystallography reveals that the Au$_{25}$(SR)$_{18}$ nanocluster comprises a Au$_{13}$ icosahedral core (Figure 5.4a, magenta) and a Au$_{12}$(SR)$_{18}$ shell (Figure 5.2a, cyan and yellow). Some of the shell atoms (triangular Au$_3$) and the icosahedral core Au atoms are shown in the space-filling model, which are accessible to reactant molecules in catalytic reactions (Figure 5.4b). The cavities or pocket-like sites may serve as the catalytically active sites. The electronic structure of Au$_{25}$(SR)$_{18}$ exhibits

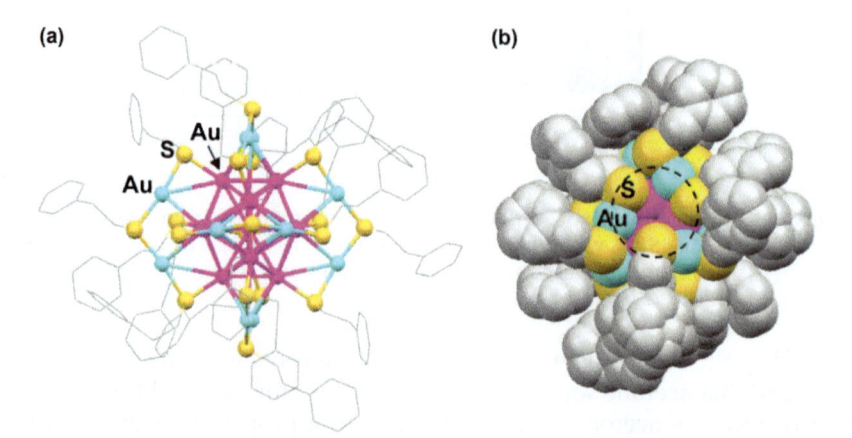

Figure 5.4 X-Ray crystal structure of Au$_{25}$(SCH$_2$CH$_2$Ph)$_{18}$ nanoclusters (magenta: gold atoms of the core, cyan: gold atoms of the shell). (a) Ball–stick model, (b) space-filling model.

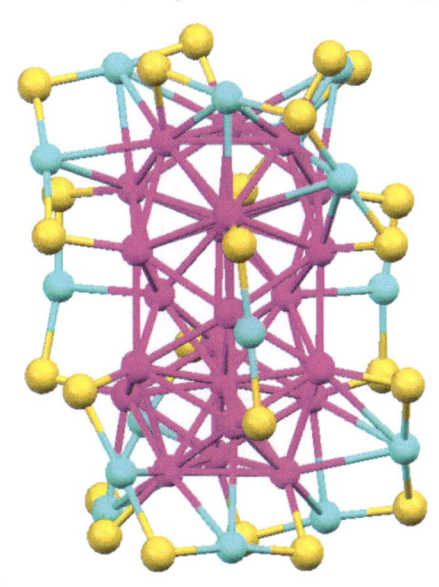

Figure 5.5 Crystal structure of $Au_{38}(SCH_2CH_2Ph)_{24}$ (magenta: gold atoms of the Au_{23} core, cyan: gold atoms of the shell, yellow: sulfur). All $-CH_2CH_2Ph$ tails are omitted for clarity.

discrete energy levels caused by the quantum-size effect. The molecular orbitals[9] can be roughly divided into the Au-core orbitals (*i.e.*, primarily contributed by Au_{13}) and the shell orbitals (*i.e.*, primarily by $Au_{12}(SR)_{18}$), as reflected in the geometric coreshell structure. It has been proved that, regardless of the thiolate type, $Au_{25}(SR)_{18}$ nanoclusters with different R groups exhibit the same geometric structure.[23]

The $Au_{38}(SCH_2CH_2Ph)_{24}$ nanocluster comprises a biicosahedral Au_{23} core and six dimeric staples $-SR-Au-SR-Au-SR-$ ($Au_2(SR)_3$ for short) and three monomeric staples $-SR-Au-SR-$ ($Au(SR)_2$ for short) protecting the Au_{23} core (Figure 5.5).[24]

5.4 Thermal Stability of $Au_n(SR)_m$ Nanoclusters

The thiolate-protected gold nanoclusters are particularly robust (*e.g.*, stable for months to years under ambient conditions). Herein we briefly discuss their thermal stability. The thermal stability of $Au_n(SR)_m$ nanoclusters is important for catalytic applications as many reactions are run under thermal conditions.

Thermogravimetric analysis (TGA) shows that $Au_n(SR)_m$ (*e.g.*, $Au_{25}(SR)_{24}$, $Au_{38}(SR)_{24}$, and $Au_{144}(SR)_{60}$) starts to lose ligands at ~ 200 °C and the ligand loss completes at ~ 250 °C (Figure 5.6a).[25] The ligand loss temperature was found to be not affected by the atmosphere (*e.g.*, N_2, air, O_2, and H_2). We further tested the isothermal stability of $Au_{25}(SR)_{18}$ nanoclusters (*i.e.*, maintained at 150 °C, in air atmosphere for 60 min), which indicates that the

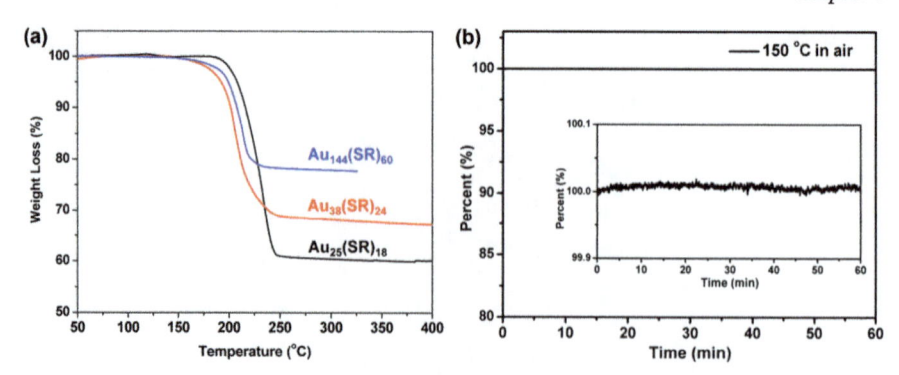

Figure 5.6 (a) Thermogravimetric analysis of Au$_n$(SR)$_m$ nanoclusters (under a N$_2$ atmosphere). (b) Isothermal stability analysis of pure Au$_{25}$(SR)$_{18}$ nanoclusters (maintained at 150 °C, in air atmosphere for 60 min). No discernible loss of ligands (starting: 100%) was observed. The inset shows a zoom-in of the TGA curve at constant 150 °C.
Adapted with permission from ref. 25 and 26.

gold nanoclusters remain stable after the annealing process, Figure 5.6b. The thermal tests confirm that the Au$_{25}$(SR)$_{18}$ nanoclusters are robust and thermally stable below the ligand loss temperature.[26]

5.5 Reactivity and Catalytic Properties of Au$_n$(SR)$_m$ Nanoclusters

5.5.1 Reversible Conversion Between [Au$_{25}$(SR)$_{18}$]^0 and [Au$_{25}$(SR)$_{18}$]$^-$

It is known that the electron transfer between metal nanocatalysts and reactants is one of the key steps in catalytic reactions. The native Au$_{25}$(SR)$_{18}$ nanoclusters from the synthesis are anionic (i.e., [Au$_{25}$(SR)$_{18}$]$^-$, counter-ion = tetraoctylammonium (TOA$^+$)). When a solution of [Au$_{25}$(SR)$_{18}$]$^-$ was exposed to air, we found that the nanoclusters were gradually converted to charge-neutral [Au$_{25}$(SR)$_{18}$]^0 clusters.[27] During the process, a discernable color change was observed, which was also reflected in the UV-visible spectral changes (Figure 5.7). This result is quite interesting, as one would normally expect that O$_2$ would first oxidize the thiolate ligands on the Au surface.[28] But our experimental results revealed that the gold core, instead of the thiolate ligands, first undergoes one-electron loss to yield neutral charge [Au$_{25}$(SCH$_2$CH$_2$Ph)$_{18}$]^0 nanoclusters. Single crystal X-ray crystallography revealed that the [Au$_{25}$(SCH$_2$CH$_2$Ph)$_{18}$]^0 structure[29] shares the same framework as that of [Au$_{25}$(SCH$_2$CH$_2$Ph)$_{18}$]$^-$. The negative charge of [Au$_{25}$(SCH$_2$CH$_2$Ph)$_{18}$]$^-$ resides on the gold core of the cluster rather than on the ligands. The redox process between [Au$_{25}$(SCH$_2$CH$_2$Ph)$_{18}$]^0 and [Au$_{25}$(SCH$_2$CH$_2$Ph)$_{18}$]$^-$ is completely reversible. If peroxide is used as the oxidant, the anion-to-neutral conversion is much faster than using O$_2$. The oxidation by O$_2$ or peroxide was

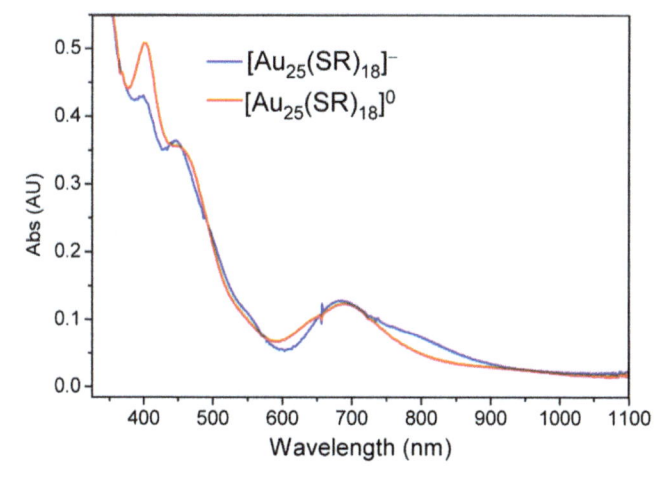

Figure 5.7 Reversible conversion between $[Au_{25}(SR)_{18}]^0$ and $[Au_{25}(SR)_{18}]^-$. Adapted with permission from ref. 29.

also found in the case of the water-soluble $Au_{25}(SG)_{18}$ nanocluster. Both the organic-soluble and the aqueous $Au_{25}(SR)_{18}$ (where R = –CH$_2$CH$_2$Ph, –C$_6$H$_{13}$ or –G (glutathione)) have been utilized in catalytic work.

5.5.2 Catalytic Oxidation

The activation of oxygen (O$_2$) plays an important role in the gold nanocluster-catalyzed selective oxidation processes.[1,30] As discussed above, $Au_n(SR)_m$ nanoclusters can interact with oxidants (e.g., O$_2$, H$_2$O$_2$, etc.) even at room temperature.[29] The oxidation reactivity of $Au_{25}(SR)_{18}$ nanoclusters inspired us to pursue their catalytic application. Below we discuss the various oxidation reactions that have been found to be catalyzed by gold nanoclusters.

5.5.2.1 Catalytic Oxidation of CO to CO$_2$

CO oxidation has been extensively investigated in nanogold catalysis.[1,31] In previous work, the support was found to largely affect the nanogold activity, and TiO$_2$ is generally the most effective support in the nanogold catalyzed oxidation of CO to CO$_2$. When it comes down to the case of $Au_n(SR)_m$ nanoclusters, surprisingly Nie et al.[32] found that the $Au_{25}(SR)_{18}$/TiO$_2$ catalyst was the least active, which had no catalytic activity even up to 200 °C (Figure 5.8a). By comparing several oxide supports, $Au_{25}(SR)_{18}$/CeO$_2$ was found to exhibit moderate activity (CO conversion onset temperature 60 °C, 62% conversion at 160 °C). These results imply that some striking differences exist between conventional (bare) Au/TiO$_2$ catalysts[31] and ligand-on $Au_{25}(SR)_{18}$/TiO$_2$ catalyst.[32]

To enhance the catalytic activity of $Au_{25}(SR)_{18}$/TiO$_2$, pretreatment was performed. Significantly, Nie et al. found that pretreatment of the

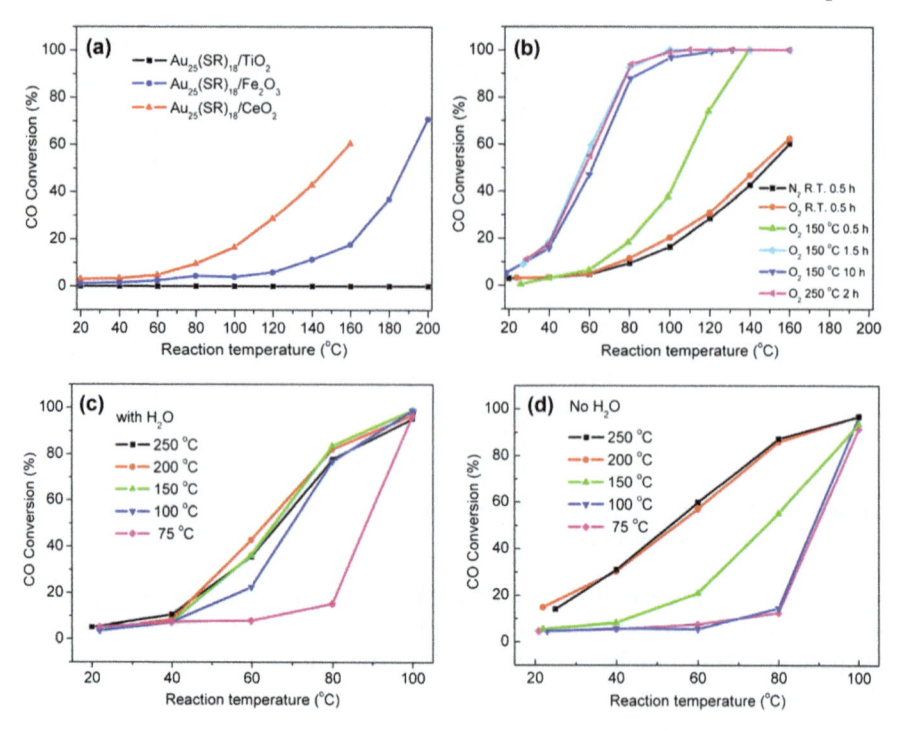

Figure 5.8 (a) Catalytic activity of different oxide-supported Au$_{25}$(SR)$_{18}$/MO$_x$ catalysts for CO oxidation (catalyst pretreatment condition: N$_2$ at room temperature (r.t.) for 0.5 h; reaction conditions: GHSV ∼7500 mL g^{-1} h^{-1}, catalyst: 0.1 g). (b) Catalytic activity of Au$_{25}$(SR)$_{18}$/CeO$_2$ after different pretreatments. (c) and (d) Effect of water vapor on CO conversion over various pretreated Au$_{25}$(SR)$_{18}$/CeO$_2$ catalysts under different pretreatment temperatures: (c) feed gases with water vapor and (d) without vapor. Adapted with permission from ref. 32.

Au$_{25}$(SR)$_{18}$/CeO$_2$ catalyst in O$_2$ for 1.5–2 h at 150 °C (T_{pre}) led to a drastic increase in catalytic activity (CO conversion onset temperature shifted to r.t.), and 100% CO conversion was obtained when the reaction temperature was as low as 100 °C (Figure 5.8b). Prolonged O$_2$ pretreatment at 150 °C (*e.g.*, >2 h) did not lead to further enhancement in catalytic activity (Figure 5.8b), and neither did it occur when the T_{pre} was increased to 250 °C (above the thiolate desorption temperature).

It is worth noting that the drastic effect of thermal O$_2$ pretreatment was not observed in the Au$_{25}$(SR)$_{18}$/TiO$_2$ system, and the Au$_{25}$(SR)$_{18}$/oxide catalysts pretreated in N$_2$ (as opposed to O$_2$) did not show any enhancement in activity. A variety of characterization of the Au$_{25}$(SR)$_{18}$/CeO$_2$ catalyst showed that the ligands remained on the cluster after the 150 °C pretreatment in O$_2$.

The presence of water vapor in the feed gases exhibited a promotional effect on the catalyst performance (Figure 5.8c and d). For example, in the case of feed-gas containing vapor, O$_2$ pretreatment at even lower temperature (e.g., $T_{pre} = 100$ °C) could lead to the same drastic enhancement in activity.[32]

In subsequent work, Nie et al.[33] further investigated the catalytic activity of $Au_{38}(SR)_{24}$ nanoclusters. The ligand-on and ligand-off CeO_2-supported Au_{38} cluster catalysts were compared. It was also found that oxygen (O_2) thermal pretreatment of $Au_{38}(SR)_{24}/CeO_2$ at a temperature between 100 and 175 °C can largely enhanced the catalytic activity. While pretreatment at higher temperatures (>200 °C) to remove the protecting thiolate ligands instead gave rise to a somewhat lower activity than that for the 175 °C pretreatment; note that the ligand-off Au_{38} clusters were unstable. With respect to the water vapor effect, it was observed that the CO conversion catalyzed by $Au_{38}(SR)_{24}/CeO_2$ in the case of wet feed-gas was appreciably higher than the case of dry feed-gas when the reaction temperature was at 60 to 80 °C, and interestingly the ligand-on and ligand-off Au_{38} catalysts exhibited opposite response to water vapor.

The above CO oxidation results imply that the interface between gold nanoclusters (e.g., $Au_{25}(SR)_{18}$ and $Au_{38}(SR)_{24}$) and CeO_2 is critical and should be the catalytic active site.[32] We speculate that O_2 should first convert to O_2^- by withdrawing an electron from the Au nanocluster, then migrates to the nanocluster/CeO_2 interface and converts to hydroperoxide species on CeO_2; the activated CO should be oxidized at the perimeter sites of the catalyst (Scheme 5.1).

To improve the high temperature (>200 °C) stability of nanoclusters, Dai and co-workers explored the stabilization of nanoclustes using mesoporous materials as supports (e.g., CuO-EP-FDU-12 and Co_3O_4-EPFDU-12).[34] The protecting ligands on $Au_{25}(SCH_2CH_2Ph)_{18}$ and $Au_{144}(SCH_2CH_2Ph)_{60}$ nanoclusters can be removed via thermal treatment at 300 °C (note: the ligand desorption onset temperature is ~200 °C) without sintering the Au_{25} and Au_{144} clusters because the ligand-off Au_{25} and Au_{144} clusters can be effectively stabilized on mesoporous material supports. The measured size of the ligand-off Au_{144} cluster was 1.67 ± 0.2 nm, which is consistent with the size of the native $Au_{144}(SCH_2CH_2Ph)_{60}$ nanoclusters and thus no aggregation of the clusters after ligands were removed. The ligand-off Au_{25} and Au_{144} clusters supported on both CuO-EP-FDU-12 and Co_3O_4-EP-FDU-12 were

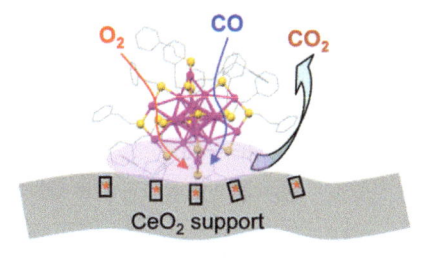

Scheme 5.1 Proposed CO oxidation at the perimeter sites of $Au_n(SR)_m/CeO_2$. Using the $Au_{25}(SR)_{18}/CeO_2$ as model catalyst.
Adapted with permission from ref. 32.

more active in CO oxidation at low temperatures (*e.g.*, 56% CO conversion at 20 °C and 100% conversion at 70 °C for the Au$_{25}$ clusters supported on the Co$_3$O$_4$-EP-FDU-12). The strategy of using functionalized mesoporous supports to stabilize nanoclusters can be applied to the nanogold catalysis in which ligand-off gold cluster catalysts are preferred.

5.5.2.2 Selective Oxidation of Styrene

Zhu et al.[35] investigated solution-phase styrene oxidation catalyzed by Au$_n$(SR)$_m$ nanocluster catalysts (free and SiO$_2$-supported gold nanoclusters) using O$_2$ as the oxidant, Scheme 5.2. The catalytic reaction was run at 80–100 °C for 12–24 h, which gave rise to major product, benzaldehyde (up to ~70% selectivity), and by-product, styrene epoxide (less than ~25% selectivity), and acetophenone (<5% selectivity). A strong size-dependence was observed in the catalytic oxidation reaction using free Au$_n$(SR)$_m$ nanocluster catalysts: the smaller Au$_n$(SR)$_m$ nanocluster catalyst exhibited much higher catalytic activity. The SiO$_2$-supported Au$_{25}$(SR)$_{18}$ catalyst gave comparable performance as that of the free nanocluster catalyst, but the advantage of Au$_{25}$(SR)$_{18}$/SiO$_2$ catalyst is its recyclability and re-use in catalytic reaction; no apparent deterioration in activity and selectivity of the reused supported catalyst was observed after multiple cycles.

Furthermore, Zhu et al.[25] compared three oxidant systems using Au$_{25}$(SR)$_{18}$/SiO$_2$ catalyst: (A) *tert*-butyl hydroperoxide (TBHP) as the oxidant; (B) TBHP as the initiator and O$_2$ as the main oxidant; (C) O$_2$ as the oxidant (without initiator). System A (only TBHP) gave a high conversion of styrene (*e.g.*, 86%) with 100% selectivity for benzaldehyde, system B (TBHP/O$_2$) gave rise to a much lower activity (25% conversion with 100% selectivity for benzaldehyde), and system C (solely O$_2$) was even lower (18% conversion with 80% selectivity for benzaldehyde). The catalytic results were consistent with the fact that TBHP is much more reactive than O$_2$. A similar phenomenon was also found in the three oxidant systems using Au$_{38}$(SR)$_{24}$/SiO$_2$ and Au$_{144}$(SR)$_{60}$/SiO$_2$ catalysts.[25] The results indicate that the ingredients of the oxidant play a critical role in the oxidation reaction, and that the oxidant (denoted as [O]) activation is a key step for achieving high conversion of styrene. The results also imply that the active oxygen species in the catalytic

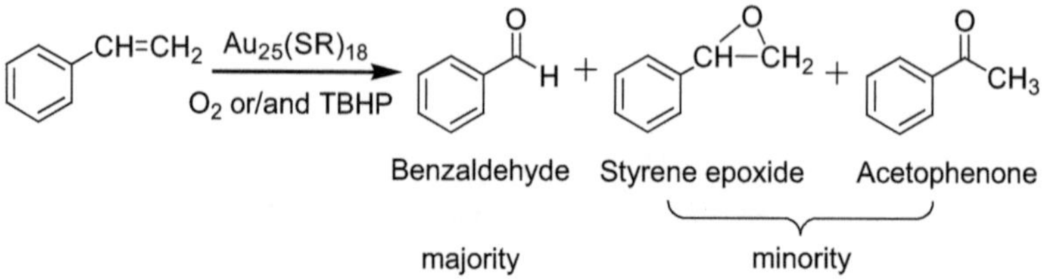

Scheme 5.2 Selective oxidation of styrene catalyzed by Au$_{25}$(SR)$_{18}$ nanocluster catalysts.

cycles may be peroxide-like or hydroperoxide-like species.[25] In addition, a distinct size dependence of $Au_n(SR)_m$ nanocluster catalysts was observed in the selective oxidation of styrene, with the order of catalytic activity of $Au_{25}(SR)_{18} > Au_{38}(SR)_{24} > Au_{144}(SR)_{60}$.

Tsukuda and co-workers[36] investigated a $Au_{25}(SG)_{18}$/HAP catalyst for selective oxidation of styrene, where HAP refers to hydroxyapatite $(Ca_{10}(PO_4)_6(OH)_2)$. The ligand-off Au_{25}/HAP catalyst was obtained by removing the ligands via 300 °C thermal treatment. The catalyst was explored in styrene oxidation (with anhydrous tert-butyl hydroperoxide (TBHP) as the oxidant). The reaction yielded styrene oxide with a selectivity of ~92%.

To compare with the catalytic activity of the $Au_{25}(SR)_{18}$, Qian et al.[37] further synthesized the monoplatinum-doped $Pt_1Au_{24}(SR)_{18}$ nanoclusters and investigated the cluster in the selective oxidation of styrene using $PhI(OAc)_2$ as the oxidant. The single platinum atom is located in the center of the core, which implies that the shell (i.e., $Au_{12}(SR)_{18}$) of the $Pt_1Au_{24}(SR)_{18}$ nanocluster is the same as that of $Au_{25}(SR)_{18}$. The catalytic reaction was carried out at 70 °C for 10 h in acetonitrile under N_2 atmosphere. $Pt_1Au_{24}(SR)_{18}/TiO_2$ catalyst gave much higher catalytic activity (i.e., conversion of styrene) and also selectivity for benzaldehyde than the $Au_{25}(SR)_{18}/TiO_2$ catalyst under the identical reaction conditions. Therefore, the different catalytic performance is apparently due to the electronic properties of the nanoclusters. The large enhancement of catalytic activity through single atom doping is remarkable, indicating the high sensitivity of nanocluster catalysts to its constituent atoms, even those not at the surface.

5.5.2.3 Selective Oxidation of Sulfides

It is known that sulfides can bind to the surface of gold nanoclusters via the interaction between the sulfur atom (–S–) and the gold surface, while sulfoxides (–S(=O)–) only weakly bind to the gold surface; hence the conversion from sulfide to sulfoxide on the clusters surface should be feasible. Gao et al.[28] investigated the oxide-supported ~1 wt% $Au_{25}(SR)_{18}$ nanocluster catalyst for selective oxidation of sulfides to sulfoxides using PhIO as oxidant (Scheme 5.3).

The TiO_2-supported $Au_{25}(SR)_{18}$ nanocluster catalyst gave rise to a high catalytic activity (e.g., ~97% conversion of sulfide with ~92% selectivity for the sulfoxide product). Meanwhile, different types of substrates have been investigated to test the effects of electronic and steric factors in the selective oxidation. Sulfides with electron-rich groups were oxidized at higher

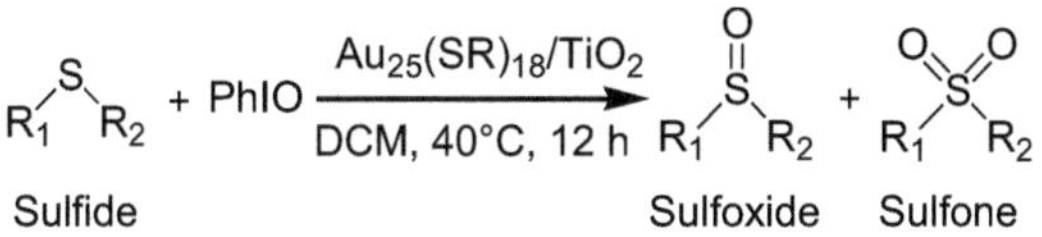

Scheme 5.3 Oxidation of sulfides using $Au_{25}(SR)_{18}/TiO_2$ catalyst.

conversion than those with electron-deficient groups, and sulfides with greater steric group showed lower conversion than those with less steric group. The support effect was also investigated, and the order of the catalyst activity of $Au_{25}(SR)_{18}$/oxide was found to be $TiO_2 > Fe_2O_3 > CeO_2 > MgO$. The oxide-supported $Au_{25}(SR)_{18}$ catalysts showed excellent recyclability in the sulfoxidation process.[28]

The above catalytic results of the oxidation by $Au_n(SR)_m$ nanoclusters allow us to gain some valuable insight at the atomic level. In the $[Au_{25}(SR)_{18}]^q$ nanocluster, the Au_{13} core possesses 8 (when $q = -1$) or 7 (when $q = 0$) delocalized valence electrons, which originate from Au(6s) and are primarily distributed within the Au_{13} core,[27] while the Au_{12} shell bears positive charges due to charge transfer from gold to sulfur atoms of the thiolate ligands. The electron-rich Au_{13} core should facilitate the activation of the oxidant [O] (e.g., O_2, TBHP, PhIO, etc.) by electron transfer to [O], accompanied by $[Au_{25}(SR)_{18}]^-$ conversion to neutral $[Au_{25}(SR)_{18}]^0$; of note, the oxidation readily occurs at elevated temperatures (e.g., under catalytic reaction conditions). For catalytic oxidation on $Au_{25}(SR)_{18}$, activation of reactants (e.g., styrene, sulfide, CO, and alcohol, etc.) should be facilitated by the partial positive charge ($Au^{\delta+}$, $\delta = \sim 0.3$)[38] on the triangular Au_3 motif of gold atoms.

5.5.3 Catalytic Selective Hydrogenation

5.5.3.1 Selective Hydrogenation of Aldehydes and Ketones

Beside the oxidation reactions, the $Au_n(SR)_m$ nanocluster catalysts were also demonstrated to be capable of catalyzing hydrogenation reactions, such as the chemoselective hydrogenation of α,β-unsaturated ketones to α,β-unsaturated alcohols under mild conditions (60 °C, in mixed solvents (1:1 toluene/acetonitrile)) (Table 5.1).[35] The $Au_{25}(SR)_{18}$ nanoclusters were found to hydrogenate preferentially the C=O bond of benzalacetone against the C=C bond, and the C=O hydrogenated product (i.e., unsaturated alcohol, UA) was obtained with 76% selectivity, whereas 14% selectivity was observed for unsaturated ketone, and 10% for saturated alcohol.[35] The UV-visible spectra and MS analysis of the free $Au_{25}(SR)_{18}$ nanocluster catalyst confirmed that the $Au_{25}(SR)_{18}$ nanoclusters remain intact during the hydrogenation reaction. Further optimization of the reaction led to a nearly complete selectivity for α,β-unsaturated alcohol in a mixed toluene/ethanol (1:1) solvent at room temperature.[39] The oxide-supported catalysts (including $Au_{25}(SR)_{18}/Fe_2O_3$, $Au_{25}(SR)_{18}/TiO_2$, and $Au_{25}(SR)_{18}/SiO_2$) were also investigated (Table 5.1). It was found that the catalytic activity was improved except the $Au_{25}(SR)_{18}/SiO_2$ catalyst, and the catalytic activity of the $Au_{25}(SR)_{18}/Fe_2O_3$ and $Au_{25}(SR)_{18}/TiO_2$ catalysts is almost equivalent (Table 5.1). In all cases, the chemoselectivity was 100%. In addition, the chemoselective hydrogenation was tested with a range of substituted

Table 5.1 Chemoselective hydrogenation reaction of α,β-unsaturated ketones to α,β-unsaturated alcohols catalyzed by Au$_{25}$(SR)$_{18}$ nanocluster catalysts.

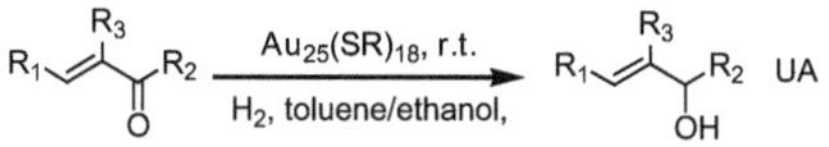

| | | | Conversion (%) | | |
Entry	Substrates	Selectivity for UA	Au$_{25}$	Au$_{25}$/Fe$_2$O$_3$	Au$_{25}$/TiO$_2$
1	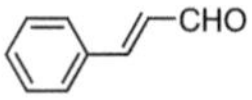 (cinnamaldehyde)	100	38	49	46
2	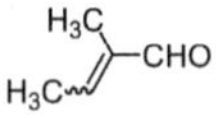 (3-methyl-2-butenal)	100	43	53	51
3	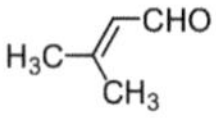 (2-methyl-2-butenal)	100	44	54	53
4	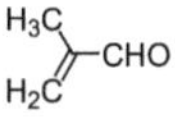 (methylenebutanal)	100	39	52	52
5	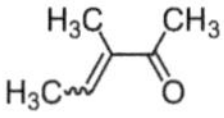 (mesityl oxide)	100	29	45	44
6	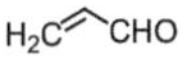 H$_2$C=CHO (acrolein)	91[a]	46	47	47

[a] The selectivity for unsaturated alcohol (UA) is 90% (free Au$_{25}$(SR)$_{18}$ catalyst, denoted as Au$_{25}$), 92% (Au$_{25}$(SR)$_{18}$/Fe$_2$O$_3$ catalyst, Au$_{25}$/Fe$_2$O$_3$), and 91% (Au$_{25}$(SR)$_{18}$/TiO$_2$ catalyst, Au$_{25}$/TiO$_2$), respectively. Adapted with permission from ref. 39.

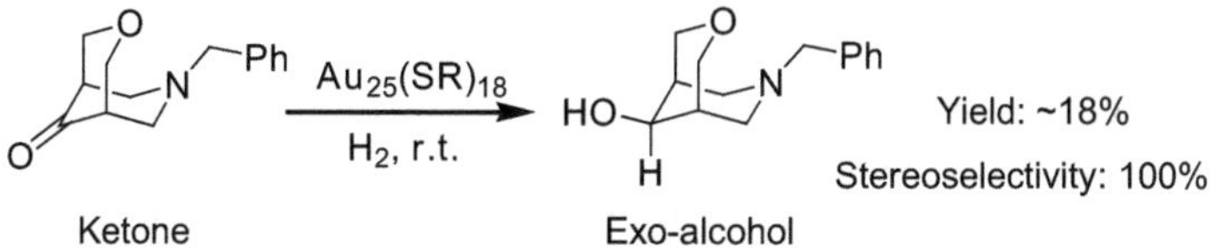

Scheme 5.4 Stereoselective hydrogenation reaction of bicyclic ketone to *exo*-alcohol catalyzed by Au$_{25}$(SR)$_{18}$ nanocluster catalysts.

α,β-unsaturated ketones and aldehydes; moderate yields and high selectivity were obtained (Table 5.1).

Zhu et al.[40] further expanded the Au$_{25}$(SR)$_{18}$ nanocluster catalyst for stereoselective hydrogenation of a bicyclic ketone (*i.e.*, 7-(phenylmethyl)-3-oxa-7-azabicyclo [3.3.1] nonan-9-one) to *exo*-alcohol (Scheme 5.4). The stereoselective hydrogenation was run at room temperature in an ethanol/toluene (5 : 1, v/v) mixed solvent using H$_2$ (1 atm) as hydrogen source; ~18% conversion of the bicyclic ketone with 100% selectivity for the *exo*-alcohol was obtained; note that the stereoselective product was determined by nuclear magnetic resonance (NMR) spectroscopy.[40] The high stereoselectivity (100%) was rationalized to be effected by the atomic spatial restraints and the unique core–shell configuration of the Au$_{25}$(SR)$_{18}$ nanocluster.

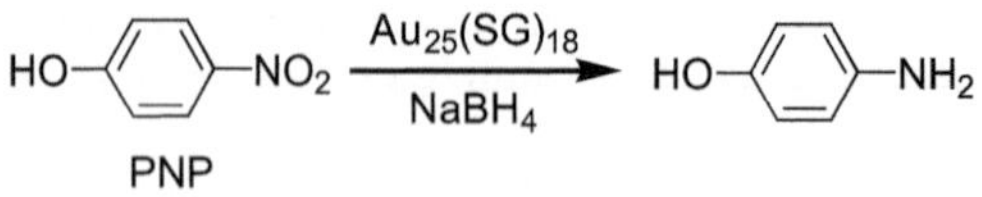

HO—⟨⟩—NO₂ $\xrightarrow[\text{NaBH}_4]{\text{Au}_{25}(\text{SG})_{18}}$ HO—⟨⟩—NH₂

PNP

Scheme 5.5 Hydrogenation reaction of 4-nitrophenol to 4-aminophenol catalyzed by Au$_{25}$(SR)$_{18}$ nanocluster catalysts.

5.5.3.2 Hydrogenation of Nitrophenol

Yamamoto et al.[41] investigated and compared the catalytic properties of Au$_{25}$(SG)$_{18}$ nanoclusters and dimethylformamide (DMF)-stabilized gold nanoclusters (not of atomic precision) in the reduction reaction of 4-nitrophenol to 4-aminophenol by NaBH$_4$ in water at room temperature (Scheme 5.5). The reduction process was detected by UV-visible spectra. The Au$_{25}$(SG)$_{18}$ nanoclusters exhibited higher catalytic activity than DMF-capped clusters in the nitrophenol reduction (pseudo-first-order rate constant 8×10^{-3} s^{-1} for Au$_{25}$(SG)$_{18}$ vs. 3×10^{-3} s^{-1} for DMF-capped Au clusters). High catalytic activity of Au$_{25}$(SG)$_{18}$ catalyst can be achieved even at a very low catalyst concentration (e.g., 1.0 µM). No induction time was observed in the case of Au$_{25}$(SG)$_{18}$ despite –SG being a strongly binding ligand due to less steric hinderance, while the DMF-stabilized Au clusters exhibited an induction time ($\sim$4000 s), which was attributed to the impedance by the surface DMF layer on the clusters when reactants access the cluster surface.[41] For Au$_{25}$(SG)$_{18}$ nanoclusters, the unique core–shell structure seems to pose less steric hinderance and thus renders the catalytic active sites more accessible to reactants.

Recently, Shivhare et al.[42] also demonstrated intact Au$_{25}$(SR)$_{18}$ (where R = C$_6$H$_{13}$, C$_8$H$_{17}$, and C$_{12}$H$_{25}$) nanoclusters for the reduction of nitrophenol by NaBH$_4$ in mixed THF/water and under N$_2$ atmosphere. The Au$_{25}$(SR)$_{18}$ nanoclusters were more stable and retained their structural integrity towards NaBH$_4$ treatment, while the larger Au$_{\sim 180}$(SC$_6$H$_{13}$)$_{\sim 100}$ nanoclusters were not.

5.5.4 Catalytic Carbon–Carbon Coupling Reaction

Gold nanoclusters also hold great promise for the carbon–carbon coupling reactions, such as Ullmann-type homo-coupling and Sonogashira cross-coupling reactions. The catalytic application in C–C coupling reaction was due to the good capability of gold nanocluster catalysts in activating the C–I bond of iodobenzene and the terminal C–H bond of phenylacetylene. Below we illustrate the C–C reactions.

5.5.4.1 Ullmann-type Homo-coupling Reaction

Gold nanoparticles are good catalysts for activating the C–I bond of the iodobenzene in the carbon–carbon homo-coupling process. Li et al.[43] investigated the catalytic activity of the oxide-supported Au$_{25}$(SR)$_{18}$ catalysts for

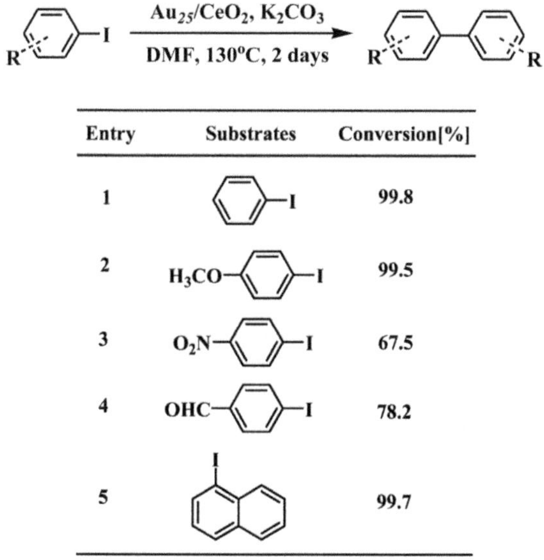

Atomically Precise Gold Nanoclusters: Synthesis and Catalytic Application   139

$$R\text{—}\bigcirc\text{—}I \xrightarrow[\text{DMF, 130°C, 2 days}]{Au_{25}/CeO_2,\ K_2CO_3} R\text{—}\bigcirc\text{—}\bigcirc\text{—}R$$

Entry	Substrates	Conversion[%]
1	⌬—I	99.8
2	H₃CO—⌬—I	99.5
3	O₂N—⌬—I	67.5
4	OHC—⌬—I	78.2
5	naphthyl–I	99.7

Scheme 5.6 Homo-coupling of aryl iodides using Au_{25}/CeO_2 catalyst.

the Ullmann-type homo-coupling reactions of aryl iodides. After optimization, the suitable conditions for homocoupling reaction were as follows: using DMF as solvent and K_2CO_3 as base at 130 °C under a N_2 atmosphere. Under such conditions, the $Au_{25}(SR)_{18}/CeO_2$ catalyst gave rise to 99.8% conversion of iodobenzene (Scheme 5.6). The application of the $Au_{25}(SR)_{18}$-catalyzed homo-coupling reaction was tested with substrates bearing a range of substituents. The electron-rich substrates performed better than electron-deficient substrates in the reaction. The support effect was also investigated, but unlike the case of CO oxidation, no distinct effect of the oxide supports was observed (i.e., CeO_2, SiO_2, TiO_2, and Al_2O_3).

5.5.4.2 Sonogashira Cross-coupling Reaction

Li et al.[26] expanded the catalytic application of the oxide-supported $Au_{25}(SR)_{18}$ nanocluster catalysts to the Sonogashira cross-coupling reaction between iodobenzene and phenylacetylene. The supported catalyst was made by impregnation of oxide powders (including CeO_2, TiO_2, MgO, and SiO_2) in a CH_2Cl_2 solution of $Au_{25}(SR)_{18}$ (~1 wt% loading), and then annealed at 150 °C (1 h) in a vacuum oven. The supported gold particle size after the annealing was confirmed by scanning transmission electron microscopy (STEM) to be identical with that of Au_{25}, and thermogravimetric analysis (TGA) confirmed that the protecting thiolate ligand remained on the surface of gold nanoclusters after the annealing. The $Au_{25}(SR)_{18}$/oxide catalysts after thermal treatment was applied in the Sonogashira cross-coupling reaction between iodobenzene and phenylacetylene.

After optimization of the conditions, the reaction was carried out under N_2 atmosphere at 160 °C using DMF as solvent and K_2CO_3 as base. A high conversion of p-iodoanisole (up to 96.1%) with excellent selectivity for cross-coupling product (up to 88.1%) was obtained when it was catalyzed by $Au_{25}(SR)_{18}/CeO_2$ catalyst (Table 5.2, entry 1). The $Au_{25}(SR)_{18}$ nanocluster catalyst performed much better than the larger sized 2–3 nm Au nanoclusters (Table 5.2, entry 1 vs. 5) and CeO_2-supported Au nanoparticles (bare, ~20 nm). Support effects were also investigated, and no distinct effect of the oxide supports was observed (i.e., CeO_2, SiO_2, TiO_2, and MgO). In the recyclability test, the conversion of p-iodoanisole exhibited no significant loss, while the selectivity was decreased from 88.1% to 64.5% after five cycles. The drop in selectivity should be caused the gradual degradation of nanoclusters in the multiple recycling tests, as larger Au clusters give a much lower selectivity. Moreover, density functional theory (DFT) calculation of the reactant adsorption shows that both reactants (i.e., iodobenzene and phenylacetylene) prefer to adsorb on the open facet with the phenyl ring facing a surface gold atom. A total adsorption energy reaches −0.90 eV when the two reactants co-adsorb on the $Au_{25}(SR)_{18}$ catalyst. The DFT results suggested that the catalytic active site is associated with the $Au_{25}(SR)_{18}$, which is consistent with the experimental results.

Table 5.2 The catalytic performance of $Au_{25}(SR)_{18}$ (R=CH_2CH_2Ph) nanoclusters supported on various oxides as catalysts for Sonogashira cross-coupling reaction of p-iodoanisole and phenylacetylene. (a) Top view of one of the two open facets of $Au_{25}(SR)_{18}$ nanoclusters where three external gold atoms are exposed; the other facet is on the back side. (b) Top view of the co-adsorption of phenylacetylene and iodobenzene on the surface of the $Au_{25}(SR)_{18}$ nanoclusters. R′ = OCH_3. Au, yellow; S, blue; C, gray; H, white; I, green.

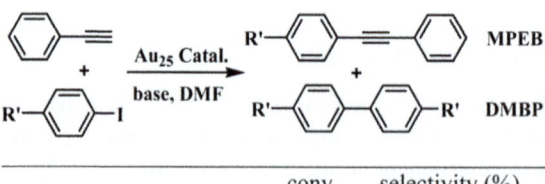

entry	catalyst	conv. (%)	selectivity (%) MPEB	DMBP
1	$Au_{25}(SR)_{18}/CeO_2$	96.1	88.1	11.9
2	$Au_{25}(SR)_{18}/TiO_2$	92.8	82.9	17.1
3	$Au_{25}(SR)_{18}/SiO_2$	90.8	79.3	20.7
4	$Au_{25}(SR)_{18}/MgO$	93.3	80.6	19.4
5	AuNC 2-3 nm $(SC_6H_{13})/CeO_2$	65.5	57.2	42.8

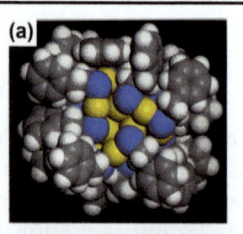

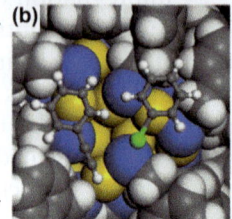

Adapted with permission from ref. 26.

5.6 Summary

In this chapter, we have discussed the size-focusing methodology for synthesizing atomically precise gold nanoclusters (such as Au$_{25}$(SR)$_{18}$ and Au$_{38}$(SR)$_{24}$ nanoclusters) and their catalytic application. These atomically precise Au$_n$(SR)$_m$ nanoclusters hold great promise in catalysis, as demonstrated in the catalytic oxidation (CO to CO$_2$, selective oxidation of styrene to benzaldehyde or styrene epoxide, and selective oxidation of sulfide to sulfoxide), selective hydrogenation (aldehyde and ketone to alcohol, and nitrophenol to aminophenol), and carbon–carbon coupling reactions (Ullmann homo-coupling of iodobenzene and Sonogashira cross-coupling of iodobenzene and phenylacetylene). The catalytic reactions strongly indicate that the catalytic activity and selectivity are associated with unique structure and electronic properties of Au$_n$(SR)$_m$ nanoclusters. As a new class of catalysts, Au$_n$(SR)$_m$ nanoclusters are expected to bridge the gap between the bulk crystal structure model catalysts and real-world catalysts. Although much work remains to be carried out, these well-defined gold nanoclusters provide exciting opportunities for achieving fundamental understanding of nanocatalysis at the atomic level, including the insight into the size dependence and deep understanding of the molecular activation, catalytic active sites, and catalytic mechanism by correlation with the structures of gold nanoclusters. We expect that future research on atomically precise nanocluster catalysts will contribute to the fundamental catalysis and the new design of highly efficient catalysts for specific chemical processes.

Acknowledgements

The work was supported by U.S. Department of Energy – Office of Basic Energy Sciences (Grant DE-FG02-12ER16354).

References

1. G. C. Bond, C. Louis and D. T. Thompson, *Catalysis by Gold*, Imperial College Press, London, 2006.
2. U. Heiz and U. Landman, *Nanocatalysis*, Springs, New York, 2007.
3. D. Astruc, F. Lu and J. R. Aranzaes, *Angew. Chem., Int. Ed.*, 2005, **44**, 7852.
4. G. Li and R. Jin, *Acc. Chem. Res.*, 2013, **46**, 1749.
5. R. Jin, *Nanotechnol. Rev.*, 2012, **1**, 31.
6. R. Jin, H. Qian, Z. Wu, Y. Zhu, M. Zhu, A. Mohanty and N. Garg, *J. Phys. Chem. Lett.*, 2010, **1**, 2903.
7. R. Jin, *Nanoscale*, 2010, **2**, 343.
8. H. Qian, M. Zhu, Z. Wu and R. Jin, *Acc. Chem. Res.*, 2012, **45**, 1470.
9. M. Zhu, C. M. Aikens, F. J. Hollander, G. C. Schatz and R. Jin, *J. Am. Chem. Soc.*, 2008, **130**, 5883.
10. D.-E. Jiang, *Nanoscale*, 2013, **5**, 7149.
11. H. Qian, C. Liu and R. Jin, *Sci. China-Chem.*, 2012, **55**, 2359.

12. C. Zeng, T. Li, A. Das, N. L. Rosi and R. Jin, *J. Am. Chem. Soc.*, 2013, **135**, 10011.
13. C. Zeng, H. Qian, T. Li, G. Li, N. L. Rosi, B. Yoon, R. N. Barnett, R. L. Whetten, U. Landman and R. Jin, *Angew. Chem., Int. Ed.*, 2012, **51**, 13114.
14. M. MacDonald, P. Zhang, N. Chen, H. Qian and R. Jin, *J. Phys. Chem. C*, 2009, **115**, 65.
15. H. Qian and R. Jin, *Nano Lett.*, 2009, **9**, 4083.
16. H. Qian, Y. Zhu and R. Jin, *Proc. Natl. Acad. Sci. U. S. A.*, 2012, **109**, 696.
17. M. Zhu, E. Lanni, N. Garg, M. E. Bier and R. Jin, *J. Am. Chem. Soc.*, 2008, **130**, 1138.
18. Z. Wu, J. Suhan and R. Jin, *J. Mater. Chem.*, 2009, **19**, 622.
19. A. C. Dharmaratne, T. Krick and A. Dass, *J. Am. Chem. Soc.*, 2009, **131**, 13604.
20. N. K. Chaki, Y. Negishi, H. Tsunoyama, Y. Shichibu and T. Tsukuda, *J. Am. Chem. Soc.*, 2008, **130**, 8608.
21. H. Qian, M. Zhu, U. N. Andersen and R. Jin, *J. Phys. Chem. A*, 2009, **113**, 4281.
22. H. Qian, Y. Zhu and R. Jin, *ACS Nano*, 2009, **3**, 3795.
23. Z. Wu, C. Gayathri, R. Gil and R. Jin, *J. Am. Chem. Soc.*, 2009, **131**, 6535.
24. H. Qian, W. T. Eckenhoff, Y. Zhu, T. Pintauer and R. Jin, *J. Am. Chem. Soc.*, 2010, **132**, 8280.
25. Y. Zhu, H. Qian and R. Jin, *Chem. – Eur. J*, 2010, **16**, 11455.
26. G. Li, D.-E. Jiang, C. Liu, C. Yu and R. Jin, *J. Catal.*, 2013, **306**, 177.
27. M. Zhu, C. M. Aikens, M. P. Hendrich, R. Gupta, H. Qian, G. C. Schatz and R. Jin, *J. Am. Chem. Soc.*, 2009, **131**, 2490.
28. G. Li, H. Qian and R. Jin, *Nanoscale*, 2012, **4**, 6714.
29. M. Zhu, W. T. Eckenhoff, T. Pintauer and R. Jin, *J. Phys. Chem. C*, 2008, **112**, 14221.
30. B. K. Min and C. M. Friend, *Chem. Rev.*, 2007, **107**, 2709.
31. M. Haruta, S. Tsubota, T. Kobayashi, H. Kageyama, M. J. Genet and B. Delmon, *J. Catal.*, 1993, **144**, 175.
32. X. Nie, H. Qian, Q. Ge, H. Xu and R. Jin, *ACS Nano*, 2012, **6**, 6014.
33. X. Nie, C. Zeng, X. Ma, H. Qian, Q. Ge, H. Xu and R. Jin, *Nanoscale*, 2013, **5**, 5912.
34. G. Ma, A. Binder, M. Chi, C. Liu, R. Jin, D.-E. Jiang, J. Fan and S. Dai, *Chem. Commun.*, 2012, **48**, 11413.
35. Y. Zhu, H. Qian, M. Zhu and R. Jin, *Adv. Mater.*, 2010, **22**, 1915.
36. Y. Liu, H. Tsunoyama, T. Akita and T. Tsukuda, *Chem. Commun.*, 2010, **46**, 550.
37. H. Qian, D.-E. Jiang, G. Li, C. Gayathri, A. Das, R. R. Gil and R. Jin, *J. Am. Chem. Soc.*, 2012, **134**, 16159.
38. D.-E. Jiang, M. L. Tiago, W. Luo and S. Dai, *J. Am. Chem. Soc.*, 2008, **130**, 2777.

39. Y. Zhu, H. Qian, B. A. Drake and R. Jin, *Angew. Chem., Int. Ed.*, 2010, **49**, 1295.
40. Y. Zhu, Z. Wu, G. C. Gayathri, H. Qian, R. R. Gil and R. Jin, *J. Catal.*, 2010, **271**, 155.
41. H. Yamamoto, H. Yano, H. Kouchi, Y. Obora, R. Arakawa and H. Kawasaki, *Nanoscale*, 2012, **4**, 4148.
42. A. Shivhare, S. J. Ambrose, H. Zhang, R. W. Purves and R. W. J. Scott, *Chem. Commun.*, 2013, **49**, 276.
43. G. Li, C. Liu, Y. Lei and R. Jin, *Chem. Commun.*, 2012, **48**, 12005.

CHAPTER 6

Electrochemical Atomic-level Controlled Syntheses of Electrocatalysts for the Oxygen Reduction Reaction

STOYAN BLIZNAKOV, MIOMIR VUKMIROVIC AND RADOSLAV ADZIC*

Chemistry Department, Brookhaven National Laboratory, Upton, NY 11973, USA
*Email: adzic@bnl.gov

6.1 Background

Polymer electrolyte membrane fuel cells (PEMFCs) are the most advanced type of fuel cells because of their high efficiency, light weight, low operational temperature, and fast-start capability. All these features define PEMFCs as a strong candidate for mobile application, especially in the electric vehicles, since the internal combustion counterparts are one of the biggest sources of air pollution. This governs an increased research interest on developing of cheap, high efficient and durable PEMFCs in the last four decades. The most challenging obstacle to the wide-spread application of the fuel cells is associated with the cathodic reduction of the oxygen. The sluggish oxygen reduction reaction (ORR) kinetics, even on Pt that is the most active electrocatalyst for the oxygen reduction reaction, causes about 400 mV over-potential losses resulting in overall reduced energy conversion efficiency in the PEMFCs. In order to compensate this loss a large amount of Pt is required, which causes their cost raising. Currently, the Pt loading in the state-of-the-art PEMFCs is 0.4 mg$_{Pt}$ cm^{-2}.[1] Reduction of the amount of the Pt contained in the PEMFCs stacks is required for large-scale automotive applications, both for reasons of cost and Pt supply limitations. In order to challenge researchers all over the world, and to boost the implementation of the PEMFCs in the electric vehicles, the United States Department of Energy (DOE) set challenging technical targets for development of fuel cell electrocatalysts for transportation applications in its Multi-Year Research, Development and Demonstration Plan.[2] The main goal of this plan is to develop and demonstrate fuel cell power system technologies for transportation, stationary and portable power applications. One of the objectives is to develop a 60% peak-efficient, 5000-h durable, direct hydrogen fuel cell power system for transportation at a cost of $30/kW by 2017. According to the DOE targets for 2020, the platinum group metal (PGM) total loading in both electrodes of the PEMFCs should be as low as 0.125 mg$_{PGM}$ cm^{-2}.[2] The mass activity is targeted to be 0.44 A mg$_{Pt}^{-1}$ at 900 mV (vs. RHE) as determined from the iR-free polarization curve measured at 80 °C H$_2$/O$_2$ in MEA, fully humidified with total outlet pressure of 150 kPa, anode and cathode stochiometry 2 and 9.5, respectively. The loss in the initial catalytic activity is targeted to be less than 40%, after 5000 accelerated stability tests, performed in a 25–50 cm^2 MEA, during triangle sweep cycles at 50 mV s^{-1} between 0.6 V and 1.0 V (vs. RHE) at 80 °C, atmospheric pressure, 100% relative humidity, H$_2$ at 200 sccm and N$_2$ at 75 sccm.[2] There are two methodologies for determining the mass activity and fundamentally more meaningful specific activity (in units of μA cm^{-2}$_{Pt}$) of the electrocatalysts for the ORR.[1] The first one is so-called thin-film rotating-disc electrode (RDE) method that is used for studding high-surface-area catalysts in aqueous electrolyte. In this method a thin layer (about 1 μm) of catalyst is deposited from catalysts ink on a glassy carbon RDE electrode and then is dryed. Then the electrode is immersed in de-aerated or oxygen saturated solution of perchloric acid and the mass and specific activities are calculated from the polarization and cyclic voltammetry (CV) curves.[3] The second method requires preparation of membrane electrode assemblies (MEAs) which consist of an anode, cathode electrodes and proton-conducting membrane (Nafion®), and testing them in a H$_2$/O$_2$-fed single cell. Both approaches are shown to yield similar activities.[1,3]

Different types of electrocatalysts have been developed in the last several decades in order to meet the DOE targets.[1,4] The general trend in the research and development of advanced electrocatalysts for the ORR is to reduce or even to eliminate the Pt and the other precious metals from the composition of the catalysts. Generally, the electrocatalysts for the ORR can be systematized in three main groups:[1,4] (i) Pt-based electrocatalysts, (ii) non-Pt catalysts, and (iii) non-precious catalysts. Among those, the Pt-based catalysts are the most promising for real fuel cell applications since, at the moment, only they satisfy all DOE requirements. This group of catalysts could be divided in three sub-groups: (i) pure Pt, (ii) Pt-alloy, and (iii) Pt monolayer (ML) type electrocatalysts for the ORR. It has recently been demonstrated that the mass and specific activities, and as well as the durability of the Pt ML type electrocatalysts exceed the DOE targets for 2020, which renders them as the most advanced type of fuel cell electrocatalysts.[5,6]

A typical chemical method for synthesis of carbon supported Pt and Pt-alloy nanoparticles includes two steps: (i) impregnation, and (ii) reduction. In the impregnation step a high surface area carbon particles are mixed and homogenized with solution of the Pt salt (K_2PtCl_4, $Pt(acac)_2$–acetylacetonate *etc.*) and/or transition metal salt, and then the reduction or co-reduction is performed by H_2 gas or other reducing agent.[4] The shape control of Pt and Pt-alloy nanoparticles (NPs) during their synthesis is achieved by controlling the NPs growth direction through binding different surfactant molecules on a crystal plane.[4] If surfactants are used to tune the shape of the NPs then the catalysts surface should be cleaned after the synthesis in a consecutive step. The cleaning procedure includes treatment of the NPs in acetone and/or alcohol, CO adsorption and stripping, plasma cleaning, ozone treatment, *etc.* Resent advances in the chemical synthesis have led to the formation of various kinds of nanoparticles with more rational control of size, shape, composition, and structure.[4]

The requirements for durability and activity of the Pt-based cathode catalysts are heightened by the need to minimize the platinum content.[2] Furthermore, durability is particularly critical for fuel-cell cathodes because of the highly aggressive conditions, namely low pH, dissolved molecular oxygen, and the high positive potentials at which they operate. The durability of the existing platinum catalysts is unsatisfactory. Platinum nanoparticles can dissolve and redeposit on other Pt NPs. In addition, Pt can deposit in the membrane by the reduction of Pt^{2+} by the hydrogen diffusing from the anode (hydrogen cross-over). The major dissolution of these catalysts occurs under potential cycling regimes, which for example happens during the stop-and-go driving of electric cars. The consequent gradual decline in performance during operation, mainly caused by the loss of the electrochemical surface area of Pt NPs at the cathode, impedes their widespread application in the PEMFCs. Thus, the durability target of 5000 h could be met only with thicker Pt layers or larger particles that can sustain sizeable Pt dissolution losses.[7] These particle types suffer from the drawbacks of high price and low Pt mass activity. In order to reduce the amount of Pt, while improving efficiency and durability of Pt-based ORR catalysts, the catalysts with a core–shell structure are designed to have a shell thickness of only a monolayer. The considerable interest for these catalysts stems from their reduced cost and improved catalyst activity and stability.[5] To produce Pt ML, a method with atomic- and molecular-level control is needed. Atomic layer deposition (ALD) methodology was first used for atomic level growth control of epitaxial metallic layers on a single crystal supports at elevated temperatures in ultra-high vacuum techniques, such as molecular beam epitaxy and chemical vapor deposition.[8] ALD is a methodology for forming nanofilms of materials, one atomic layer at a time, using surface limited reactions (SLRs).[9] SLRs occur only at the deposit surface, and once the surface (the limiting reagent) is covered the reaction stops. The term 'atomic layer' refers to monolayer coverage, a ML being a unit of coverage particular to the deposit being formed. From a surface science point of view, a full ML is formed when there is one deposit atom for each surface atom. The electrochemical atomic layer deposition, also known in the literature as a surface limited redox replacement (SLRR), is an ALD in an electrochemical environment. The electrochemical methods for Pt ML deposition are well suited and atomic-level control of Pt deposition can be achieved either by precisely controlling the amount of reactants or quenching the reaction. In addition, they can be used to produce a core on a suitable electrode. Thus, core–shell catalysts can be synthesized by applying only electrochemical methods. Since they are inexpensive, flexible, and easily scalable, the electrochemical methods can provide a possible way to reduce manufacturing cost of fuel cells.

In the first part of this chapter we review the methods for the syntheses of Pt monolayer on different single crystal supports, using electrochemical deposition strategies with an atomic level control. These methodologies include: (i) galvanic displacement of under-potentially pre-deposited (UPD) Cu, Pb or H monolayer by Pt that allows a precise control at a fraction of a monolayer of deposited Pt amount; and (ii) electro-deposition of Pt and limiting its coverage to a ML by quenching from strongly adsorbed layer of non-metallic elements or molecules (H_{ads} or CO) that hinders the growth of second ML. In the second part of the chapter we outline the advantages of the electro-deposition techniques over the conventional chemical methods for synthesis of electrocatalysts for the oxygen reduction reaction, and demonstrate the applicability of the SLRR strategy for synthesis of Pt ML on electro-deposited Pd, PdAu, PdIr, PdNiW nanostructures. As obtained Pt ML core–shell fuel cells electrocatalysts are with ultra-low total precious metal loadings, and have very high activity and excellent performance stability as evidenced from their polarization curves.

6.2 Electrochemical Deposition of Monolayers of Precious Metals on Different Transition Metal Supports

The electrochemical methods for deposition of a monolayer of precious metal on different metal supports could be divided in two groups, based on the surface specific reaction that is used to limit the deposition of metal of interest only to monolayer coverage. In an electrochemical environment the gas molecule adsorption and the metal under-potential deposition are the only two phenomena that could cause surface limited reactions. Figure 6.1 illustrates both 'displacement driven' and 'adsorption driven' strategies for a monolayer deposition. In the 'displacement driven' methodology the amount of the precious metal is limited to a monolayer from a pre-deposited

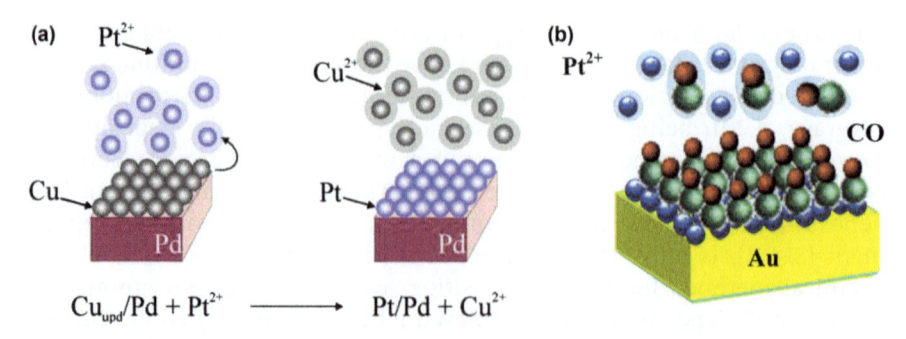

Figure 6.1 Cartoons of the 'displacement driven' electrochemical atomic layer deposition strategy (a) (Reproduced by permission of ECS – The Electrochemical Society, Interface, Summer 2011, pp. 33–40. Copyright 2011), and 'adsorption driven' strategy for Pt monolayer deposition (b).

monolayer at under-potentials of less noble metal that is displaced by more noble metal in a consecutive step.[10–12] This strategy is based on the UPD phenomenon. As Figure 6.1a shows, the strategy is realized in two steps. In the first step, that is pure electrochemical step, a complete Cu ML layer is formed by immersing the electrode in a solution containing Cu^{2+} ions and applying potential more positive than the equilibrium potential for bulk Cu deposition. In the second step, that is an electro-less step, the electrode is transferred in a second container containing Pt^{2+} ions. Herein, as formed Cu UPD ML is galvanically displaced by Pt. The galvanic displacement is a process where the ions of more noble species oxidize less noble species and replace them. In the 'adsorption driven' strategy the growth of second monolayer is prevented by adsorption of layer of non-metallic element or molecule that adsorbs strongly on the formed ML and thus hinders further growth process.[13,14] Figure 6.1b, illustrates how the CO molecules are adsorbed on the Pt ML, electro-deposited on the Au surface, and inhibit further Pt ions reduction and prevent bulk Pt formation.

6.2.1 Adsorption-driven Surface-limited Reactions for Atomic Monolayer Deposition

It has been recently reported[14] that the formation of a saturated H_{UPD} layer at high negative over-potentials in thetra-chloro-platinate $[PtCl_4]^{2-}$ solutions with pH ranging from 2.5 to 4, exerts a quenching or self-terminating effect on Pt deposition on Au, restricting it to a high coverage of two-dimensional (2D) Pt islands (sub-monolayer coverage). When repeated, by using pulse potential waveform to periodically oxidize the H_{UPD} layer, sequential deposition of discrete Pt layers can be achieved. The process is thus analogous to the atomic layer deposition, but with a rapid potential cycle replacing the time-consuming displacement of the ambient reactant. According to the authors, the self-termination of the Pt deposition reaction arose from perturbation of the double-layer structure that accompanies H_{ads} saturation of the Pt surface. Thus, the H-covered Pt surface impacts the adjacent water structure, which leads to minimum in coupling between the electrode and electrolyte. In addition, the quenching of the metal deposition reaction occurs at potentials negative to the potential of zero charge of Pt, where all anions are desorbed. The combination of these two phenomena exerts remarkable effect whereby $[PtCl_4]^{2-}$ reduction is completely quenched while diffusion-limited proton reduction continues unabated.

Alternatively, an idea for Pt ML deposition on polycrystalline Au, based on adsorption driven surface specific reaction, has been most recently proposed.[13] In this work the authors studied the electro-deposition of Pt on Au from CO saturated K_2PtCl_4 plus H_2SO_4 solution into thin layer flow cell under potential control. The potential chosen is far below the Nernst redox potential for the Pt^{2+}/Pt couple, and bulk Pt deposition would be expected. However, a distinct reduction peak in the current transient curve corresponding to reduction of Pt ions to Pt metal on Au surface is observed, followed by fast decay of current to zero as soon as saturation Pt coverage is reached. Thus, according to the authors, immediately after Pt ML deposition the layer is covered by strongly adsorbed CO molecules that inhibit further Pt ions reduction and prevent bulk Pt formation. The monolayer geometry is confirmed by electrochemical and *in situ* IR spectroscopic data. The IR spectra indicate that adsorbed CO revealed distinct electronic ligand and strain effects, characteristic for an electronic modification of the Pt ML as compared with bulk Pt(111). Also, it is reported that oxidative removal of the CO adlayer results in a distinct restructuring of the Pt films, yielding three-dimensional (3D) Pt nanostructures.[13] This finding puts into question the applicability of this strategy for synthesis of highly efficient and durable Pt ML type electrocatalysts for the ORR.

We note that these both strategies, self-termination of Pt ML by hydrogen adsorption at high redox over-potentials and self-limited Pt ML deposition by CO adsorption, have been demonstrated for Pt ML deposition on Au substrates and have not been explored for synthesis of Pt ML core–shell catalysts.

6.2.2 Displacement-driven Surface-limited Reactions for Atomic Monolayer Deposition

Under-potential deposition is a type of electrochemical SLR in which up to a monolayer of metal is deposited on another metal at a potential prior to (under) that needed to deposit the first metal on itself.[15] UPD is a thermodynamic phenomenon, where the interaction energy between the two elements is larger than the interaction of an element with itself, thus resulting in the formation of a surface compound or alloy.[15–17]

Generally, UPD involves deposition of less noble metal on a more noble metal surface. The driving force for UPD is formation of a compound, which is energetically favored relative to the bulk elements whit a stoichiometry defined by the surface chemistry of the substrate. With metals, a two-dimensional bimetallic compound is formed. This process avoids formation of three-dimensional crystallites and Stranski–Krastanov growth. Cu UPD, Pb UPD and H UPD are three reactions used widely for deposition of Pt ML on different supports by the SLRR strategy.[11,18,19] The SLRR methodology is realized in two steps. In the first step a monolayer of Cu (Pb or H) is deposited at under-potentials on a substrate metal of interest, and in the second step the UPD monolayer is displaced by Pt ML (Figure 6.1a). Technically, the SLRR protocol has been initially implemented either by shuttling of the electrode between separate cells (multiple immersion) with different solutions[10,20] or by subjecting it to an alternating flux of different solutions in a flow cell.[12] Thus, the potentially controlled UPD layer formation and the spontaneous redox replacement steps take place in different solution.

Most recently an alternative method for SLRR, known as 'one-cell configuration' approach has been proposed.[11] It has been initially introduced as a 'one-pot' configuration and first implemented in the growth of Cu on a Au(111) single crystal surface.[21] In this approach the electrode is held all the time in one cell, in contact with a solution containing both ions/complexes of UPD and growing metals in a concentration ratio that would result in only marginal (no more than 1 at.%) electro-deposition of the growing metal during the sacrificial layer formation step.

6.2.2.1 Surface-limited Redox Replacement of Cu UPD Monolayer: a Highly Efficient Method for Synthesis of Pt ML Core–Shell Type Electrocatalysts

The electrochemical atomic layer deposition strategy coincided with demonstration of the Pt monolayer type electrocatalysts, which is considered as the most advanced class of electrocatalysts for the ORR. The strategy has been demonstrated for first time by deposition of a Pt ML on single crystal Au(111) electrode.[10,18] The SLRR protocol has been performed in multiple immersion configuration, where a Cu UPD ML is first deposited under potential control from $CuSO_4$ plus H_2SO_4 solution, and then is displaced in $[PtCl_6]^{2-}$ plus H_2SO_4 solution. It is observed that the Au single crystal is uniformly covered by one ML high Pt nanoclusters that does not form a continuous layer.[10] The amount of Pt deposited by the displacement of a full Cu UPD ML is limited to coverage of a 1/2 monolayer because two Cu^0 atoms are oxidized in order to reduce one Pt^{4+} ion. Thus, the stoichiometry of the displacement is 2:1, and a maximum coverage of 50% could be achieved when Pt^{4+} salt solution is used in the displacement step of the SLRR protocol. Subsequent studies used Pt^{2+} and it has been confirmed[22] that if a Pt^{2+} solution is used in the displacement step, the stochiometry of the displacement is 1:1 and a uniform and pinhole-free Pt ML on Au surface is formed. The stoichiometry of Pt deposition via surface-limited redox replacement of the under-potentially deposited Cu monolayer on Au(111) from hexa-chloro-platinate $[PtCl_6]^{2-}$ plus $HClO_4$ solution, has recently been comprehensively studied.[23] In this work the authors combined the statistical scanning tunneling microscopy (STM) data analysis and conventional electrochemical techniques, and studied the relation between the Pt deposit coverage and the amount of the replaced Cu UPD ad-layer. It is reported that as obtained Pt sub-monolayers possess 2D morphology and linear dependence of the Pt coverage on the amount of the Cu UPD ML. Interestingly, the authors found that four Cu UPD ad-atoms are replaced by one Pt ad-atom. As observed 4 : 1 stoichiometry has been attributed to the specific experimental conditions where the Cu UPD ad-atoms oxidation to the Cu(I) ions is thermodynamically more favorable than their oxidation to Cu(II). This occurs as result of the direct ligand transfer process from depositing $[PtCl_6]^{2-}$ to Cu ions, which is the reaction mechanism that should be generally applicable to any situation where supporting electrolyte contains anions with no complexing ability toward Cu. This work stresses the general importance of the anions for metal deposition *via* surface-limited redox replacement of a UPD ML.

The scanning tunneling microscopy images of Pt ML deposited by SLRR of Cu UPD ML from Pt^{2+} and Pt^{4+} solutions are presented in Figure 6.2. In both images formation of 2D Pt islands is observed. It is seen from Figure 6.2A, that the coverage of the Au(111) terraces by Pt is only 0.25 ML

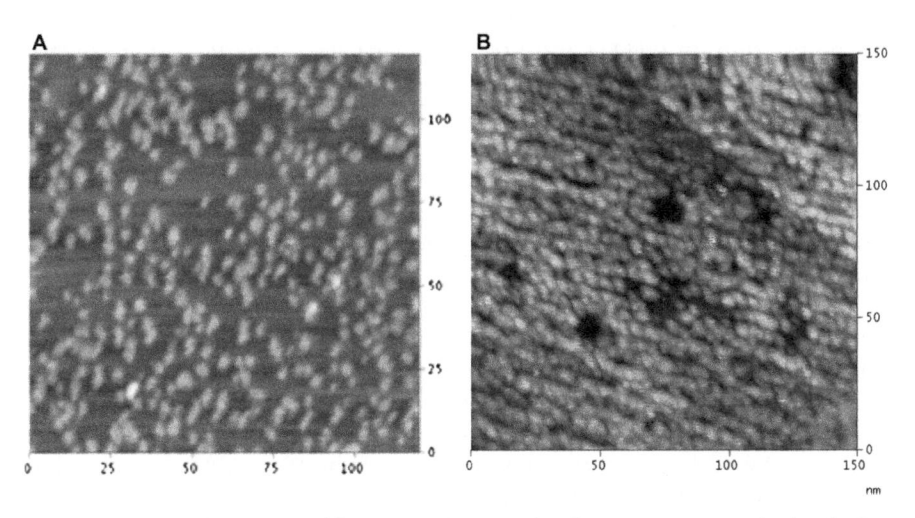

Figure 6.2 STM images of: (a) Pt deposit on Au(111), 125×125 nm, obtained after galvanic displacement of a Cu UPD ML deposited at under-potentials from 1 mM Cu^{2+} + 0.1 M $HClO_4$ by Pt^{4+} from 1 mM $[PtCl_6]^{2-}$ + 0.1 M $HClO_4$ solution (Adapted from ref. 23, reproduced by permission of ECS – The Electrochemical Society. Copyright 2010). (b) Pt ML deposited on Au(111), 150×150 nm, by galvanic displacement of Cu UPD ML under-potentially deposited from 50 mM Cu^{2+} + 50 mM H_2SO_4 solution by Pt^{2+} from 1 mM $[PtCl_4]^{2-}$ + 50 mM H_2SO_4 solution.

when the displacement step of the SLRR protocol is realized in Pt^{4+} solution. In contrast, Figure 6.2B shows that when the displacement is conducted in Pt^{2+} solution almost complete Pt ML is accomplished.

Further studies addressed the role of support in determining the activity of Pt monolayer as catalysts for the oxygen reduction reaction. Several single crystal surfaces, including Pd(111), Ru(0001), Ir(111), Rh(111), and Au(111) have been investigated as supports for Pt ML catalysts for the ORR. It is reported that the ORR activity of Pt ML on Au(111), Ir(111), Pd(111), Rh(111), and Ru(0001) surfaces showed volcano-type dependence *versus* the d-band center of the Pt ML.[24–26] Surprisingly, it was found that Pt ML supported on Pd(111) single crystal surface possessed better activity toward the ORR than the best known up to date Pt(111) single crystal surface in perchloric acid solution.[26] An improved activity of the Pt_{ML}/Pd(111) for the ORR is explained by geometric and electronic (ligand) effects.[25] The geometric effect accounts about the degree of compression/tension of the Pt ML as a result of the atomic size difference between Pt and supporting metal substrate. Thus, due to the close similarity in the atomic size of Pt and Pd, the Pt ML supported on Pd(111) surface, is slightly compressed. The compression results in down-shift of the d-band center of Pt ML, the electronic effect which is beneficial for the ORR kinetics.[26] According to the DFT calculations, an electronic interaction (ligand effect) between Pt ML and Pd support results in superior ORR catalytic activity of Pt_{ML}/Pd(111) due to the reduced OH coverage.[24,25]

Significant enhancement of the kinetics of the ORR has also been confirmed for Pt ML on Pd carbon supported nanoparticles (NPs) in comparison to the reaction on Pt(111) and Pt/C nanoparticles.[27] The Pt_{ML}/Pd/C electrocatalysts are prepared by using the SLRR strategy, where a Cu UPD ML is first deposited on the Pd NPs surface and then is galvanically displaced by Pt from $[PtCl_4]^{2+}$ containing solution. It is reported that the four-electron reduction mechanism, with a first-charge transfer-rate determining step, is operative on both Pt_{ML}/Pd(111) and Pt_{ML}/Pd/C surfaces, with a very small amount of H_2O_2 detected on the ring electrode in the hydrogen-adsorption potential region.[27]

Initially, the synthesis of the Pt ML elctrocatalysts is carried out on a glassy carbon electrode with diameter of 5 mm and the quantity of the electrocatalysts synthesized by one experiment is limited to a few tens of micrograms. Later a scale-up synthesis method using a new electrochemical cell that allows a synthesis of gram quantities of Pt monolayer electrocatalysts has been developed.[28] The core–shell structure of the Pt_{ML}/Pd/C electrocatalyst, synthesized by the scaled-up method, has been verified using high-angle annular dark field (HAADF) scanning transmission electron microscopy (STEM) Z-contrast images, STEM-EELS (electron energy-loss spectroscopy), and STEM/EDS (energy-dispersive X-ray spectrometry) line profile analysis.

Figure 6.3, shows a HAADF–STEM image of a single carbon supported Pt_{ML}/Pd nanoparticle, synthesized by the scaled-up method based on the SLRR of Cu UPD ML by Pt from Pt^{2+} solution, and a line profile analysis using STEM-EDS.

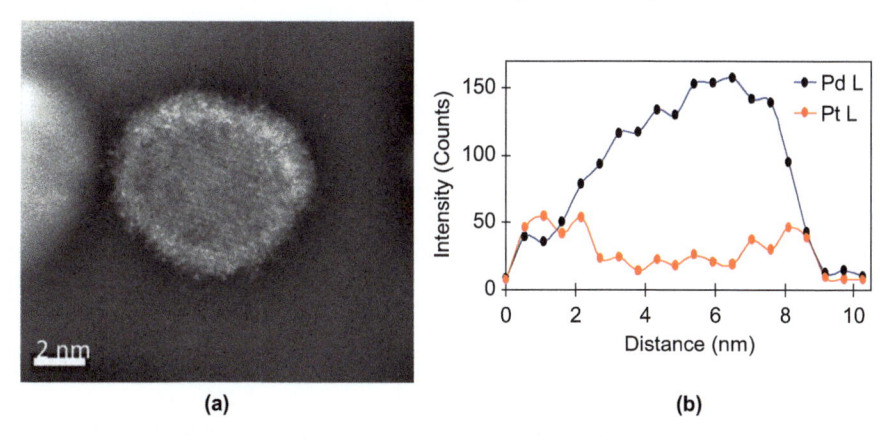

Electrochemical Atomic-level Controlled Syntheses of Electrocatalysts 153

Figure 6.3 Microscopic and compositional observations of a single Pt ML on Pd/C nanoparticles. (a) HAADF/STEM Z-contrast image of carbon-supported Pt$_{ML}$/Pd nanoparticle. (b) Distribution of the elements in the Pt$_{ML}$/Pd nanoparticle, obtained by the line profile analysis using STEM/EDS. Reprinted from ref. 28 (copyright 2010 with permission from Elsevier).

The HAADF technique can give the observations of interfaces/locations between different elements since the image intensity from an element (atomic number Z) follows approximately Z^2 dependence. The Z-contrast image in Figure 6.3a shows bright shells on a relatively darker nanoparticle core, which suggests the formation of a core–shell structure, *i.e.*, a Pt ($Z=78$) monolayer or bilayer (shell) on a Pd ($Z=46$) nanoparticle (core). Individual bright dots are observed around the nanoparticle. They are presumably due to Pt atoms bombarded by the high voltage electron beam during the observation. It should be noted that some HAADF images showed only partial coverage of shells on core nanoparticles, or gave no clear contrast of core-shell structures. Some HAADF images observed under zone axis conditions showed contrast patterns due to strain fields but no contrasts due to the difference in atomic weight. When the samples are tilted from the zone axis, the image contrast becomes one predicted by atomic number variation (Z-contrast). In STEM analysis the nanoparticles lie in completely random orientations with respect to the electron beam and every single particle is thus not amenable to the analysis. This is likely to be one of the reasons why some pictures show incomplete Pt monolayers.

Figure 6.3b illustrates the STEM/EDS line profile analysis of a single nanoparticle and shows the distribution of Pt and Pd components. It is seen that the width of the Pd core is 8 nm. The Pt profile has the following three marked features: (i) Pt atoms fully cover the Pd nanoparticle surface, (ii) the Pt intensity is fairly constant along the center of Pd nanoparticle, and (iii) at the both edges of Pd nanoparticle the Pt intensity is approximately doubled in comparison with that at center. These features demonstrate the formation of a core (Pd)–shell (Pt) structure. From the atomic ratio analysis of Pt and Pd, and assuming that the height of Pd nanoparticle (normal to the picture)

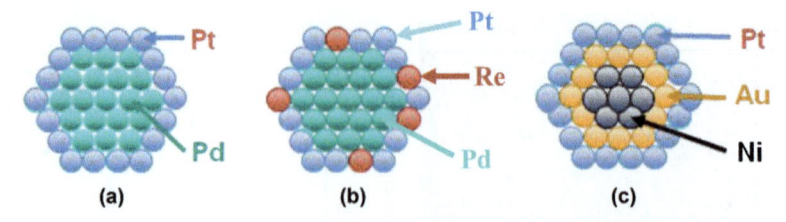

Figure 6.4 Illustrations of Pt ML shells on different nanoparticle cores: (a) Pt ML shell on Pd nanoparticle core; (b) Mixed Pt–Re ML shell on Pd nanoparticle core; (c) Pt ML shell on Ni core covered by ML of Au.

is equal to the width (8 nm), the thickness of the Pt shell is consistent with the presence of a monolayer.

Furthermore, the core–shell structure of the $Pt_{ML}/Pd/C$ electrocatalyst have also been characterized by *in situ* extended X-ray absorption fine structure (EXAFS), yielding information on atomic structures averaged from thousands of nanoparticles. The coordination numbers of Pt–Pt and Pt–Pd obtained from the EXFAS analysis verified the formation of a Pt monolayer on the surfaces of the Pd nanoparticles.[28]

It has been shown that Pt ML on Pd nanoparticles electrocatalysts possessed higher activity and stability compared to the state-of-the-art Pt/C electrocatalysts.[27] Pt specific mass activity for $Pt_{ML}/Pd/C$ electrocatalysts increases up to 10 times in comparison to the commercial Pt/C catalyst.[25] Further increasing of the Pt mass activity by factor of 20 is achieved on mixed-metal Pt ML on Pd(111) and Pd/C nanoparticles,[29] where part of the Pt atoms (20%) in the ML are replaced by the atoms of a second noble metal (*e.g.*, Ru, Ir, Os, Re). An additional decreasing of the noble metal loading in the catalysts is proposed by preparing a Pt ML on non-noble-core noble-shell electrocatalysts. In this case a Co or Ni core is covered by monolayer of Au or Pd and ML of Pt or mixed monolayer is deposited on this core–shell structure by galvanic displacement of Cu UPD monolayer.[24] The cartoons of the three types of above discussed Pt ML shells on Pd or non-noble cores are presented in Figure 6.4.

6.2.2.2 Surface-limited Redox Replacement of Pb UPD Monolayer for Synthesis of Pt Monolayer Core–Shell Type Electrocatalysts

Copper is not a single metal suitable for forming a UPD monolayer to be displaced by more noble metals. Pb UPD monolayer displacement by Pt has been proposed as another SLRR possibility for processing of Pt ML core–shell electrocatalysts and epitaxially grown thin films.[11,22] Fayette *et al.*,[11] have been comprehensively studied Pt displacement of both Cu UPD and Pb UPD monolayers on mono- and poly-crystal Au and Pt surfaces by various techniques. The SLRR protocol that includes Pb UPD deposition and Pt displacement from tetra-chloro-platinate $[PtCl_4]^{2-}$ solution has been performed in a 'one-cell configuration'. According to the authors, the advantages of this protocol over the other SLRR protocols are: (i) easier handling; (ii) better control and uniformity of oxygen-free environment; (iii) better uniformity of the deposit, and (iv) easier adaptation and scaling up of systems with practical importance. It is demonstrated that with only one SLRR event of Pb UPD it is possible to form Pt ML film on Au that behaves like pure Pt in the H UPD reaction. The Pt film grows in 2D mode with no roughness increase up to 10 redox replacement cycles. An efficiency of 1 Pt ML per cycle (1 : 1 exchange Pb:Pt ratio) is achieved. As obtained Pt films consist of small clusters that show very similar size and distribution for different thicknesses. Interestingly, an amount of Pt above stoichiometry depositing per replacement event has been detected once the bare Au surface is platinized. This anomaly is attributed to the presence of extra reduction power that could be associated with the oxidative adsorption of OH_{ads} and/or to H adsorption/desorption phenomena, both reported to take place along with the UPD of Pb layers on Pt surfaces. In addition, a compositional analysis of the Pt layers grown by SLRR suggested incorporation of a minimum 4 atom% Pb and 13 atom% Cu when these metals are used as sacrificial ones in the deposition process.

6.2.2.3 Hydrogen Adsorption and Absorption as a Surface-limited Reaction for Synthesis of Pt ML Catalysts

A viable SLRR protocol for the growth of Pt MLs, and as well as thin films on Pt and Pd (including Pd nanocubes) substrates, based on adsorbed H or also called under-potentially deposited H (H-UPD), has been most recently proposed.[19] In contrast to the previously reported studies on SLRR deposition, this work presents the first application of the SLRR protocol in a one-cell configuration using a nonmetal UPD system. Thus, any possible mediator metal incorporation, reported in standard Pt SLRR protocols, is precluded herein. The growth of Pt using H UPD via SLRR method is conducted in tetra-chloro-platinate $[PtCl_4]^{2-}$ solution containing perchlorite ions as a supporting electrolyte. The protocol consist of repeating elementary steps of H UPD formation by potential impulse for 1 s followed by redox replacement step that takes place at OCP. It is reported that as grown Pt films feature steady roughness evolution and the growth is in the realm of quasi-2D growth mode that generates uniform Pt films. In addition, it is confirmed that no side reactions are taking place during the redox exchange, and the exchange Pt:H ratio is 1 : 2. The generality of the proposed approach is validated by growth of Pt films on two types of Pd surfaces: Pd ultra-thin films on Au (Pd/Au) and Pd nanocubes. The Pt over-layers grown by 30 replacements on Pd structures have been characterized by H-UPD and their catalytic activity examined by formic acid oxidation reaction.

A method for producing Pt surface monolayer on palladium or palladium alloy nanoparticles having low Pt loading coupled with catalytic activity for the ORR has been proposed in a patent assigned to Wang and Adzic.[30] The method includes contacting hydrogen adsorbed palladium or palladium alloy (PdNi, PdCo, PdFe, PdAu) particles with one or more metal salts, particularly Pt, to produce a sub-monolayer or mono-atomic metal or metal–alloy coating on the surface of the hydrogen-absorbed Pd or Pd alloy particles. The method does not require electro-deposition and does not generate waste. Hydrogen absorbed Pd or Pd alloy particles are produced by exposing the particles to hydrogen gas for a period of time sufficient to bring stoichiometric coefficient x in PdH_x to value of 0.6. After that the hydrogen absorbed Pd or Pd alloy particles are exposed to a solution containing Pt salt (or mixture of Pt salt with other metal salts) and an atomic monolayer of Pt or mixed Pt alloy monolayer-shell is deposited on the Pd or Pd alloy-core.

6.3 Platinum Monolayer on Electro-deposited Mono- and Bi-metallic (Pd, PdAu, PdIr, NiW) Nanostructures: Highly Efficient Electrocatalysts for the ORR

Among all strategies for Pt ML core–shell electrocatalyst processing, the SLRR of Cu UPD ML by Pt ML is the most viable for fuel cell catalysts synthesis. The catalysts obtained by this strategy use as a core Pd or Pd alloys nanoparticles supported on carbon, which are synthesized by conventional chemical methods, and then are cleaned from the surfactants that are commonly used during the synthesis. This type of Pt ML catalysts is well studied in the literature and several reviews have been published.[5,6,24,28,31] In addition, the method for the Pt ML shell deposition on Pd and PdAu nanoparticles cores has been scaled up. Furthermore, these electrocatalysts are considered as the most promising candidate for the second generation of the PEMFCs for the electric vehicles.

In this chapter we report on the next generation fuel cell electrocatalysts where the cores of mono-, bimetallic- or multi-component Pd alloys (including refractory alloys) nanostructures are electro-deposited on functionalized carbon (C_{ox}) or directly on the gas diffusion layer (GDL), which serve also as a current collector, and then Pt ML is deposited on their surface by SLRR of Cu UPD monolayer. The advantages of the electro-deposition over the conventional chemical methods for synthesis of electrocatalysts are as follows: (i) ultra-precise quantitative control of the deposited nanostructures; (ii) fine tuning of the microstructure and the morphology of the deposits without the need for the presence of any surfactants in the electrolyte; (iii) versatility in metals or alloys deposition; (iv) high utilization rate of the row materials and low materials waste; (v) low energy consumption; (vi) little capital investment; and (vii) ease of development of large-scale manufacturing processes.

In the Pt ML shell catalyst on electro-deposited nanostructures cores the total precious metal loading could be reduced to 50 µg cm^{-2}, which is a half of the U.S. Department of Energy's target for 2020,[2] and they possess excellent activity and performance stability. Furthermore, these catalysts have a 100% Pt utilization, because of the nature of their processing, which means that each Pt atom is accessible and is participating in the ORR.

6.3.1 Electro-deposited Pd Nanostructures and Bimetallic Pd Alloys on Functionalized Carbon Substrates: Advanced Cores for Pt ML Shell Electrocatalysts

It has been reported recently that the morphology of the electro-deposited Pd nanostructures on functionalized carbon nanoparticles from 1 mM $PdCl_2$ and 0.1 M NaCl solution, strongly depends from the applied over-potential.[32] Thus, Pd nanoparticles (Pd_{NP}) or nanorods (Pd_{NR}) or nanowires (Pd_{NW}) can be easily electro-deposited directly on functionalized carbon support at different over-potentials that have to be established.[20,32,33] Figure 6.5 presents the TEM images of electro-deposited Pd_{NP}, Pd_{NR} and Pd_{NW} at different potentials. The elongated nanostructures obtained are excellent cores for Pt ML shells electrocatalysts, because they possess smooth and well-ordered facets with reduced number of low coordinating sites and defects that are usually source of dissolution at fuel cell operation conditions. The anisotropic growth of the one-dimensional Pd nanowires is ascribed to the H under-potential deposition-mediated, layer-by-layer deposition of Pd. The growth mode operates on the cluster's facets covered by adsorbed hydrogen atoms, while, at these potentials, the chloride ions adsorbed only on the edges of the Pd clusters, precluding growth in other directions.[32] In addition, the surface structure of deposited nanowires is composed predominantly of (111)-oriented facets that facilitate the enhanced ORR activity and stability of the $Pt_{ML}/Pd_{NW}/C_{OX}$ electrocatalysts.

In order to prepare Pt monolayer type catalysts, the surface of as electro-deposited Pd nanostructures is modified by a monolayer of Pt, using the Cu

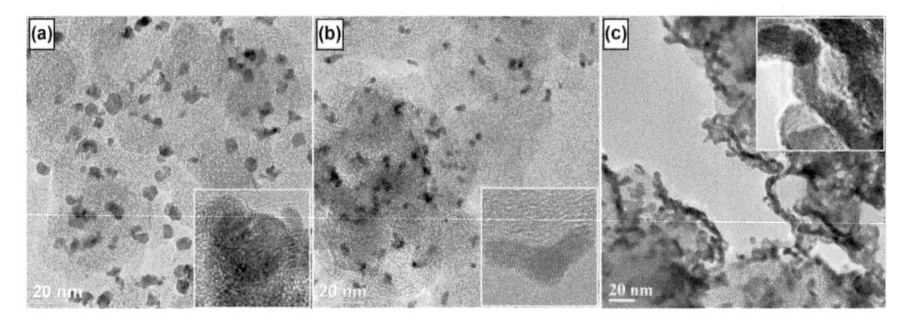

Figure 6.5 TEM images of electro-deposited Pd nanostructures at different over-potentials: (a) at 0.25 V, (b) at 0.1 V, and (c) at 0.05 V. The HRTEM images of nanoparticles, nanorods and nanowires are presented as insets. The second inset in (c) shows the fast Fourier transform pattern of selected area from the nanowires.

UPD displacement strategy described above. The electrocatalytic activity and stability of Pt_{ML}-shell on Pd_{NP}, Pd_{NR}, and Pd_{NW}-core supported on functionalized carbon ($Pt_{ML}/Pd_{NP,NR,NW}/C_{OX}$) electrocatalysts for the ORR is studied by measuring their polarization curves in oxygen-saturated 0.1 M $HClO_4$ at room temperature, and at rotation rate of 1600 rpm.[32]

In Figure 6.6, the Pt mass- and area-specific activities, as well as the total platinum group metal (PGM) mass-activities of Pt_{ML}-shell on Pd nanorods/nanowires-core supported on functionalized carbon electrocatalysts for the ORR, are compared with those of the state-of-the-art commercial Pt/C catalyst[3] and the DOE's targets for 2020.[2] Generally, the mass- or area-specific activity of the commercial Pt/C catalysts are well below the DOE's requirements. The Pt mass-specific activity values for the Pt_{ML} on Pd nanowires and nanorods electrocatalysts are 4.1 and 3.7 times higher than the DOE's targets, respectively. Also, it is seen that the values for the total PGM mass activity values for the Pt_{ML} on Pd_{NR} and Pd_{NW} electrocatalysts are lower than the targets, but comparable and even higher than the state-of-the-art Pt/C catalyst. In addition, Figure 6.6 reveals that the area-specific activity of the $Pt_{ML}/Pd_{NR}/C_{OX}$ electrocatalysts is very close to the DOE's target, while that of the $Pt_{ML}/Pd_{NW}/C_{OX}$ electrocatalysts exceeds it. Furthermore, the stability of both Pt_{ML}-shell on

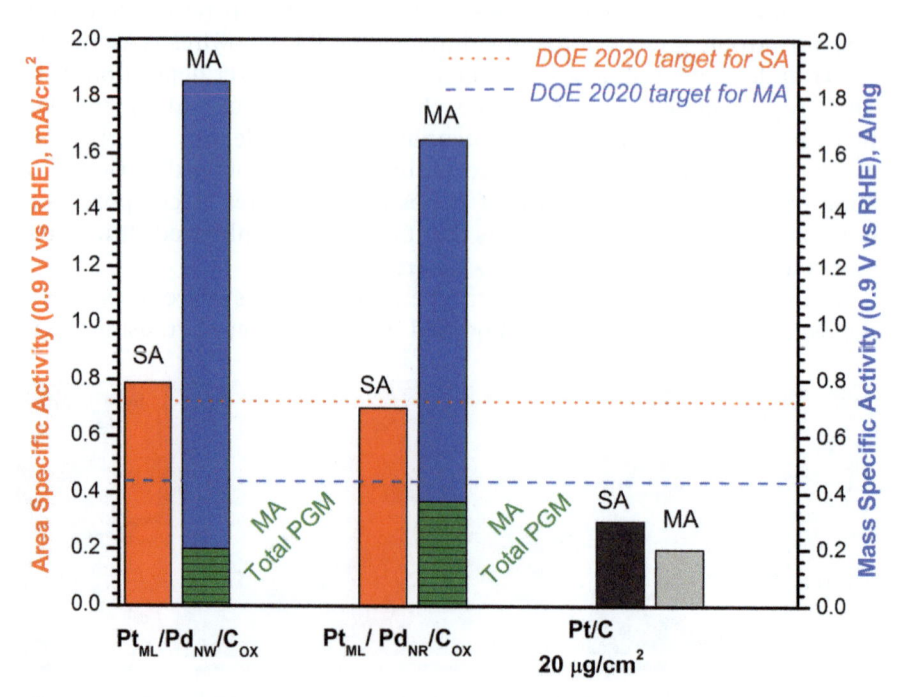

Figure 6.6 Comparison of the Pt mass- and area-specific activities of Pt_{ML}-shell on Pd nanorods/nanowires-core supported on functionalized carbon electrocatalysts for the ORR with the activities of commercial Pt/C catalysts and the DOE targets for 2020. The total platinum group metal (PGM) content values account for the total amount of the precious metals (Pt + Pd) in the catalysts.

Pd nanorods/nanowires-core electrocatalysts for the ORR is improved significantly in comparison to the one of the state-of-the-art Pt/C catalyst.[32]

Formation of a Pd 'band' in the Nafion membrane after 100 000 potential cycles has been reported for MEAs prepared with Pt_{ML}/Pd/C catalysts in the cathode.[7] During the cycling a considerable amount of Pd is dissolved from the catalyst core and Pd^{2+} ions diffuse out of the catalysts layer into the membrane. Part of the Pd^{2+} ions are reduced by the H_2 permeating from the anode in the membrane, and part of them are deposited on the anode. Negligible amounts of Pt in the membrane are also detected, but Pt shell integrity and the core–shell structure of the catalysts remained after the stability tests.[7] After prolonged potential cycling and significant dissolution of Pd, the Pt monolayer shell undergoes a small contraction that makes it less reactive and raises its existing high stability and dissolution resistance. Such a contraction causes a decrease in the particle size to form a more stable structure. As a result, the excess of Pt atoms from a monolayer shell of larger particles form a partial bi-layered structure. Also, a substantial dissolution of Pd causes formation of hollow particles, and induces additional contraction in the Pt ML shell. Thus, the Pd core protects the Pt shell from dissolution, and improves the activity and the stability of the Pt ML catalysts.[7]

Additional improvement of the stability of the Pt_{ML}/Pd/C electrocatalysts is possible by alloying the Pd core with another transition metal, such as Au and Ir.[7] Sasaki et al.,[6] reported a new class of highly durable and active electrocatalysts that comprise Pt ML on $Pd_{0.9}Au_{0.1}$ alloy nanoparticles. This electrocatalysts showed minimal degradation of the initial activity (only 8%) over 100 000 accelerated stability test cycles at real fuel cell operating conditions. The improved stability is attributed to increased oxidation resistance of the Pd core as a result of the alloying with Au. Alloying of Pd with Au causes a positive shift of the Pd oxidation potential that prevents the core from dissolution at fuel cell operating conditions. Therefore, in these $Pd_{0.9}Au_{0.1}$-supported Pt monolayer core–shell nanoparticles, the core increases the stability of the shell by shifting positively its oxidation potential, and by preventing the cathode potential reaching values at which Pt dissolution takes place. In addition, the Pd atoms from the core may dissolve through the imperfection cites in the Pt ML shell, and Au atoms from the PdAu solid solution could segregate on the surface of the nanoparticles. The segregation of the Au atoms (or the anti-segregation of the Pd atoms) is examined by DFT calculations.[6] The calculations make it clear that Au atoms preferentially remain on the surface by the segregation process, and support the experimental observations that preferential replacement of the Au atoms takes place at the vertex or corner sites on the Pd nanoparticles surface. This mends the defective sites on the Pt ML and, thus, markedly inhibits the Pd and Pt dissolution. This 'healing' effect by Au atoms is considered as a second factor for the improved stability of Pt_{ML}/$Pd_{0.9}Au_{0.1}$ electrocatalysts.

Improvement of the ORR activity, and as well as enhancement of the Pd core stability is also reported for the Pt_{ML}/PdIr/C catalysts.[34] The Pt-specific activity for Pt_{ML}/PdIr/C electrocatalyst is three times and 25% higher than that of Pt/C and Pt_{ML}/Pd/C, respectively. Also, the Pt mass activity of this catalyst is more than 20 times and 25% higher than that of Pt/C and Pt$_{ML}$/Pd/C, respectively. The electrochemical measurements pointed out that the enhanced ORR kinetics observed on Pt$_{ML}$/PdIr/C electrocatalysts originated from a decreased Pt–OH coverage.[34] This has been rationalized by self-consistent DFT calculations of the binding energy of OH on various relevant model systems. Clearly, the existence of a PdIr$_2$ monolayer located right below a Pt$_{ML}$/Pd$_{ML}$ bilayer and above the Pd(111) crystal leads to the destabilization of the surface OH compared to that already shown on a Pt$_{ML}$/Pd(111).

Pd$_{0.9}$Au$_{0.1}$ and Pd$_{0.9}$Ir$_{0.1}$ alloys nanoparticles are co-electro-deposited at constant potential from solution containing PdCl$_2$ and HAuCl$_4$ or IrCl$_3$ salts in presence of 0.1 M NaCl. The ratio between the salts in the solutions is chosen to give a deposit containing 90 at.% Pd and 10 at.% of Au or Ir. Although the deposition potential has been applied in the range where anisotropic growth of Pd nanowires is taking place, Pd$_{0.9}$Au$_{0.1}$ and Pd$_{0.9}$Ir$_{0.1}$ nanorods are obtained. It seems that due to the atomic size difference between Pd and Au or Ir atoms the proposed mechanism is not enough efficient to grow nanowires. After the deposition the surface of the Pd$_{0.9}$Au$_{0.1}$ and Pd$_{0.9}$Ir$_{0.1}$ nonorods is modified by Pt ML, and the activity and stability of these electrocatalysts is studied by measuring their polarization curves. In Figure 6.7, the TEM image of the electro-deposited Pd$_{0.9}$Au$_{0.1}$ nanoroads is presented along with the polarization curves for the ORR of the Pt$_{ML}$/Pd$_{0.9}$Au$_{0.1}$/C electrocatalysts measured in 0.1 M HClO$_4$ solution at 1600 rpm.

The mass activity of this electrocatalyst, as calculated from the polarization curves, exceeds by an order of magnitude the DOE's targets for mass activity of electrocatalysts for the ORR for 2020. Also, it is seen from Figure 6.7a that the polarization curve measured after 15,000 accelerated

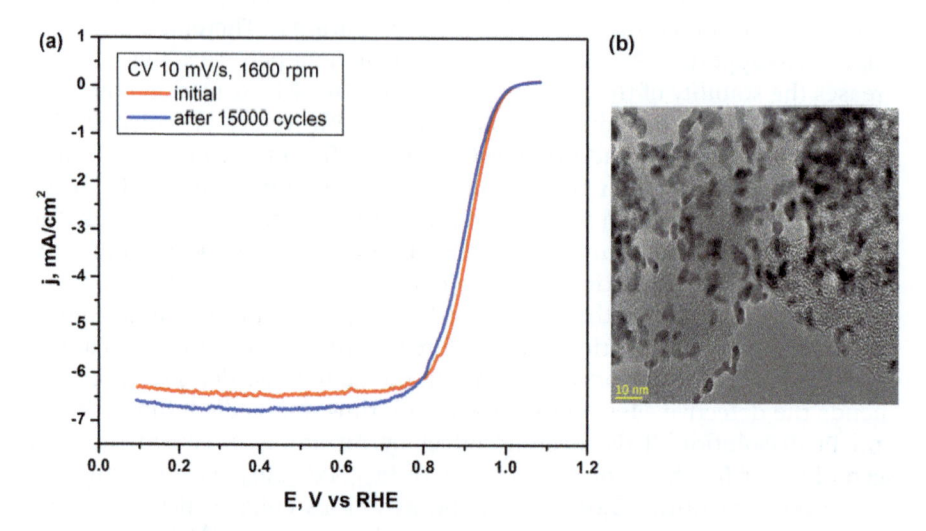

Figure 6.7 (a) ORR polarization curves, measured in 0.1 M HClO$_4$ acid solution at 1600 rpm, and (b) TEM image of the Pd$_{0.9}$Au$_{0.1}$ nanorods.

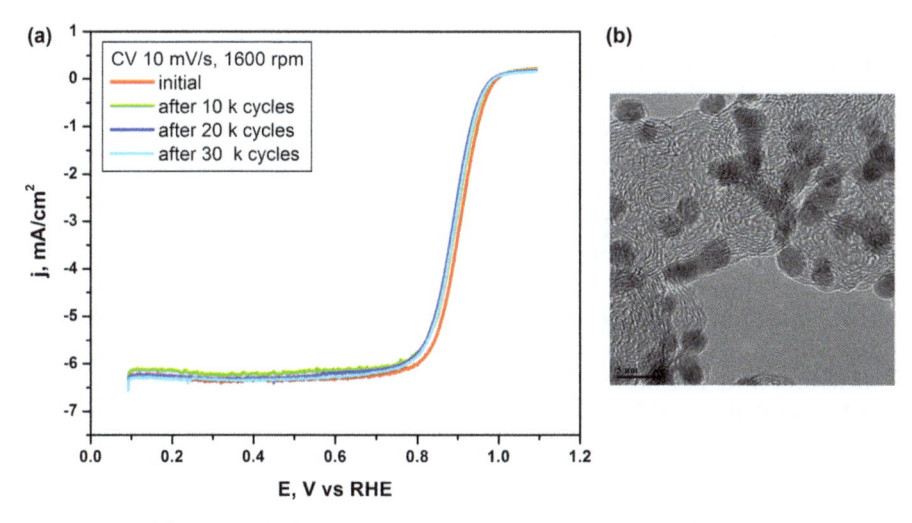

Figure 6.8 (a) ORR polarization curves, measured in 0.1 M HClO$_4$ acid solution at 1600 rpm, and (b) TEM image of the Pd$_{0.9}$Ir$_{0.1}$ nanostructures.

stability test cycles is shifted negatively only about 10 mV, which indicates an excellent stability of the catalysts of interest.

Figure 6.8 shows the polarization curves for the ORR measured on Pt$_{ML}$/Pd$_{0.9}$Ir$_{0.1}$/C electrocatalysts and the TEM image of the Pd$_{0.9}$Ir$_{0.1}$ nanostructures obtained after electro-deposition on functionalized carbon.

This electrocatalysts possessed five times higher Pt mass activity than the targeted value for 2020 and excellent stability as evident from the polarization curves measured after 30 000 accelerated stability tests.

6.3.2 Pt ML on Electro-deposited Pd/WNi Refractory Alloys: Advanced Fuel Cell Electrocatalysts

Refractory metals, such as W and Mo or their alloys, are of particular interest as cores for the Pt ML type electrocatalysts for the ORR. Such refractory cores will reduce the precious metal loading in the Pt ML electrocatalysts and will improve their stability, because the refractory metals are known for their excellent corrosion resistance.

It is well known that tungsten metal itself cannot be electro-deposited in the pure metal state from any aqueous or non-aqueous solutions. However, tungsten is readily co-deposited with iron group transition metals such as nickel or cobalt to form an alloy from aqueous solutions containing the tungstate ions.[35–37] Electro-deposition of W–Ni alloys is a typical example of induced co-deposition. A typical bath for the deposition of WNi alloys consist of NiSO$_4$, Na$_2$WO$_4$ and an organic polyacid, such as citric acid, acting as a complexing agent, for both Ni and W ions. Also, it is known that an addition of ammonia improves the quality of the deposits and increases the Faradaic efficiency, but it limits the tungsten content of the alloy significantly.[37] Ammonia-free solutions permitted the achievement of 1:1 W–Ni alloys, while the presence of ammonia caused a decrease in the W content in the alloys to about 15 at.%. The mechanism of co-deposition of W–Ni alloys has been studied in citrate-based electrolytes,[36] and the authors concluded that the precursor for the anomalous co-deposition of W is a mixed complex like $[(Ni)(WO_4)(H)(Cit)]^{2-}$.

In this chapter we report on electro-deposition of W–Ni nanocomposite alloys directly on functionalized carbon support (GDL) from citrate solution with 0.05 M $NiSO_4$ + 0.4 M Na_2WO_4 + 0.6 M Na_3Cit composition (pH = 7.5). The deposition is performed at constant current of 15 mA cm^{-2}. As obtained deposits have very fine composite amorphous–nanocrystalline microstructure, and 1:1 atomic ratio between W and Ni, as confirmed from the TEM and EDS analysis. The deposition is performed in a specially designed cell that allows exchanging of the electrolyte with an external peristaltic pump, and holding an inert (Ar) atmosphere in the cell. After finishing the deposition step, the electrolyte is replaced by 1 mM $PdCl_2$ and 50 mM H_2SO_4

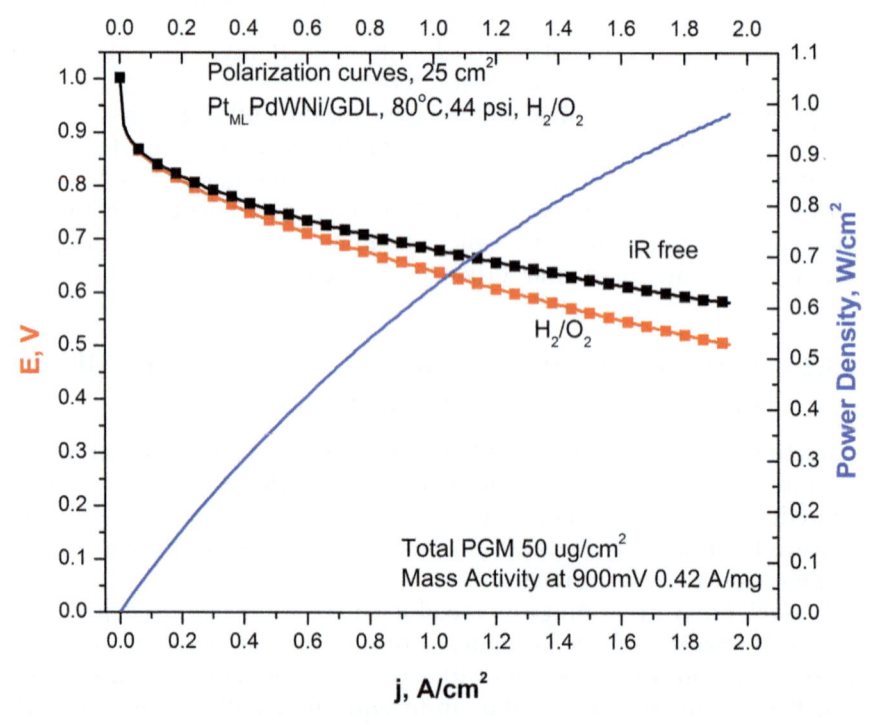

Figure 6.9 Polarization curve of MEA assembled from a cathode, prepared from electro-deposited Pt_{ML}/Pd/WNi catalyst on GDL, Nafion 211 membrane and an anode prepared by spraying ink from standard Pt/C nanoparticles and Nafion 5% solution. The total PGM loading of the cathode is 50 μg cm^{-2}, while the Pt loading in the anode is 100 μg cm^{-2}. The size of the electrodes is 25 cm^2, and the curve is measured at 80 °C in H_2/O_2 atmosphere.

solution, where part of the Ni atoms from the alloy are galvanically displaced by Pd and a Pd reach shell is formed on the WNi core. In the next steps, performed in the same cell, a Pt ML layer is deposited on the top of as obtained Pd/WNi nanostructures by using the Cu UPD strategy. As obtained Pt$_{ML}$/Pd/WNi electrocatalysts are deposited directly on the gas diffusion layer and after rinsing the electrode is dried and assembled with a Nafion 211 (25 μm thick) membrane and an anode with Pt loading of 100 μg cm^{-2}, by hot pressing at 130 °C for 1 min. As prepared membrane electrode assembly is first activated and then tested at real fuel cell operating conditions. Figure 6.9, presents the polarization curves measured on the MEA with size of the electrodes of 25 cm^2.

The Pt mass activity and the total PGM mass activity, as estimated from the iR compensated curve (plotted in black in Figure 6.9) at 0.9 V, are 1.1 A mg^{-1} and 0.42 A mg^{-1}, respectively. These values indicate that the electrocatalysts of interest possessed activity that meets the DOE targets. In addition, the accelerated stability test cycles (not shown here), performed on the same MEA at 80 °C in H$_2$/N$_2$ atmosphere, showed no degradation and even better performance after 15 000 cycles.

6.4 Conclusions

The Pt ML core–shell type electrocatalysts for the ORR are the most advanced class of fuel cell catalysts that possess a potential to boost the implementation of the PEMFCs in the electric vehicles. This type of electrocatalysts offer ultra-low Pt content, complete Pt utilization, very high activity and excellent performance stability surpassing the US DOE targets for 2020. In this chapter we reviewed for first time the electrochemical strategies for atomic layer deposition of Pt ML on different transition metal surfaces, including nanoparticles. The electrochemical ALD strategies are divided in two groups, based on the surface specific reaction that is used to limit the deposition of Pt only to maximum monolayer coverage. In the 'displacement driven' strategy a UPD ML of less noble element (Cu, Pb or H) is first deposited on the surface of interest, and then is galvanically displaced by Pt. Based on the stoichiometry of the displacement reaction the amount of Pt can be controlled at sub-monolayer level. This strategy is also known in the literature as surface limited redox replacement. In the second group, named 'adsorption driven', once a Pt ML is electro-deposited on the Au substrate a layer of strongly adsorbed non-metallic elements or molecules (H$_{ads}$ or CO) is formed and hinders the growth of a second ML. This electrochemical ALD strategy has been introduced in the literature in the last couple years, and has been demonstrated to work only for deposition of Pt ML on single- and poly-crystalline Au surfaces.

The advantages of the electro-deposition techniques over conventional chemical methods for synthesis of electrocatalysts for the oxygen reduction reaction are also discussed herein. It is demonstrated that the SLRR of Cu UPD ML, pre-deposited on electro-deposited Pd, PdAu, PdIr, NiW nanostructures, by Pt ML is a viable methodology for synthesis of highly efficient electrocatalysts for the ORR. The excellent performance of the catalysts of interest is evidenced from their polarization curves, measured either by rotating disc electrode techniques in perchloric acid solution or at real fuel cell operating conditions.

Pt$_{ML}$ core–shell catalysts are highly effective catalytic system having several unique properties. One of these properties is associated with an ease and directness of tuning their catalytic activity and stability by core–shell interactions. This can be accomplished with an atomic-level precision at a fraction of a monolayer coverage. Designing cores with desired composition, shape, size, surface sites coordination, can be made to high degree of perfection. We described a number of ways for controlled deposition of Pt monolayer on such surfaces that adds to the versatility of this approach and opens up broad area for their applications. Tailoring Pt$_{ML}$ properties by controlling core particle properties presents a great potential of Pt$_{ML}$ electrocatalysts for resolving problems of availability of Pt in future catalysis and electrocatalysis and removing the main obstacles to the broad scale application of fuel cell technology.

Acknowledgements

This work is supported by the U.S. Department of Energy, Division of Chemical Sciences, Geosciences and Biosciences Division, under the Contract No. DE-AC02-98CH10886.

References

1. H. A. Gasteiger, S. S. Kocha, B. Sompalli and F. T. Wagner, *Appl. Catal., B*, 2005, **56**, 9–35.
2. http://www1.eere.energy.gov/hydrogenandfuelcells/mypp/pdfs/fuel_cells.pdf
3. Y. Garsany, O. A. Baturina, K. E. Swider-Lyons and S. S. Kocha, *Anal. Chem.*, 2010, **82**, 6321–6328.
4. S. Guo, S. Zhang and S. Sun, *Angew. Chem., Int. Ed.*, 2013, **52**, 8526–8544.
5. R. R. Adzic, *Electrocatalysis*, 2012, **3**, 163–169.
6. K. Sasaki, H. Naohara, Y. Choi, Y. Cai, W.-F. Chen, P. Liu and R. R. Adzic, *Nat. Commun.*, 2012, **3**.
7. K. Sasaki, H. Naohara, Y. Cai, Y. M. Choi, P. Liu, M. B. Vukmirovic, J. X. Wang and R. R. Adzic, *Angew. Chem., Int. Ed.*, 2010, **49**, 8602–8607.
8. T. Suntola and J. Atson, *United States Patent*, 4,058,430, 1977.
9. J. Stickney, *Electrochem. Soc. Interface*, 2011, **20**, 3.
10. S. R. Brankovic, J. X. Wang and R. R. Adzic, *Surf. Sci.*, 2001, **474**, L173–L179.

11. M. Fayette, Y. Liu, D. Bertrand, J. Nutariya, N. Vasiljevic and N. Dimitrov, *Langmuir*, 2011, **27**, 5650–5658.
12. Y. Kim, J. Kim, D. Vairavapandian and J. Stickney, *J. Phys. Chem. B*, 2006, **110**, 17998–18006.
13. S. Brimaud and R. Behm, *J. Am. Chem. Soc.*, 2013, **135**, 11716–11719.
14. Y. Liu, D. Gokcen, U. Bertocci and T. Moffat, *Science*, 2012, **338**, 1327–1330.
15. K. Junter and W. Lorenz, *Zeitschrift Fur Physikalische Chemie*, 1980, **122**, 163–185.
16. B. Gregory and J. Stickney, *J. Electroanal. Chem.*, 1991, **300**, 543–561.
17. D. Kolb, M. Przasnys and H. Gerische, *J. Electroanal. Chem.*, 1974, **54**, 25–38.
18. S. R. Brankovic, J. X. Wang and R. R. Adzic, *J. Serb. Chem. Soc.*, 2001, **66**, 887–898.
19. J. Nutariya, M. Fayette, N. Dimitrov and N. Vasiljevic, *Electrochim. Acta*, 2013, **112**, 11.
20. M. B. Vukmirovic, S. T. Bliznakov, K. Sasaki, J. X. Wang and R. R. Adzic, *Electrochem. Soc. Interface*, 2011, **20**, 33–40.
21. L. Viyannalage, R. Vasilic and N. Dimitrov, *J. Phys. Chem. C*, 2007, **111**, 4036–4041.
22. M. Mrozek, Y. Xie and M. Weaver, *Anal. Chem.*, 2001, **73**, 5953–5960.
23. D. Gokcen, S. Bae and S. Brankovic, *J. Electrochem. Soc.*, 2010, **157**, D582–D587.
24. R. R. Adzic, J. Zhang, K. Sasaki, M. B. Vukmirovic, M. Shao, J. X. Wang, A. U. Nilekar, M. Mavrikakis, J. A. Valerio and F. Uribe, *Top. Catal.*, 2007, **46**, 249–262.
25. A. U. Nilekar, Y. Xu, J. L. Zhang, M. B. Vukmirovic, K. Sasaki, R. R. Adzic and M. Mavrikakis, *Top. Catal.*, 2007, **46**, 276–284.
26. J. L. Zhang, M. B. Vukmirovic, Y. Xu, M. Mavrikakis and R. R. Adzic, *Angew. Chem., Int. Ed.*, 2005, **44**, 2132–2135.
27. J. Zhang, Y. Mo, M. B. Vukmirovic, R. Klie, K. Sasaki and R. R. Adzic, *J. Phys. Chem. B*, 2004, **108**, 10955–10964.
28. K. Sasaki, J. X. Wang, H. Naohara, N. Marinkovic, K. More, H. Inada and R. R. Adzic, *Electrochim. Acta*, 2010, **55**, 2645–2652.
29. J. L. Zhang, M. B. Vukmirovic, K. Sasaki, A. U. Nilekar, M. Mavrikakis and R. R. Adzic, *J. Am. Chem. Soc.*, 2005, **127**, 12480–12481.
30. J. Wang and R. R. Adzic, *United States Pat.*, US 20060134505A1, 2006.
31. K. Sasaki, M. B. Vukmirovic, J. X. Wang and R. R. Adzic, in *Fuel Cell Science: Theory, Fundamentals, and Biocatalysis*, ed. A. Wieckwski and J. K. Nørskov, John Wiley & Sons, Inc., Hoboken, New Jersey, 2010 p. 215–237.
32. S. T. Bliznakov, M. B. Vukmirovic, L. Yang, E. A. Sutter and R. R. Adzic, *J. Electrochem. Soc.*, 2012, **159**, F501–F506.

33. S. Bliznakov, M. Vukmirovic, E. Sutter and R. Adzic, *Maced. J. Chem. Chem. Eng.*, 2011, **30**, 19–27.
34. S. L. Knupp, M. B. Vukmirovic, P. Haldar, J. A. Herron, M. Mavrikakis and R. R. Adzic, *Electrocatalysis*, 2010, **1**, 213–223.
35. O. Younes, L. Zhu, Y. Rosenberg, Y. Shacham-Diamand and E. Gileadi, *Langmuir*, 2001, **17**, 8270–8275.
36. O. Younes-Metzler, L. Zhu and E. Gileadi, *Electrochim. Acta*, 2003, **48**, 2551–2562.
37. S. Franz, A. Marlot, P. Cavallotti and D. Landolt, *Trans. Inst. Met. Finish.*, 2008, **86**, 92–97.

CHAPTER 7

Atomic Layer Deposition in Nanoporous Catalyst Materials

JOLIEN DENDOOVEN

Ghent University, Department of Solid State Sciences, COCOON, Krijgslaan 281/S1, Gent B-9000, Belgium
Email: Jolien.Dendooven@UGent.be

7.1 Introduction

Atomic layer deposition (ALD) is a thin-film growth method that is characterized by alternating self-limiting chemisorption reactions of gas phase precursor molecules with a solid surface. The unique surface-controlled chemistry of ALD yields a sub-nanometer thickness control and an excellent conformality on complex three-dimensional (3D) substrates. As will become clear throughout this chapter, these key advantages render ALD an enabling technology for the controlled atomic-scale design of supported catalysts.

This introductory section starts with a brief history and basic principles of the ALD technique. Next, the focus moves to how ALD proceeds in complex 3D porous networks with pore diameters in the range 1–30 nm. The conformal coating of nanometer-sized mesopores and the infiltration of metal oxide ALD in nanoporous powder particles are discussed by means of experimental studies. The section then concludes with an outlook on the versatile opportunities of ALD for catalytic applications.

RSC Catalysis Series No. 22
Atomically-Precise Methods for Synthesis of Solid Catalysts
Edited by Sophie Hermans and Thierry Visart de Bocarmé
© The Royal Society of Chemistry 2015
Published by the Royal Society of Chemistry, www.rsc.org

7.1.1 Atomic Layer Deposition

The ALD technique, formerly known as ALE (atomic layer epitaxy) or ALCVD (atomic layer chemical vapor deposition), was invented[†] and patented by Suntola and co-workers in Finland in the 1970s.[2] Its application was initially limited to the deposition of materials used in electroluminescent flat panel displays. Driven by the decreasing device dimensions, ALD research intensified in the 1990s, which has led to a major commercial breakthrough of ALD in the semiconductor industry for growing high-k gate oxides. Already in the 1990s, several authors also recognized the potential impact of ALD in the area of catalyst preparation.[3–8] Since the early 2000s, an increasing number of researchers are exploring ALD as a generic coating technique for all sorts of nanostructures, as illustrated in a number of recent review papers.[9–14] Catalysis is an important potential application field, but also gas separation, sensors, batteries, capacitors, fuel cells, photovoltaics and photonics are explored by the growing ALD community.

Various materials that are relevant to catalysis including several oxides, nitrides, sulfides and (noble) metals can be deposited *via* ALD. For an extensive overview of existing ALD processes, the reader is referred to a recent review paper.[15] A classic example of an ALD process is the growth of Al_2O_3 by alternating exposure of the substrate to trimethylaluminium (TMA) and H_2O vapor.[16,17] The so-called ALD cycle is depicted in Figure 7.1. In part A of the reaction cycle, the surface is exposed to TMA molecules which react with the OH surface groups, resulting in strong Al–O bonds and a CH_3-terminated surface. Because TMA is inert towards the CH_3 surface groups, the reaction will stop once all accessible[‡] OH groups have been consumed. The TMA pulse is followed by an evacuation of the growth chamber through pumping or purging with an inert gas. In part B of the reaction cycle, H_2O vapor hydrolyses the residual CH_3 groups on the surface resulting in the formation of a (sub)monolayer of Al_2O_3. The reaction is again self-limiting because H_2O does not react with the newly generated OH surface groups. Instead, after a second evacuation of the chamber, these surface groups allow for the repetition of the ALD cycle to continue Al_2O_3 growth in a self-limited (sub)monolayer-by-(sub)monolayer manner.

The basic chemical mechanism that offers ALD an exceptional growth control is the self-saturation of the surface reactions. It is therefore of critical importance that the precursors do not decompose upon adsorption on the surface. This requirement sets an upper limit to the growth temperature of ALD processes. On the other hand, the sample temperature must be sufficiently high to provide the activation energy required for the surface reactions and to avoid condensation of precursor molecules as this would

---

[†] A similar type of process based on self-limited gas–solid reactions called 'molecular layering' was independently developed by Aleskovskii and co-workers in the former Soviet Union in the late 1960s.[1]

[‡] Some OH groups are no longer available because of steric hindrance from chemisorbed neighboring TMA molecules.

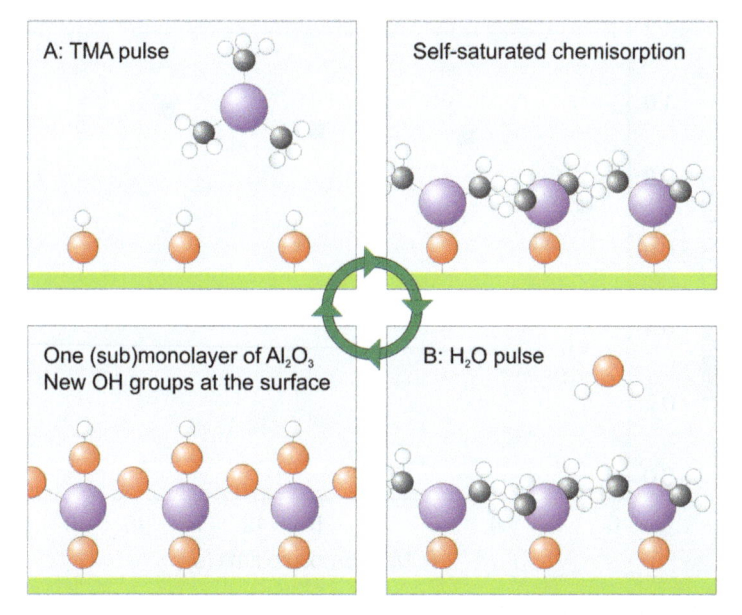

Figure 7.1 Principle of ALD, illustrated by the process for deposition of Al_2O_3 using TMA and H_2O. Small white spheres represent hydrogen, black spheres carbon, red spheres oxygen and large violet spheres aluminium atoms.

also destroy the self-limitation of the ALD half-reaction. For processes using metal–organic precursors, the temperature window for which 'ideal ALD' occurs, is typically in the range 100–350 °C.

Figure 7.2 shows the self-limiting behavior of the first half-reaction in the tetrakis(dimethylamino)titanium (TDMAT)/H_2O process for the deposition of TiO_2 at 200 °C.[18] After reaching saturated coverage of the surface, additional TDMAT precursor molecules do not result in deposition of more material. During steady-state growth, each saturated ALD cycle thus results in the same amount of material on the surface, even when the precursors remain in the growth chamber for a very long time. An immediate and advantageous consequence is that the deposited film thickness can accurately be controlled on a (sub)monolayer scale by the number of repeated ALD cycles (Figure 7.3).

A second consequence, and also the key advantage of ALD, is that the deposited films are highly conformal. The self-limited growth mechanism guarantees that ALD is insensitive to differences in precursor flux, meaning that the growth rate is the same everywhere in the ALD reaction chamber and thus also within the pores of nanoporous materials. This is in contrast to flux-controlled deposition methods, such as chemical vapor deposition (CVD), where material is mainly deposited near the pore entrances making it difficult to avoid pore clogging. Also in ALD, the regions near the pore openings will experience a larger flux of precursor vapor and, therefore, will be covered much sooner than the interior surfaces. Once saturated, however,

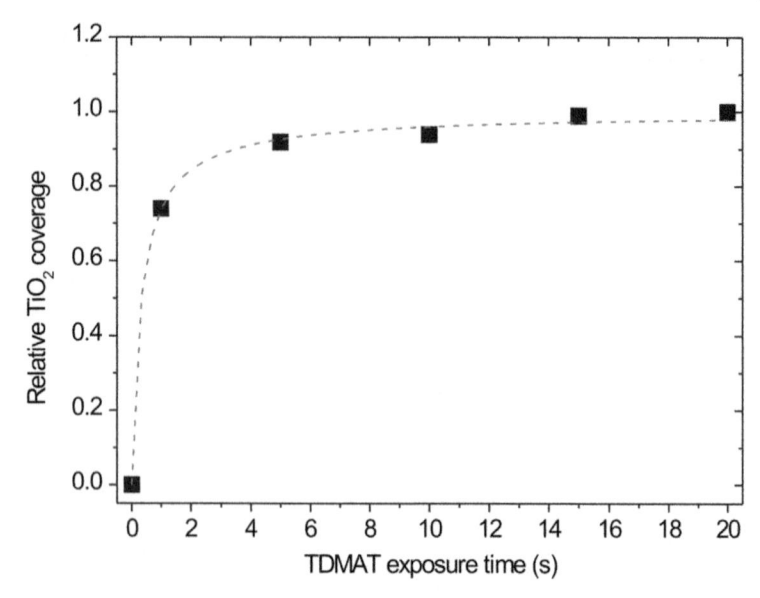

Figure 7.2 Saturation of the TiO$_2$ coverage against the exposure time of the Ti precursor (TDMAT) at 200 °C. The dashed line serves as a guide to the eye. The relative TiO$_2$ coverage was obtained from X-ray fluorescence measurements.

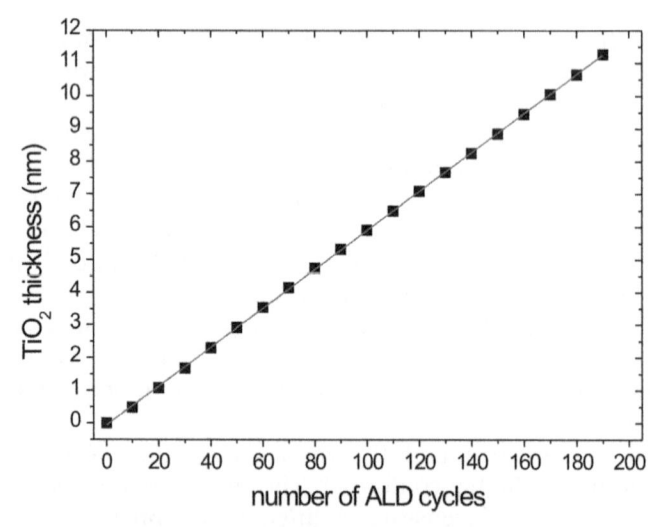

Figure 7.3 Linear growth of the TDMAT/H$_2$O process on a SiO$_2$ surface. The gray line is a linear fit to the data points. Every ALD cycle results in the deposition of 0.06 nm of TiO$_2$. The film thickness was obtained from spectroscopic ellipsometry measurements.

no further deposition can take place near the pore mouths and precursor molecules will diffuse into the structure to also coat the internal surface of the pores.

ALD has proven to be an effective technique for the deposition of conformal coatings in Si trenches[19,20] and anodic alumina nanopores.[21,22] Nevertheless, it is clear that, because of the diffusion limited gas transport, achieving a good conformality requires careful optimization of the ALD process parameters.[23,24] This is especially true when ALD is applied to nanoporous materials that are typically used as high surface area catalyst supports. Conformal ALD requires the precursor molecules to penetrate into the bulk of these materials often comprising a complex 3D network of interconnected extremely narrow channels (pore sizes <10 nm). This presents a challenge in itself and, therefore, the next section is dedicated to the conformality of ALD in nanoporous materials.

7.1.2 Conformality of ALD in Nanoporous Materials

A good understanding of the conformality of ALD in nanoporous materials is essential to the optimization of the deposition parameters for ALD processing of supported catalysts. Section 7.1.2.1 addresses the ability of ALD to conformally coat the interior surface of pores with diameters in the low mesoporous regime. Section 7.1.2.2 focuses on the transport of precursor molecules in nanoporous powder particles exhibiting a high surface area.

7.1.2.1 ALD in Nanometer-sized Mesopores

It is not straightforward to obtain direct information about the conformality of ALD coatings deposited in nanometer-sized mesopores. In a recent study, porous titania thin films were used as substrate for the detailed investigation of HfO_2 ALD in ink-bottle shaped mesopores by means of advanced characterization techniques such as quantitative electron tomography and synchrotron-based X-ray fluorescence (XRF).[25] The main results of this work are reviewed in the next paragraphs.

A mesoporous titania film was synthesized by spin coating a solution of titanium precursor and polymeric templates on a Si substrate.[26] As revealed by ellipsometric porosimetry (EP),[27,28] the titania film contained ink-bottle shaped mesopores with an average pore diameter of 6.8 nm and an average neck size of 4.8 nm. The mesoporous film was coated with HfO_2 using 30 cycles of the tetrakis(ethylmethylamino)hafnium (TEMAH)/H_2O ALD process at 200 °C.[29]

Quantitative electron tomography enabled a direct and local 3D characterization of the ultra-thin HfO_2 coating deposited in the titania mesopores. Where conventional transmission electron microscopy (TEM) 'only' provides a two-dimensional (2D) projection of a 3D object, electron tomography aims at reconstructing the 3D structure of the object from a series of 2D TEM images acquired at different tilt angles.[30–32] In order to obtain a full tilt range of 2D projection images a nanopillar was cut from the sample and a dedicated on-axis tomography holder was used.[33,34] Figure 7.4a shows a 3D volume rendering of the porous film obtained using the total variation

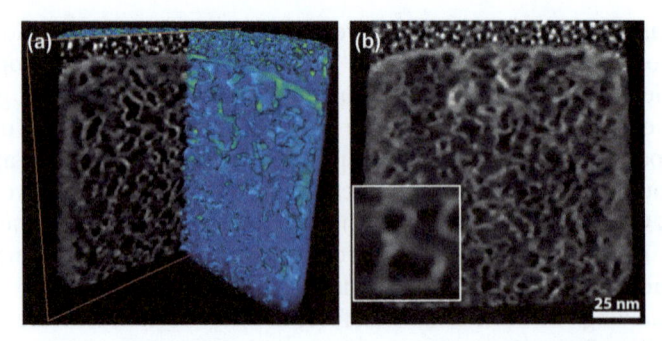

Figure 7.4 Electron tomography reconstructions of a thin HfO$_2$ ALD layer deposited in a mesoporous titania film (initial porosity: 30%, thickness: 120 nm). (a) 3D visualization of the reconstruction. (b) XZ-orthoslice through the 3D reconstruction with HfO$_2$ coating in light gray, titania matrix in dark gray and voids in black. The detail shows conformally coated ink-bottle shaped mesopores.
Adapted with permission from ref. 25. Copyright 2012 American Chemical Society.

minimization (TVM) reconstruction algorithm.[35] To examine the inner structure in more detail, slices were made *through* the 3D reconstruction (Figure 7.4b). Dark areas represent voids, light gray zones correspond to the HfO$_2$ coating and dark gray zones to the titania pore walls. The conformal deposition of an ultra-thin HfO$_2$ layer (2.2 ± 0.5 nm) was confirmed throughout the whole mesoporous film.

More insights on the conformal coating of the mesoporous titania film with HfO$_2$ were obtained from *in situ* synchrotron-based XRF measurements.[36,37] Figure 7.5 shows the Hf XRF intensity against the number of ALD cycles for deposition on the porous film as well as on a planar SiO$_2$ reference substrate. During the first *ca.* 19 ALD cycles, the XRF intensity increased faster in the mesoporous film than on the planar substrate, proving that HfO$_2$ was deposited on the internal surface of the porous thin film.[§] With progressing growth, the slope of the Hf XRF intensity curve decreased and became more similar to the slope obtained for ALD on the planar reference sample. This suggests that the pore necks became too narrow and were no longer accessible for the TEMAH precursor, and that ALD continued on top of the coated mesoporous film.

An X-ray reflectivity measurement after 70 ALD cycles on the planar reference substrate yielded a growth rate of 0.12 nm per cycle for the TEMAH/H$_2$O process. Assuming a similar growth rate in the titania ink-bottle pores, 19 ALD cycles would result in a 2.3 nm thick coating, which is in agreement

---

[§]While a gradually decreasing slope was expected during the first *ca.* 19 ALD cycles, due to a decreasing surface area with progressing growth, the slope of the Hf XRF intensity curve increased during the first *ca.* 13 cycles and only then started to decrease. This could possibly be explained by substrate-inhibited growth of HfO$_2$ on the mesoporous titania film, implying an increasing growth rate during the first ALD cycles (ref. 17).

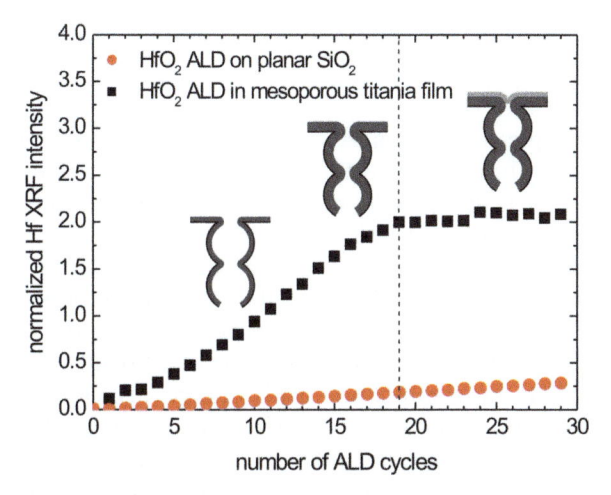

Figure 7.5 *In situ* X-ray fluorescence during HfO$_2$ ALD on a planar SiO$_2$ substrate and on a mesoporous titania film containing ink-bottle shaped mesopores: Hf Lα (7.9 keV) peak area (normalized to the incoming beam intensity) against the number of ALD cycles. The Hf XRF intensity is proportional to the number of Hf atoms deposited. The schematics show conformal deposition on the pore walls (in dark gray) followed by deposition on the exterior surface (in light gray).
Adapted with permission from ref. 25. Copyright 2012 American Chemical Society.

with the film thickness obtained from the electron tomography study. The size of the ink-bottle necks is then reduced to 0.3 nm, below the estimated kinetic diameter of the TEMAH molecule (*ca.* 0.7 nm). As such, the *in situ* XRF data show the feasibility of ALD to tune pore sizes down to the molecular level without clogging the pore mouths (as TEMAH can still enter the pores up to cycle 19).

In conclusion, the ALD technique was successfully applied to conformally deposit an ultra-thin HfO$_2$ layer in ink-bottle mesopores that were only accessible *via* extremely narrow channels (<5 nm). This result demonstrates the unique suitability of ALD for atomic level tuning of pore sizes in the low mesoporous regime.

7.1.2.2 ALD on High Surface Area Powder Particles

Considering a mesoporous thin film as a model system for the porous high surface area materials typically used as catalyst supports, the previous section demonstrated that ALD precursor molecules can access and coat nanometer-sized mesopores. In this section, the focus moves to the penetration of ALD precursors into the interior portions of mesoporous powder particles. Two experimental studies are presented discussing the infiltration of metal oxide ALD in the unordered pore structure of mesoporous silica gel and in the ordered 3D mesoporous network of Zeotile-4 material, respectively.

7.1.2.2.1 Isotropic Infiltration of Al$_2$O$_3$ ALD in Silica Gel.
Elam *et al.* investigated the infiltration of Al$_2$O$_3$ ALD in mesoporous silica gel spheres with a particle size of 75–200 μm.[38] The silica gel exhibited an internal surface area of 99.6 m^2 g^{-1} and an average pore size of 30 nm. The conditions for reaching saturation in 1 g of this mesoporous material were determined *via* weight-gain experiments. At a TMA partial pressure of *ca.* 20 Pa, 90 s of exposure were sufficient to fully cover the interior surface of the spherical silica particles. Using cross-sectional scanning electron microscopy (SEM) in combination with energy dispersive X-ray analysis (EDX), the spreading of the Al$_2$O$_3$ material in a silica gel sphere was investigated for unsaturated exposure times of 5, 15 and 30 s (Figure 7.6). The EDX elemental maps of Al revealed the diffusion-limited nature of the TMA/H$_2$O ALD process in mesoporous silica gel. Coating of the interior surface occurred progressively from the outside of the particles to the core. The authors further reported that the rate of Al$_2$O$_3$ uptake in the silica gel was not only governed by diffusional limitations but also by the rate of precursor delivery in their experiments.

7.1.2.2.2 Anisotropic Infiltration of TiO$_2$ ALD in Zeotile-4 Material.
Sree *et al.* studied the infiltration of TiO$_2$ ALD in Zeotile-4 powder with an average particle size of 0.5 μm and a total surface area of 1224 m^2 g^{-1}.[39] Zeotile-4 material is characterized by a 3D mesoporosity that is achieved *via* stacking and annealing of silicalite nanoslabs.[40,41] The specific organization of the nanoslabs measuring *ca.* 8×4×2 nm^3 creates about 8–10 nm wide hexagonal parallel channels connected to each other *via* 2–4 nm wide slits (Figure 7.7). The nanoslabs themselves have a zeolitic character and present internal microporosity.

0.1 g of Zeotile-4 powder was highly dispersed onto the *ca.* 25 cm^2 bottom of an aluminium tray and exposed to 30 cycles of the TDMAT/H$_2$O process

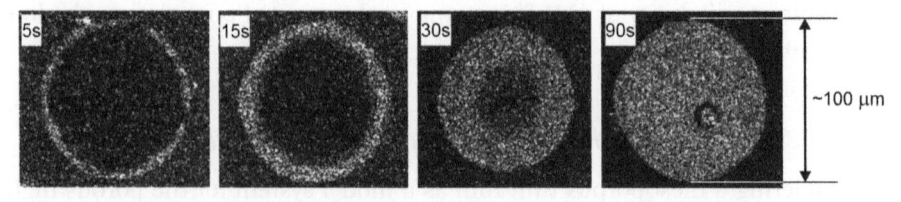

Figure 7.6 Cross-sectional EDX elemental maps revealing the spreading of Al atoms deposited in a spherical silica gel particle using five TMA/H$_2$O ALD cycles and TMA exposure times of 5, 15, 30 and 90 s. With progressing deposition, an isotropic deposition front is penetrating to the core of the unordered silica gel particle.
Adapted with permission from ref. 38. Copyright 2010 American Chemical Society.

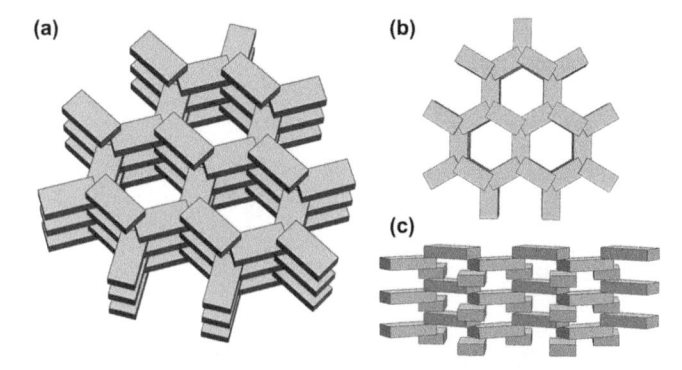

Atomic Layer Deposition in Nanoporous Catalyst Materials 175

Figure 7.7 (a) Structure of Zeotile-4 material as established by discrete electron tomography.[41] (b) Structure viewed along the main channels (pore size 8–10 nm). (c) Sideways view on the slit-like pores (pore size 2–4 nm).

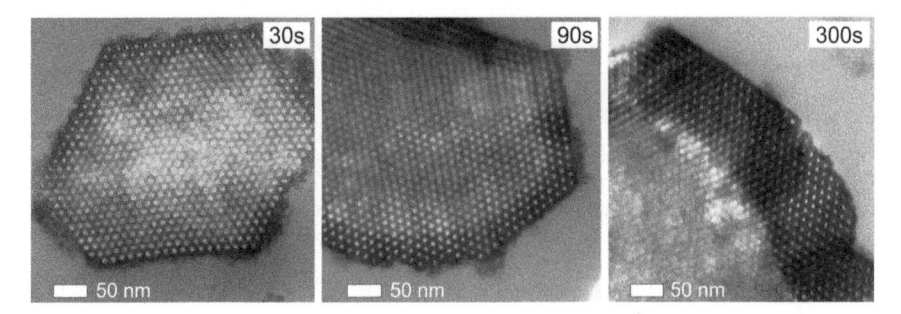

Figure 7.8 Cross-sectional TEM images revealing the spreading of TiO$_2$ deposited in a Zeotile-4 powder particle using 30 TDMAT/H$_2$O ALD cycles and TDMAT exposure times of 30, 90 and 300 s.
Adapted with permission from ref. 39. Copyright 2012 American Chemical Society.

for ALD of TiO$_2$ using 30, 90 or 300 s exposure times. During the TDMAT exposures, Ar was used as a carrier gas and the pressure in the ALD chamber was *ca.* 0.15 Pa. In the first experiments using 30 and 90 s TDMAT exposure times, the deposition temperature was 200 °C. For the 300 s sample, it was necessary to reduce the deposition temperature to 120 °C in order to avoid substantial decomposition of the Ti precursor. XRF characterization of the samples confirmed a higher TiO$_2$ content with longer exposure times. TEM images of coated powder particles oriented along the hexagonal main channels of the Zeotile-4 structure are shown in Figure 7.8. The dark outer rim observed in the TEM images is ascribed to the infiltration of TiO$_2$ material in the slit-like mesopores of the Zeotile-4 structure (Figure 7.7). As in the silica gel particle, the TiO$_2$ deposition front moved progressively deeper into the powder particles with increasing TDMAT exposure time. The dark rim measured *ca.* 20, 45 and 120 nm for 30, 90 and 300 s exposure, respectively.

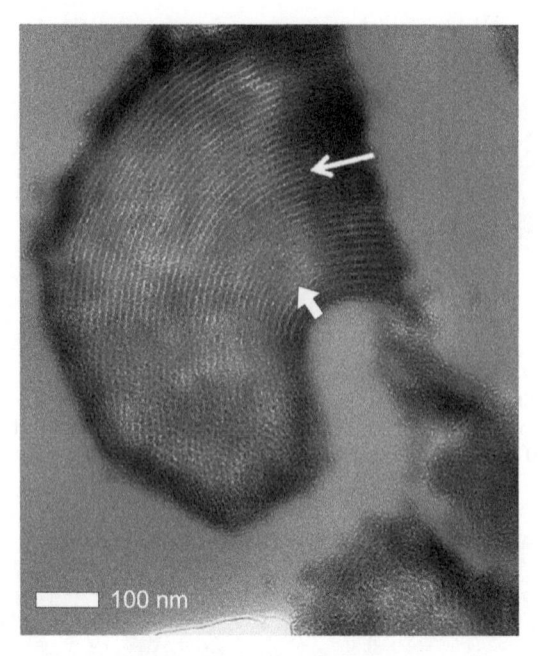

Figure 7.9 Cross-sectional TEM image revealing anisotropic penetration of the TiO$_2$ deposition in the hexagonal mesopores (thin white arrow) and in the smaller slit-like mesopores (thick white arrow). The Zeotile-4 material was subjected to 30 TDMAT/H$_2$O ALD cycles with a TDMAT exposure time of 90 s.
Adapted with permission from ref. 39. Copyright 2012 American Chemical Society.

Given the specific pore structure of Zeotile-4, the TDMAT molecules can penetrate the powder particles either *via* the slit-like mesopores or *via* the larger hexagonal channels. In TEM, occasionally we observed a Zeotile-4 particle that was cut parallel to the main channels (Figure 7.9). This image revealed that the deposition front penetrated deeper in the direction along the main mesopores than in the perpendicular directions. The formation of anisotropic deposition profiles within the Zeotile-4 material can be understood based on its specific structure containing differently sized mesopores.

7.1.2.2.3 Conclusions. The aforementioned experimental studies prove the ability of ALD to coat the interior surface of nanoporous materials. In the silica gel, full coverage of the internal surface area with Al$_2$O$_3$ was achieved using an exposure time of 90 s for the Al precursor. On the other hand, in the infiltration study of TiO$_2$ in Zeotile-4, even an exposure time of 300 s for the Ti precursor was insufficient to reach full coverage of the bulk of the material. Reasons for this are the very large surface area of Zeotile-4 (1224 m^2 g^{-1}) and the low precursor pressures used in the experiments. This work also illustrates the practical relevance of real-time sensing tools that monitor the surface reaction dynamics and saturation

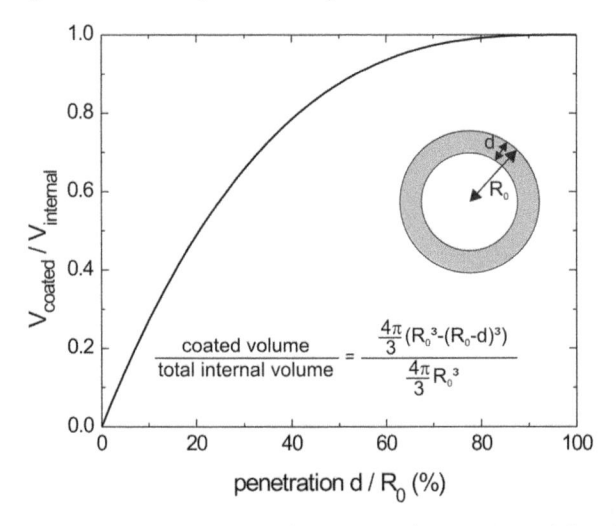

Atomic Layer Deposition in Nanoporous Catalyst Materials 177

Figure 7.10 Percentage ALD coating of a spherical porous particle with radius R_0 against penetration depth d of the precursor.
Reproduced from ref. 11.

during ALD on powders.[42–44] It should, however, also be noted that already a significant fraction of the interior surface of a mesoporous particle is reached when only an outer rim is coated.[45] For a simple model of a spherical porous particle, deposition to a depth of 100 nm in a particle with a radius of 0.5 μm suffices to cover already 50% of the internal porous volume (Figure 7.10). Moreover, also in catalytic processes, the reactants have to diffuse into the pores of the catalysts, and, as for the ALD precursors, only a minority of them will reach the core of the nanoporous catalyst particles. With this in mind, it is valid to conclude that the conformality of ALD can be exploited to apply conformal coatings on nanoporous materials for catalytic applications.

7.1.3 Opportunities of ALD in Supported Catalyst Preparation

The versatility of ALD as a method to design supported catalysts is illustrated in Figure 7.11. In this figure, the nanoporous support is schematically represented by means of a cross-section covering three nanopores. Figure 7.11a represents the situation where ALD is used as a vapor phase grafting method to engineer the presence of active sites on the pore walls.[4,46] Very few (or even only one) ALD cycles are typically applied in this case.

In Figure 7.11b, ALD is used to synthesize catalytic active nanoparticles nicely dispersed on the large surface area support. For nucleation-controlled ALD processes, islands are formed at the start of the deposition instead of a continuous layer.[17] This is, for example, the case for many noble metal ALD processes on metal oxide surfaces. By carefully controlling the nucleation

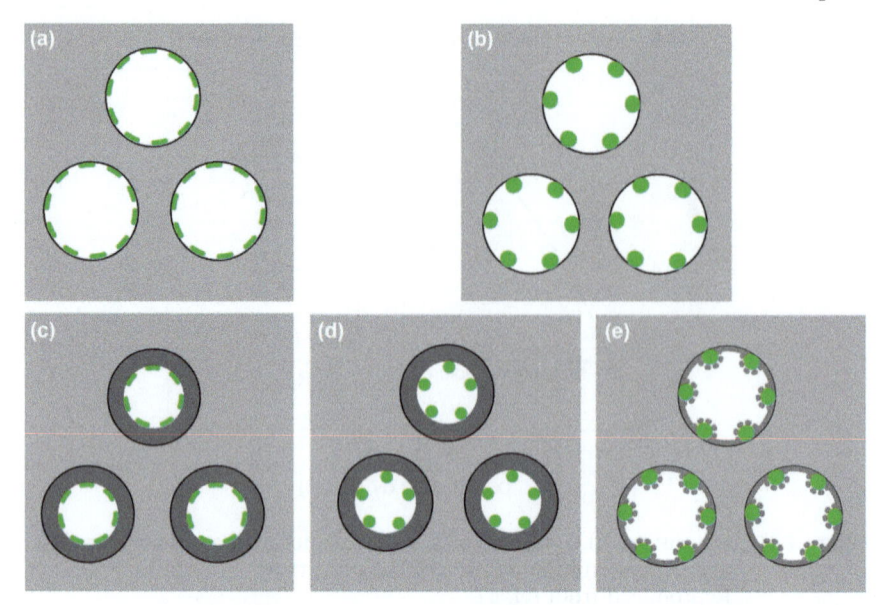

Figure 7.11 ALD for catalysis: (a) Introducing catalytic sites on the pore walls. (b) Introducing catalytic nanoparticles on the pore walls. (c) and (d) Tuning the pore size of the support prior to the introduction of catalytic sites (c) or nanoparticles (d). (e) Stabilizing catalytic nanoparticles with an ALD over-layer. The coating is dense on the oxide support material; a porous layer is formed on top of the metal nanoparticles.

stage of these processes, ALD can be used for the conformal deposition of catalytic noble metal particles with the desired size and composition.[47–50] Alternatively, nanoparticles can also be formed by the controlled break-up of an ALD-deposited layer through calcination.[51]

Besides, for the deposition of the catalytically active material, ALD can be used to modify the support material prior to the introduction of the active species or particles. In Figures 7.11c and d, the atomic level thickness control of ALD is exploited to precisely tune the pore size of the support material.[52] In this way, the selectivity of the catalyst can be tailored to the specific application.

In Figure 7.11e, a thin oxide coating is applied on top of the catalytic metal nanoparticles to stabilize them.[53–55] The over-layer is dense and continuous on the oxide support, while it contains some porosity on top of the metallic particles (formed during the ALD process or induced by thermal treatment). In this way, an ultra-thin ALD coating can act as a physical barrier to prevent migration and sintering of the particles while keeping the surface of the active nanoparticles accessible for the chemical species that are to be catalytically converted.

In the following sections, examples are provided for the different ALD approaches depicted in the schematic picture. Section 7.2 briefly reviews a selection of catalysis-related ALD papers, while Section 7.3 elaborates on two

Atomic Layer Deposition in Nanoporous Catalyst Materials

specific case studies focusing on the functionalization of mesoporous silica with acid sites and photo-active nanoparticles, respectively.

7.2 ALD for Catalysis – a Literature Overview

Figure 7.12 shows the annual number of ALD publications related to catalysis[¶] and reveals a growing interest in ALD for catalytic applications. This section presents an overview of selected papers on this topic. It does not intend to provide a complete literature review, but rather aims to give a general flavor of the various ALD processes, supports and catalytic reactions that have been considered. It also provides a context for the case studies presented in Section 7.3. Early studies that were conducted from the mid 1990s until the early 2000s are discussed in Section 7.2.1, while Sections 7.2.2 to 7.2.5 cover more recent work.

7.2.1 Early Work on Catalyst Preparation by ALD

The suitability of ALD for coating porous high surface area materials was realized by Suntola and co-workers in the early 1990s.[3,8,46,56–58] Since then, ALD has been reported to produce active catalysts for a variety of purposes. Extensive work in this field has been carried out by the group of Prof. Outi Krause at the Laboratory of Industrial Chemistry, Aalto University School of Chemical Technology.[‖] Co catalysts were, for example, synthesized on

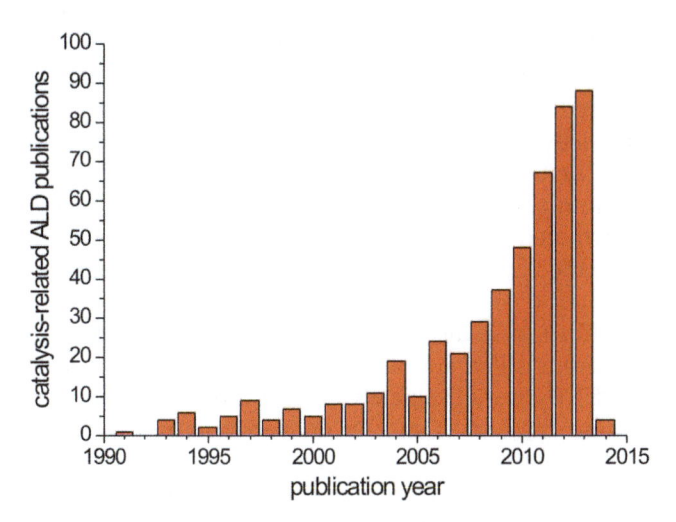

Figure 7.12 Number of catalysis-related ALD publications per year.

[¶]Search made on 12 February 2014. ISI Web of Science, Topic = ("atomic layer deposition" OR "atomic layer epitaxy" OR "atomic layer chemical vapor deposition" OR "atomic layer chemical vapor deposition" OR "molecular layering") AND Topic = (catalysis OR catalyst OR catalysts).
[‖]Formerly, Helsinki University of Technology.

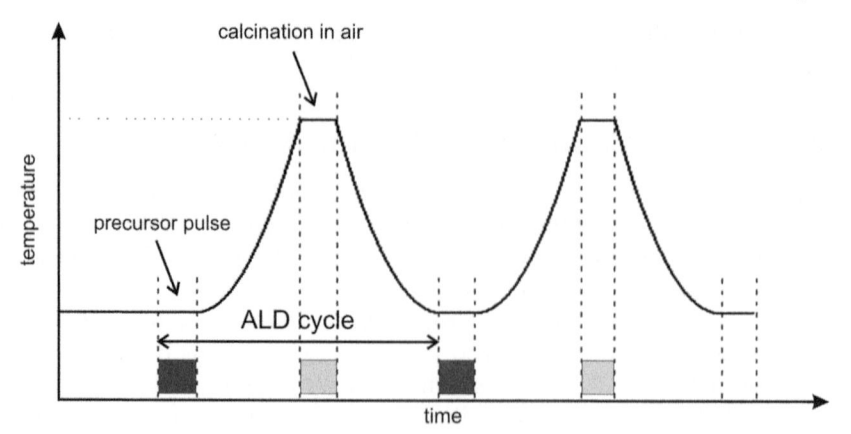

Figure 7.13 For the preparation of catalysts by ALD, temperature variations are sometimes introduced in the ALD cycle.

mesoporous silica**,6,7,59 and alumina††,60 powder for the hydroformylation of ethene[7] and the hydrogenation of toluene.[6,59,60] After preheating of the support, the Co-precursor, cobalt(III)acetylacetonate or Co(acac)$_3$, was chemisorbed on the surface at a temperature of 180 °C. Removal of the remaining ligands occurred by calcination in synthetic air at 450 °C leading to CoO$_x$ species (Figure 7.13). This ALD cycle was repeated up to five times, after which the CoO$_x$ species were partially reduced to Co in flowing H$_2$ at 500–600 °C. It was shown that omitting the calcination step after the last Co(acac)$_3$ reaction step led to an enhanced reducibility of the CoO$_x$ species. The ALD-synthesized Co catalysts showed improved activities compared to catalysts prepared by impregnation from Co(NO$_3$)$_2$, due to the higher dispersion of the active species.[6] A similar ALD approach was applied to fabricate catalysts for alkane dehydrogenation by Cr(acac)$_3$/air reactions on the mesoporous alumina support.[5,61] Although the active species were more uniformly distributed in catalysts prepared by ALD than by impregnation from Cr(NO$_3$)$_3$, the catalysts were equally active in the dehydrogenation of isobutane.[61]

Significant efforts were also made to modify the support prior to catalyst loading.[62–64] Porous silica and alumina powders covered with AlN by ALD from TMA and NH$_3$ were evaluated as supports for the preparation of Co hydroformylation catalysts[65] and Cr dehydrogenation catalysts,[66] respectively. The active species were generated by the chemisorption and subsequent reduction in H$_2$ of Co(acac)$_3$ or Cr(acac)$_3$ precursor molecules. In the considered reactions, however, the AlN-coated supports did not offer additional benefits for the catalytic activity.

**Grace 432 Silica: particle size 0.5–1 mm, surface area 320 m^2 g^{-1}, pore volume 1.2 cm^3 g^{-1}, and average pore diameter 10–15 nm.
††AKZO γ-alumina: particle size 0.2–0.4 mm, surface area 120–195 m^2 g^{-1}, pore volume 0.45–0.5 cm^3 g^{-1}, and average pore diameter 10–15 nm.

Keränen et al. published several papers on the synthesis of vanadium-based catalysts.[67–70] In their early work, they used the VO(acac)$_2$ precursor to modify non-porous silica[‡‡] and alumina[§§] powders.[67,68] One chemisorption step at 180 °C and subsequent air treatment at 350 °C resulted in highly dispersed V$_2$O$_5$ species on the support materials. Compared to catalysts prepared by impregnation, the ALD catalysts were found to have stronger acid sites and a higher activity in the dehydrogenation of propane.[67,68] Later, by exposing mesoporous silica[¶¶] to titanium isopropoxide and calcining it under O$_2$ flow, Keränen et al. synthesized a highly dispersed titania/silica material.[69] The acidic character of this material was further enhanced by a vanadyl isopropoxide/O$_2$ ALD treatment. The generated isolated vanadium species were shown to exhibit higher acidic strength than the less dispersed and more crystalline sites prepared with solutions of NH$_4$VO$_3$ and oxalic acid.

The one-step VO(acac)$_2$/air reaction was also used to introduce VO$_x$ species in ordered mesoporous MCM-48 material.[71,72] Prior to the ALD reaction, the MCM-48 support was treated in the liquid phase with a bifunctional silane and subsequently stirred in water. This treatment rendered the material hydrophobic while also creating OH groups for the reaction with the VO(acac)$_2$ precursor. Due to their partial hydrophobicity, these materials were found to be highly stable towards structural collapse and leaching. Furthermore, the VO$_x$/MCM-48 catalysts were active in the gas phase oxidation of methanol and the liquid phase selective oxidation of toluene.[72]

7.2.2 Supported Noble Metal Catalysts by ALD

The increase in catalysis-related ALD research from 2004 onward (Figure 7.12) can be partly ascribed to the successful development of ALD processes for the growth of noble metals that started in 2003.[73,74] As illustrated in the next paragraphs, the island growth mode that is typically present at the start of these ALD processes can be exploited for the fabrication of catalytic nanoparticles (Figure 7.14).

Krause and co-workers published several papers on the synthesis of noble metal catalysts using beta-diketonate[‖‖] ALD precursors, e.g., Pd(thd)$_2$, Ru(thd)$_3$, and Ir(acac)$_3$.[47,75–79] In one of these studies, H-beta zeolites were used as the support material for Ir catalysts prepared by one Ir(acac)$_3$ chemisorption step at 190 °C and subsequent ligand removal through calcination or reduction in H$_2$.[78] The deposition of Ir(acac)$_3$ was shown to be limited to the mesoporous part of the zeolites. Two H-beta zeolites with an effective surface area of 580 m^2 g^{-1} but with a different mesoporosity

---

[‡‡]Aerosil 200 silica (Degussa): particle size 12 nm and surface area 208 m^2 g^{-1}.
[§§]Oxid C γ-alumina (Degussa): particle size 20 nm and surface area 108 m^2 g^{-1}.
[¶¶]EP10 Silica (Crosfield): surface area 300 m^2 g^{-1}, pore volume 1.2 cm^3 g^{-1}, and average pore size 20 nm.
[‖‖]thd: 2,2,6,6-tetramethyl-3,5-heptanedione, acac: pentane-2,4-dionate or acetylacetonate.

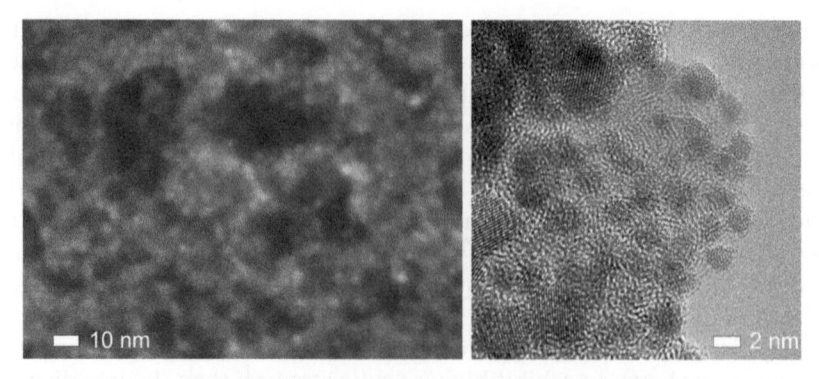

Figure 7.14 High resolution SEM (left) and TEM (right) images of ALD-grown Pt nanoparticles (size 2–3 nm) on a mesoporous silica support.

(corresponding to 90 m^2 g^{-1} and 220 m^2 g^{-1}, respectively) were loaded with the Ir precursor. Higher saturated loadings were achieved in the zeolite with the higher mesoporosity. It was also shown that the Ir content was reduced almost by half when the zeolitic support was treated with Hacac prior to the Ir(acac)$_3$ chemisorption step. TEM images revealed the presence of 2–3 nm Ir particles. The ALD catalysts were tested in the decalin ring opening reaction and were found to be more active and selective than samples prepared by wet impregnation of the zeolites using IrCl$_3$·3H$_2$O in water.

King et al. demonstrated the feasibility of ALD for the fabrication of Pt nanoparticles on the inner surfaces of carbon aerogels.[48] The aerogel support with a surface area of ~480 m^2 g^{-1} exhibited a macroporous network formed by interconnected microporous carbon particles. Pt was deposited using 2, 5 or 10 ALD cycles of the (methylcyclopentadienyl)trimethyl platinum (MeCpPtMe$_3$)/O$_2$ process at 320 °C. TEM images revealed the presence of evenly dispersed Pt nanoparticles (<5 nm). Remarkably, despite the lower amount of Pt, the catalyst prepared with 2 ALD cycles was as active as the 10 cycle catalyst for the oxidation of CO.

Li et al. used the MeCpPtMe$_3$/O$_2$ process to disperse Pt particles on the interior surface of mesoporous silica gel.***,[49] The application of 3, 5 and 10 ALD cycles resulted in 1.2, 1.9 and 2.3 nm nanoparticles, respectively. The three-cycle sample was subjected to a thermal stability test. After 4 h of heat treatment at 450 °C in air, the particle size was increased from 1.2 to 2 nm, indicating sintering of the particles. After following heating cycles, however, no additional sintering was observed. The Pt catalysts showed high activity for the oxidation of CO.

Pt nanoparticles were also deposited on nonporous, single-crystalline, cubic-shaped SrTiO$_3$ (STO) particles with an average edge length of 60 nm and a surface area of 20 m^2 g^{-1}.[50] STO particles treated with one MeCpPtMe$_3$/O$_2$ ALD cycle were evaluated in the oxidation of propane. The

---

***Silica gel (Aldrich): particle size 30–75 μm, surface area 513 m^2 g^{-1}, pore volume 0.75 cm^3 g^{-1}, and average pore size 6 nm.

ALD-synthesized catalyst converted propane at a 50 °C lower temperature than a conventional alumina-supported catalyst with comparable Pt loading, and showed better resistance to deactivation. The improved catalytic behavior was attributed to the strong epitaxy between the crystalline STO cubes and the Pt particles, preventing full oxidation of the Pt.

Feng et al. synthesized catalysts for methanol decomposition using ALD of Pd into mesoporous silica gel.[†††,80] Prior to Pd ALD, the support was covered with a ca. 0.5 nm thick Al_2O_3 or ZnO ALD coating. Pd nanoparticles (1 to 2 nm in diameter) were obtained by one or two cycles of the $Pd(hfac)_2$/formalin process.[‡‡‡] While the Pd/Al_2O_3 catalysts showed quite high and stable activity over time, the Pd/ZnO catalysts deactivated quickly at temperatures higher than 230 °C. The poor performance of the Pd/ZnO catalysts was related to the facilitated formation of Pd–Zn alloys under the conditions required for the decomposition of methanol. The authors showed that an ultra-thin Al_2O_3 ALD over-layer could improve the stability and the catalytic performance of the Pd particles on the ZnO surface, while keeping the active Pd species accessible for the methanol reactant.

An Al_2O_3 ALD overcoat was also reported to protect Pd catalysts from coking[55] and/or sintering[53,54] in high-temperature reactions. Feng et al. showed the effectiveness of a sub-nanometer thick Al_2O_3 over-layer to prevent sintering of supported Pd nanoparticles up to 500 °C.[53] The protective layer was found to preferentially nucleate at corners, steps and edges of the Pd nanoparticles, while keeping the more active Pd(111) facets accessible for methanol conversion. Lu et al. used thicker Al_2O_3 ALD overcoats (∼8 nm) to reduce coke formation and improve the thermal stability of Pd nanoparticle catalysts at temperatures as high as 675 °C.[55] Accessibility for reactant gases to the surface of the active nanoparticles was achieved by high-temperature treatments inducing microporosity in the Al_2O_3 ALD layer. It was furthermore shown that the overcoat had a positive effect on the selectivity of Pd catalysts in the oxidative dehydrogenation of ethane.

Christensen et al. demonstrated the potential of ALD for synthesizing bimetallic nanoparticles.[81] Ru–Pt particles with an average size of 1.2 nm were synthesized on nonporous alumina powder[§§§] using two Ru-ALD[¶¶¶] cycles, one Pt-ALD cycle, and another Ru-ALD cycle. TEM images showed that the particles were crystalline and had a crystal structure similar to bulk Pt. At temperatures above 210 °C, the bimetallic catalyst showed higher methanol conversion than a physical mixture of ALD-synthesized Pt/Al_2O_3 and Ru/Al_2O_3 catalysts with a similar loading.

[†††] Silicycle S10040M: particle size 75–200 μm, surface area 99.6 $m^2\ g^{-1}$, and pore diameter 30 nm.
[‡‡‡] hfac = hexafluoroacetylacetonate.
[§§§] Al_2O_3 nanospheres (NanoDur, Alfa Aesar): particle size 40–60 nm and surface area 35 $m^2\ g^{-1}$.
[¶¶¶] 2,4-(Dimethylpentadienyl)(ethylcyclopentadienyl)Ru was used in combination with O_2 to grow Ru, while Pt was deposited from $MeCpPtMe_3$ and O_2.

7.2.3 ALD for Photocatalysis

ALD has been used to synthesize photoactive nanostructures by conformal deposition of TiO$_2$ on nanoporous anodic aluminium oxide (AAO) membranes and arrays of Ni nanorods.[82,83] In other studies, AAO membranes were used as template for the fabrication of TiO$_2$ nanotube arrays by ALD.[84,85] Formation of the active anatase phase was achieved during the deposition process[82] or *via* post-deposition annealing.[84,85] Recent work, detailed in the next section (Case 2), used ALD of TiO$_2$ followed by a thermal treatment for the synthesis of anatase nanoparticles supported on a mesoporous silica film.[51] All nanostructured TiO$_2$ materials exhibited higher photocatalytic activity than planar TiO$_2$ reference samples, due their larger surface area.[51,82–85]

Adding noble metal particles to the surface of a TiO$_2$ photocatalyst can improve the photocatalytic performance. This effect is often attributed to the noble metal acting as an electron trap, thus suppressing electron–hole recombination.[86] Kemell *et al.* prepared a photocatalyst by first applying TiO$_2$ ALD and then Ir ALD on cellulose fibers.[87] The Ir process using Ir(acac)$_3$ resulted in Ir islands on the TiO$_2$ surface, which caused an enhancement of the photocatalytic activity of the TiO$_2$/cellulose composite, as evaluated in the photo-degradation of methylene blue.

More recently, Zhou *et al.* showed that the photoactivity of TiO$_2$ nanopowder‖‖ could be enhanced by loading it with highly dispersed Pt nanoparticles using ALD.[88] The authors systematically studied the effect of the number of ALD cycles and the deposition temperature on the Pt deposition and the photocatalytic activity, and concluded that one single MeCpPtMe$_3$/O$_2$ cycle deposited at 350 C° was optimal for the photodegradation of methylene blue. In a second study of the same authors, a magnetic photocatalyst was created by ALD of TiO$_2$ thin films on iron-based magnetic nanoparticles.[89]

7.2.4 Synthesis of Catalytic Membranes by ALD

Using AAO membranes as supports, Stair and co-workers demonstrated the suitability of ALD for synthesizing catalytic membranes with precise control of pore wall composition and pore size (10 to 100 nm).[52,90–92] AAO membranes supported by an aluminium ring were infiltrated with 1 nm Al$_2$O$_3$ and subsequently loaded with VO$_x$ by either wet impregnation or ALD.[92] The prepared catalysts were tested in the oxidative dehydrogenation of cyclohexane. In order to force the reactants to flow through the nanopores of the membrane, the coated AAO structures were placed in a standard Swagelok VCR fitting. This sealed structure was called a nanolith. It was found that the effective catalytic activity of the ALD-loaded catalysts was substantially

---

‖‖Aeroxide P25 TiO$_2$ (Evonik): surface area 50 m^2 g^{-1}.

higher compared to the impregnated samples, most likely due to a better dispersion of the active sites in case of ALD. Moreover, the selectivity to the desired product cyclohexene was higher for ALD samples with lower VO$_x$ loadings.

More recently, Deng *et al.* synthesized cobalt-based membrane catalysts for the oxidative decomposition of 1-methoxy-2-methyl-2-propanol, which is a model compound for cellulose.[93] The catalytic breakdown of cellulose, a structural material in plants, has attracted strong scientific interest for the production of ethanol as an alternative (and renewable) transportation fuel to petroleum fuels.

7.2.5 Use of Ordered Mesoporous Supports in Catalyst Preparation by ALD

Despite their attractive properties, ordered mesoporous materials have been used relatively rarely in catalysis-related ALD research. Recently, two papers reported the use of SBA-15 as ordered support material for ALD-prepared catalysts.[94,95] SBA-15 is a high surface area material with well-ordered hexagonal mesopores which are interconnected by irregular micropores. In the work by Pagàn-Torres *et al.*, the mesopore walls were uniformly coated by ALD of Nb$_2$O$_5$ from niobium ethoxide and H$_2$O.[94] N$_2$ adsorption measurements recorded after 0, 10, 19 and 30 ALD cycles revealed the progressive reduction in pore size from 6.5 to 3.7 nm. The microporosity was completely eliminated after 10 ALD cycles. While the uncoated SBA-15 material collapsed upon treatment in liquid water at 200 °C, the coated materials exhibited excellent hydrothermal stability. It was furthermore shown that the synthesized catalysts were active in the dehydration of 2-propanol, even after the water treatment.

In the second study, SBA-15 powders were coated with 1 ALD cycle of the TDMAT/H$_2$O process and/or 1 ALD cycle of the vanadyl isopropoxide/H$_2$O process.[95] The metal loading was varied by the exposure time of the metal-containing precursor. The synthesized catalysts were tested in the liquid phase epoxidation of cyclohexene to cyclohexene oxide, using *tert*-butyl hydroperoxide as oxidizing agent. In general, both the activity and selectivity towards cyclohexene oxide increased with increasing metal loading. While the VO$_x$ catalysts showed higher conversion rates for similar metal loading, the TiO$_x$ catalysts exhibited higher selectivity towards the desired product in the reaction. A maximum selectivity of 85% was found for a VO$_x$/TiO$_x$ catalyst. Moreover, a mechanism for the epoxidation reaction in these materials was proposed.

The applicability of ALD to ordered mesoporous materials is further addressed in the next section. Case 1 demonstrates the potential of ALD for generating acid catalytic activity in the ordered 3D silica network of Zeotile-4 material (Figure 7.7).[11,96]

7.3 ALD for Catalysis: Case Studies

7.3.1 Case 1: Introducing Acid Sites in Ordered Mesoporous Materials

Zeolites are microporous crystalline oxide materials that are widely used as industrial catalysts, in particular in the oil refining and petrochemical industry. The isomorphic substitution of Si^{4+} with a trivalent element such as Al^{3+} in the tetrahedral oxide framework of zeolites introduces negative framework charges that are compensated with protons (or exchangeable cations), thus giving these materials high Brønsted acidity (and ion exchange capacity).[97] The zeolite micropores offer excellent shape selectivity properties. However, in specific applications such as hydrocracking, the narrow pores cause diffusional limitations due to the similarities in size of the zeolite micropores and the hydrocarbons involved in the reaction. Therefore, acidic materials with pores in the mesoporous regime are of great interest for cracking molecules that are too bulky to fit within the narrow zeolite pores.

A promising class of materials in this respect are the ordered mesoporous silica materials characterized by large surface areas and regular pore arrangements. Their use in acid catalysis requires the generation of Brønsted sites, often attempted by the incorporation of Al in the silica pore walls. The introduction of Al directly during the synthesis of ordered mesoporous silicas has proven to be difficult.[98,99] The acid strength of the Al sites is typically very low and the order in the material becomes less well defined than for the pure-silica form. Another problem is the relatively low thermal stability at the elevated temperatures used in catalytic processes. Therefore, several methods have been proposed to introduce Al atoms in a post-synthesis operation,[99–105] e.g., by treatment of the silicate framework with $Al(NO_3)_3$,[102,103] anhydrous $AlCl_3$,[104] or aluminium chlorohydrate[101] solutions.

This case study presents an alternative ALD-based method for post-synthesis alumination of mesoporous silicate materials.[11,96] Incorporation of Al species in Zeotile-4, a silica material containing micropores as well as a 3D ordered network of mesopores,[40,41] was achieved in an elegant way by exposing the powder to 10 cycles of the TMA/H_2O ALD process.

The amount and coordination of the Al species added by the ALD process were studied using ^{27}Al magic angle spinning nuclear magnetic resonance (MAS NMR) spectroscopy. The as-synthesized Zeotile-4 material is a pure silica and did not show any signal in ^{27}Al MAS NMR. The spectrum measured after the ALD treatment is shown in Figure 7.15 and presents three signals with intensity maxima around 54, 30 and 0 ppm. The largest signal around 0 pmm is due to octahedral Al species located on the surface of the pore walls. The signals around 54 and 30 ppm are ascribed to tetrahedral Al atoms incorporated in the Zeotile-4 pore walls and distorted tetrahedral or pentacoordinated Al atoms, respectively.[106] The total Al content estimated from the signal intensities was found to be 1.815 mol kg^{-1}.

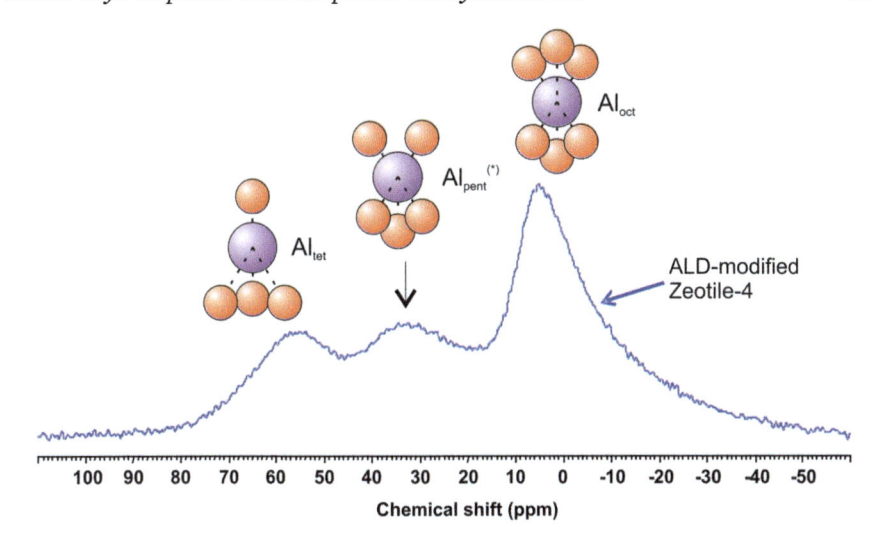

Figure 7.15 ^{27}Al MAS NMR spectrum for the Zeotile-4 powder after ALD treatment with 10 cycles of the TMA/H$_2$O process. Red and violet spheres represent oxygen and aluminium atoms, respectively. (*) Note that the peak at 30 ppm could also correspond to distorted tetrahedral Al atoms.[106] Adapted from ref. 11.

N$_2$ adsorption measurements provided an additional proof for the deposition of Al species in the pores of the Zeotile-4 material. Adsorption/desorption isotherms of N$_2$ gas at 77 K measured on the parent and ALD modified Zeotile-4 material are shown in Figure 7.16. Both isotherms are characterized by a marked hysteresis loop caused by capillary condensation in mesopores. A decreased N$_2$ adsorption capacity is observed for the ALD modified sample. Quantification revealed a decrease in surface area and mesopore volume due to the deposition of Al$_2$O$_3$ in the mesopores (Figure 7.16). The micropores were still largely accessible to N$_2$ after the ALD treatment, meaning that they were not closed off by the deposition process.

The catalytic performance of the parent and aluminated Zeotile-4 powder was evaluated in the hydroconversion of *n*-decane. The powders were impregnated with an aqueous solution of Pt(NH$_4$)$_3$Cl$_2$ to obtain a Pt loading of 0.5 wt%. The Pt is responsible for the dehydrogenation and hydrogenation of alkanes and alkenes, respectively, while the Al sites catalyze two competing reactions: hydroisomerization (converting linear alkanes into branched isomers) and hydro-cracking (converting the molecules into shorter fragments). Figure 7.17 shows the *n*-decane conversion yield against the reaction temperature. The as-synthesized Zeotile-4 material was inactive at temperatures below 270 °C. After incorporation of Al by ALD, the Zeotile-4 powder became active starting at 220 °C and reached a conversion yield of nearly 100% at 300 °C.

In conclusion, this case study demonstrated that the ALD process using TMA and H$_2$O is suited for the incorporation of Al species and the

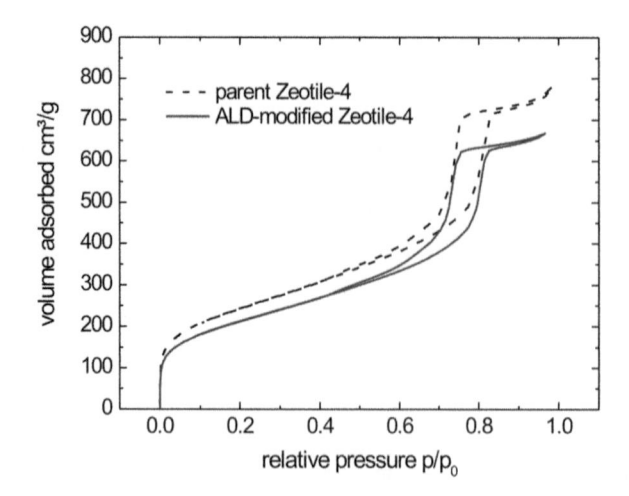

ALD cycles	Total surface area (m²/g)	Mesopore volume (ml/g)	Micropore volume (ml/g)
0	873	0.889	0.090
10	754	0.783	0.076

Figure 7.16 Adsorption/desorption isotherms of N_2 at 77 K measured on Zeotile-4 powder before and after ALD treatment with 10 cycles of the TMA/H_2O process. The total surface area and the micro- and mesopore volume of the samples estimated from t-plots are listed in the table. Adapted from ref. 11.

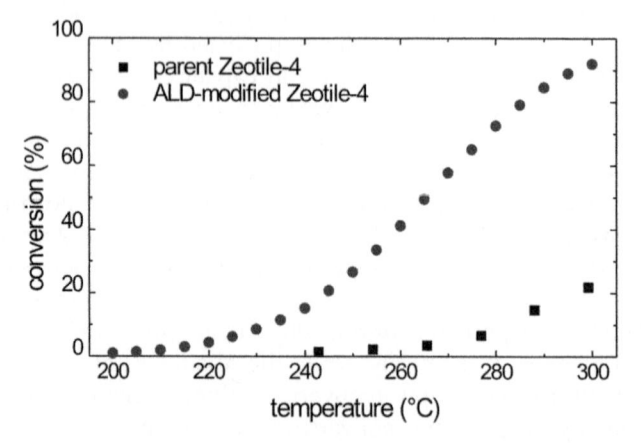

Figure 7.17 Decane conversion over parent and ALD-modified Zeotile 4 powder. Adapted from ref. 11.

enhancement of catalytic activity in ordered micro/mesoporous Zeotile-4 material. This method is furthermore extendable to other mesoporous materials, as illustrated by Sree *et al.*[96] for large pore ultra-stable Y zeolites. The presented ALD approach is also of potential practical interest for the reinsertion of Al in regenerated catalysts that are de-aluminated due to removal of the deposited coke *via* calcination.

7.3.2 Case 2: Introducing Photo-Active Nanoparticles in Mesoporous Films

TiO_2 nanomaterials play an important role in photocatalytic air and water remediation and in solar energy conversion.[86,107,108] Nanosized powders are among the best photocatalysts but their separation after a water purification process is time consuming. Nanopowders also present a risk to human health making their immobilization of vital importance in gas phase applications. Another issue is the loss in exposed surface area due to agglomeration of powder particles. These problems can be circumvented by direct synthesis of surface-immobilized TiO_2 nanostructures. Because anatase TiO_2 nanoparticles exhibit enhanced photocatalytic activity for very small particle sizes (<10 nm),[107,109] the TiO_2 nanostructures should ideally consist of defect-free nanometer-sized anatase particles stacked in a porous arrangement or well dispersed on a large surface area support.

Mesoporous silica films are often used for the preparation of photoactive surfaces. Incorporation of TiO_2 in mesoporous silica films is mostly realized during synthesis of the film,[110–116] *e.g.*, by spin coating a suspension of silica precursor, polymeric templates and colloidal TiO_2 nanoparticles on a substrate, followed by thermal treatment.[114–116] These direct approaches typically result in TiO_2–SiO_2 composite films with anatase nanocrystals contained within the walls of the porous structure. Only limited papers have used a post-synthesis method for introducing TiO_2 nanoparticles in the pore channels of mesoporous silica thin films; in the studies by Hua *et al.*[117] and Kohno *et al.*[118] nanometer-sized TiO_2 crystals were synthesized in the mesopores of an ordered silica network through impregnation of a Ti precursor and subsequent calcination.

This case study presents an alternative post-synthesis approach comprising ALD followed by a thermal treatment to introduce anatase particles into mesopores.[51] Mesoporous silica films with an exceptional high porosity (75%) and an average pore size of 12 nm were selected as support material.[119] An amorphous TiO_2 layer was conformally deposited using 50 ALD cycles of the $TDMAT/H_2O$ process at 200 °C. According to ellipsometric porosimetry measurements the average pore size decreased to 6 nm, indicating a TiO_2 film thickness of *ca.* 3 nm. After ALD, the amorphous as-deposited TiO_2 layer was converted into crystalline anatase by calcination of the sample in air for 5 h at 550 °C starting at room temperature and heating at a rate of 1 °C min^{-1}.

Quantitative electron tomography was used to investigate the morphology of the crystallized TiO_2 coating. Figure 7.18a shows an *XZ*-orthoslice *through* the 3D reconstruction of the porous film obtained using conventional reconstruction algorithms.[120] The white dots correspond to TiO_2

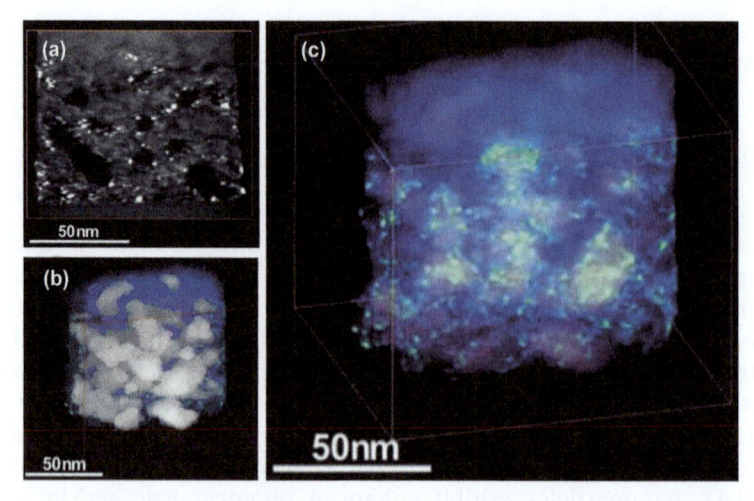

Figure 7.18 Electron tomography reconstructions of TiO$_2$ nanoparticles in a mesoporous silica film (initial porosity: 75%, thickness: 150 nm) synthesized by ALD of TiO$_2$ and subsequent annealing. (a) *XZ*-orthoslice through the 3D reconstruction. (b) and (c) 3D visualization of the reconstruction with the silica matrix material in blue and mesopores highlighted in white in (b), and anatase particles in green in (c).
Adapted from ref. 51.

nanoparticles, the gray zones to the silica pore walls, and the black zones to the voids. During annealing, the TiO$_2$ ALD film clearly broke up into anatase nanoparticles. To allow for quantitative image analysis, a 3D reconstruction was also carried out using the advanced total variation minimization algorithm.[35] This study yielded an average anatase particle size of 4.0 ± 2.6 nm. Figure 7.18b and c shows 3D voltex visualizations of the reconstructed film emphasizing voids and nanoparticles, respectively. The TiO$_2$ nanoparticles are represented as green dots, the silica matrix material is blue and the voids are white.

The photoactivity of the supported TiO$_2$ nanoparticles was tested in the degradation of methylene blue (MB) in aqueous solution. Figure 7.19 shows the MB degradation against the irradiation time for the mesoporous silica film with TiO$_2$ nanoparticles as well as for a mesoporous reference film without any TiO$_2$. As expected, the presence of anatase nanoparticles clearly enhanced the photodegradation. The third curve in Figure 7.19 was measured for a crystallized TiO$_2$ layer on a planar Si substrate that contained the same amount of Ti as the mesoporous film with TiO$_2$ nanoparticles according to XRF. The superior photocatalytic activity of the mesoporous film can be ascribed to the excellent dispersion of the anatase nanoparticles on a highly porous and accessible support (Figure 7.18).

As a final note it is worth mentioning that the size of the TiO$_2$ nanoparticles can be controlled by the thickness of the amorphous TiO$_2$ coating or, in other words, the number of ALD cycles. As detailed by Sree *et al.*,[51]

Atomic Layer Deposition in Nanoporous Catalyst Materials 191

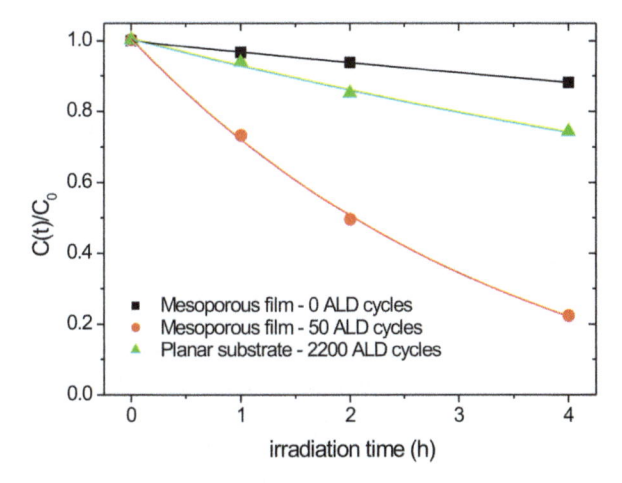

Figure 7.19 Photocatalytic degradation of methylene blue on a blank mesoporous silica film, a mesoporous silica film loaded with anatase nanoparticles, and a planar TiO_2/Si sample.
Adapted from ref. 51.

particle size estimations from XRD patterns indicated an increasing trend in anatase particle size with increasing TiO_2 deposition in similar mesoporous silica films.

7.4 Conclusions

Atomic layer deposition (ALD) is discussed as a promising method for supported catalyst preparation. The self-saturation of the chemical surface reactions in ALD enables the conformal coating of high surface area nanoporous materials and provides atomic-level control over the coating thickness. These unique advantages offer ALD the ability to precisely control the pore size and chemical surface composition of nanoporous materials, and therefore render ALD an effective method for the nanoscale design of catalysts. As illustrated with selected examples from literature and two case studies, a wide variety of oxides, metals and other materials can be deposited on the interior surface of nanoporous supports, resulting in catalytic active materials. This can be either as an ultra-thin conformal layer or, if growth is inhibited during nucleation, as dispersed nanoparticles. Nanosized particles can also be fabricated by the controlled breakup of an ALD-deposited layer through calcination, as demonstrated here for the formation of anatase nanoparticles. Because of its versatility and exceptional growth control, the ALD technology has attracted attention from researchers in various fields such as photocatalysis and acid catalysis, and is also particularly relevant to the synthesis of model catalysts and microchannel reactors used for screening purposes. While gram quantities of catalyst support material can be generally treated in ALD research labs, the optimization and scale-up of the technique for industrial catalysis requires novel and dedicated reactor concepts for ALD powder coating, beyond the current commercial designs that are mainly developed for wafer manufacturing. Encouraged by the increasing efforts on this topic,[121] we believe that the ALD technique has significant potential for continued exploration and ultimately breakthrough in catalysis on an industrial scale.

Acknowledgements

The research leading to these results was supported by the Flemish IWT in the frame of the strategic basic research (SBO) project, the Flemish FWO, and the European Research Council (ERC Starting Grant 239865). I gratefully acknowledge the Flemish FWO for a postdoctoral fellowship. Past and present ALD colleagues at Ghent University are thanked for contributions to the experiments and generous sharing of knowledge and experience. Thanks go especially to Prof. Christophe Detavernier, Dr Davy Deduytsche, Dr Jan Musschoot and Kilian Devloo-Casier. I thank Prof. Johan Martens, Dr Sree and colleagues (COK, Catholic University of Leuven) for significant contributions to this work and pleasant collaborations on many of the topics discussed here. Prof. Sara Bals and colleagues (EMAT, University of Antwerp) are acknowledged for the electron tomography characterizations, Prof. Jan D'Haen (Hasselt University) for the TEM characterization of Zeotile-4, Dr Hilde Poelman (Ghent University) for providing the SEM and TEM picture of Pt nanoparticles and Prof. Karl Ludwig (Boston University) for collaboration on the *in situ* XRF experiments.

References

1. V. B. Aleskovskii, *J. Appl. Chem. USSR*, 1974, **47**, 2207.
2. T. Suntola and J. Antson, U.S. Patent No. 4 058 430, 1977.
3. S. Haukka, E. L. Lakomaa and T. Suntola, *Thin Solid Films*, 1993, **225**, 280.
4. M. Lindblad, L. P. Lindfors and T. Suntola, *Catal. Lett.*, 1994, **27**, 323.
5. A. Kytökivi, J. P. Jacobs, A. Hakuli, J. Meriläinen and H. H. Brongersma, *J. Catal.*, 1996, **162**, 190.
6. L. B. Backman, A. Rautiainen, A. O. I. Krause and M. Lindblad, *Catal. Today*, 1998, **43**, 11.
7. T. A. Kainulainen, M. K. Niemelä and A. O. I. Krause, *Catal. Lett.*, 1998, **53**, 97.
8. S. Haukka, E. L. Lakomaa and T. Suntola, *Studies in Surface Science and Catalysis*, ed. A. Dabrowski, Elsevier, Amsterdam, 1999, vol. 120, Part A, pp. 715–750.
9. M. Knez, K. Nielsch and L. Niinistö, *Adv. Mater.*, 2007, **19**, 3425.
10. H. Kim, H.-B.-R. Lee and W.-J. Maeng, *Thin Solid Films*, 2009, **517**, 2563.
11. C. Detavernier, J. Dendooven, S. P. Sree, K. F. Ludwig and J. A. Martens, *Chem. Soc. Rev.*, 2011, **40**, 5242.

12. C. Bae, H. Shin and K. Nielsch, *MRS Bull.*, 2011, **36**, 877.
13. J. W. Elam, N. P. Dasgupta and F. B. Prinz, *MRS Bull.*, 2011, **36**, 899.
14. C. Marichy, M. Bechelany and N. Pinna, *Adv. Mater.*, 2012, **24**, 1017.
15. V. Miikkulainen, M. Leskelä, M. Ritala and R. L. Puurunen, *J. Appl. Phys.*, 2013, **113**, 021301.
16. G. S. Higashi and C. G. Fleming, *Appl. Phys. Lett.*, 1989, **55**, 1963.
17. R. L. Puurunen, *J. Appl. Phys.*, 2005, **97**, 121301.
18. Q. Xie, Y.-L. Jiang, C. Detavernier, D. Deduytsche, R. L. Van Meirhaeghe, G.-P. Ru, B.-Z. Li and X.-P. Qu, *J. Appl. Phys.*, 2007, **102**, 083521.
19. C. Dücsö, N. Q. Khanh, Z. Horváth, I. Bársony, M. Utriainen, S. Lehto, M. Nieminen and L. Niinistö, *J. Electrochem. Soc.*, 1996, **143**, 683.
20. M. Ritala, M. Leskelä, J.-P. Dekker, C. Mutsaers, P. J. Soininen and J. Skarp, *Chem. Vap. Deposition*, 1999, **5**, 7.
21. A. W. Ott, J. W. Klaus, J. M. Johnson, S. M. George, K. C. McCarley and J. D. Way, *Chem. Mater.*, 1997, **9**, 707.
22. J. W. Elam, D. Routkevitch, P. P. Mardilovich and S. M. George, *Chem. Mater.*, 2003, **15**, 3507.
23. R. G. Gordon, D. Hausmann, E. Kim and J. Shepard, *Chem. Vap. Deposition*, 2003, **9**, 73.
24. J. Dendooven, D. Deduytsche, J. Musschoot, R. L. Vanmeirhaeghe and C. Detavernier, *J. Electrochem. Soc.*, 2009, **156**, P63.
25. J. Dendooven, B. Goris, K. Devloo-Casier, E. Levrau, E. Biermans, M. R. Baklanov, K. F. Ludwig, P. Van Der Voort, S. Bals and C. Detavernier, *Chem. Mater.*, 2012, **24**, 1992.
26. J. H. Pan and W. I. Lee, *New J. Chem.*, 2005, **29**, 841.
27. M. R. Baklanov, K. P. Mogilnikov, V. G. Polovinkin and F. N. Dultsev, *J. Vac. Sci. Technol., B*, 2000, **18**, 1385.
28. J. Dendooven, K. Devloo-Casier, E. Levrau, R. Van Hove, S. P. Sree, M. R. Baklanov, J. A. Martens and C. Detavernier, *Langmuir*, 2012, **28**, 3852.
29. K. Kukli, M. Ritala, T. Sajavaara, J. Keinonen and M. Leskelä, *Chem. Vap. Deposition*, 2002, **8**, 199.
30. P. A. Midgley and M. Weyland, *Ultramicroscopy*, 2003, **96**, 413.
31. P. A. Midgley and R. E. Dunin-Borkowski, *Nat. Mater.*, 2009, **8**, 271.
32. K. J. Batenburg, S. Bals, J. Sijbers, C. Kübel, P. A. Midgley, J. C. Hernandez, U. Kaiser, E. R. Encina, E. A. Coronado and G. Van Tendeloo, *Ultramicroscopy*, 2009, **109**, 730.
33. E. Biermans, L. Molina, K. J. Batenburg, S. Bals and G. Van Tendeloo, *Nano Lett.*, 2010, **10**, 5014.
34. X. Ke, S. Bals, D. Cott, T. Hantschel, H. Bender and G. Van Tendeloo, *Microsc. Microanal.*, 2010, **16**, 210.
35. B. Goris, W. Van den Broek, K. J. Batenbrug, H. H. Mezerji and S. Bals, *Ultramicroscopy*, 2012, **113**, 120.
36. J. Dendooven, S. P. Sree, K. De Keyser, D. Deduytsche, J. A. Martens, K. F. Ludwig and C. Detavernier, *J. Phys. Chem. C*, 2011, **115**, 6605.

37. K. Devloo-Casier, K. F. Ludwig, C. Detavernier and J. Dendooven, *J. Vac. Sci. Technol., B*, 2014, **32**, 010801.
38. J. W. Elam, J. A. Libera, T. H. Huynh, H. Feng and M. J. Pellin, *J. Phys. Chem. C*, 2010, **114**, 17286.
39. S. P. Sree, J. Dendooven, J. Jammaer, K. Masschaele, D. Deduytsche, J. D'Haen, C. E. A. Kirschhock, J. A. Martens and C. Detavernier, *Chem. Mater.*, 2012, **24**, 2775.
40. S. P. B. Kremer, C. E. A. Kirschhock, A. Aerts, K. Villani, J. A. Martens, O. I. Lebedev and G. Van Tendeloo, *Adv. Mater.*, 2003, **15**, 1705.
41. S. Bals, K. J. Batenburg, D. Liang, O. Lebedev, G. Van Tendeloo, A. Aerts, J. A. Martens and C. E. A. Kirschhock, *J. Am. Chem. Soc.*, 2009, **131**, 4769.
42. J. Hyvärinen, M Sonninen and R. Törnqvist, *J. Cryst. Growth*, 1988, **86**, 695.
43. D. Longrie, D. Deduytsche, J. Haemers, K. Driesen and C. Detavernier, *Surf. Coat. Technol.*, 2012, **213**, 183.
44. J. E. Maslar, W. A. Kimes, B. A. Sperling, P. F. Ma, J. Anthis, J. R. Bakke and R. Kanjolia, *In Situ Optical Characterization of Solid Precursor Delivery for ALD Processes*. Presented at the AVS Topical Conference on ALD, San Diego, USA, July 28–31, 2013.
45. J. A. Libera, J. W. Elam and M. J. Pellin, *Thin Solid Films*, 2008, **516**, 6158.
46. S. Haukka and T. Suntola, *Interface Sci.*, 1997, **5**, 119.
47. M. Lashdaf, T. Hatanpää, A. O. I. Krause, J. Lahtinen, M. Lindblad and M. Tiitta, *Appl. Catal., A*, 2003, **241**, 51.
48. J. S. King, A. Wittstock, J. Biener, S. O. Kucheyev, Y. M. Wang, T. F. Baumann, S. K. Giri, A. V. Hamza, M. Baeumer and S. F. Bent, *Nano Lett.*, 2009, **8**, 2405.
49. J. Li, X. Liang, D. M. King, Y.-B. Jiang and A. W. Weimer, *Appl. Catal., B*, 2010, **97**, 220.
50. J. A. Enterkin, W. Setthapun, J. W. Elam, S. T. Christensen, F. A. Rabuffetti, L. D. Marks, P. C. Stair, K. R. Poeppelmeier and C. L. Marshall, *ACS Catal.*, 2011, **1**, 629.
51. S. P. Sree, J. Dendooven, K. Masschaele, H. M. Hamed, S. Deng, S. Bals, C. Detavernier and J. A. Martens, *Nanoscale*, 2013, **5**, 5001.
52. M. J. Pellin, P. C. Stair, G. Xiong, J. W. Elam, J. Birrell, L. Curtiss, S. M. George, C. Y. Han, L. Iton, H. Kung, M. Kung and H. H. Wang, *Catal. Lett.*, 2005, **102**, 127.
53. H. Feng, J. Lu, P. C. Stair and J. W. Elam, *Catal. Lett.*, 2011, **141**, 512.
54. X. Liang, J. Li, M. Yu, C. N. McMurray, J. M. Falconer and A. W. Weimer, *ACS Catal.*, 2011, **1**, 1162.
55. J. Lu, B. Fu, M. C. Kung, G. Xiao, J. W. Elam, H. H. Kung and P. C. Stair, *Science*, 2012, **335**, 1205.
56. E. L. Lakomaa, S. Haukka and T. Suntola, *Appl. Surf. Sci.*, 1992, **60-1**, 742.
57. E. L. Lakomaa, *Appl. Surf. Sci.*, 1994, **75**, 185.

58. M. Lindblad, S. Haukka, A. Kytökivi, E. L. Lakomaa, A. Rautiainen and T. Suntola, *Appl. Surf. Sci.*, 1997, **121**, 286.
59. L. B. Backman, A. Rautiainen, M. Lindblad, O. Jylhä and A. O. I. Krause, *Appl. Catal., A*, 2001, **208**, 223.
60. L. B. Backman, A. Rautiainen, M. Lindblad and A. O. I. Krause, *Appl. Catal., A*, 2000, **191**, 55.
61. A. Hakuli, A. Kytökivi and A. Krause, *Appl. Catal., A*, 2000, **190**, 219.
62. R. L. Puurunen, A. Root, S. Haukka, E. I. Iiskola, M. Lindblad and A. O. I. Krause, *J. Phys. Chem. B*, 2000, **104**, 6599.
63. R. L. Puurunen, A. Root, P. Sarv, S. Haukka, E. I. Iiskola, M. Lindblad and A. O. I. Krause, *Appl. Surf. Sci.*, 2000, **165**, 193.
64. R. L. Puurunen, A. Root, P. Sarv, M. M. Viitanen, H. H. Brongersma, M. Lindblad and A. O. I. Krause, *Chem. Mater.*, 2002, **14**, 720.
65. R. L. Puurunen, T. A. Zeelie and A. O. I. Krause, *Catal. Lett.*, 2002, **83**, 27.
66. R. L. Puurunen, S. M. K. Airaksinen and A. O. I. Krause, *J. Catal.*, 2003, **213**, 281.
67. J. Keränen, A. Auroux, S. Ek-Härkönen and L. Niinistö, *Thermochim. Acta*, 2001, **379**, 233.
68. J. Keränen, A. Auroux, S. Ek and L. Niinistö, *Appl. Catal., A*, 2002, **228**, 213.
69. J. Keränen, C. Guimon, E. Iiskola, A. Auroux and L. Niinistö, *Catal. Today*, 2003, **78**, 149.
70. J. Keränen, P. Carniti, A. Gervasini, E. Iiskola, A. Auroux and L. Niinistö, *Catal. Today*, 2004, **91–2**, 67.
71. P. Van Der Voort, M. Morey, G. D. Stucky, M. Mathieu and E. F. Vansant, *J. Phys. Chem. B*, 1998, **102**, 585.
72. P. Van Der Voort, M. Baltes and E. F. Vansant, *Catal. Today*, 2001, **68**, 119.
73. T. Aaltonen, P. Alen, M. Ritala and M. Leskelä, *Chem. Vap. Deposition*, 2003, **9**, 45.
74. T. Aaltonen, M. Ritala, T. Sajavaara, J. Keinonen and M. Leskelä, *Chem. Mater.*, 2003, **15**, 1924.
75. M. Lashdaf, A. O. I. Krause, M. Lindblad, A. Tiitta and T. Venäläinen, *Appl. Catal., A*, 2003, **241**, 65.
76. R. J. Silvennoinen, O. J. T. Jylhä, M. Lindblad, H. Österholm and A. O. I. Krause, *Catal. Lett.*, 2007, **114**, 135.
77. R. J. Silvennoinen, O. J. T. Jylhä, M. Lindblad, J. P. Sainio, R. L. Puurunen and A. O. I. Krause, *Appl. Surf. Sci.*, 2007, **253**, 4103.
78. H. Vuori, R. J. Silvennoinen, M. Lindblad, H. Österholm and A. O. I. Krause, *Catal. Lett.*, 2009, **131**, 7.
79. E. Rikkinen, A. Santasalo-Aarnio, S. Airaksinen, M. Borghei, V. Viitanen, J. Sainio, E. I. Kauppinen, T. Kallio and A. O. I. Krause, *J. Phys. Chem. C*, 2011, **115**, 23067.
80. H. Feng, J. W. Elam, J. A. Libera, W. Setthapun and P. C. Stair, *Chem. Mater.*, 2010, **22**, 3133.

81. S. T. Christensen, H. Feng, J. L. Libera, N. Guo, J. T. Miller, P. C. Stair and J. W. Elam, *Nano Lett.*, 2010, **10**, 3047.
82. M. Kemell, V. Pore, J. Tupala, M. Ritala and M. Leskelä, *Chem. Mater.*, 2007, **19**, 1816.
83. M. Kemell, E. Härkönen, V. Pore, M. Ritala and M. Leskelä, *Nanotechnology*, 2010, **21**, 035301.
84. C. J. W. Ng, H. Gao and T. T. Y. Tan, *Nanotechnology*, 2008, **19**, 445604.
85. Y.-C. Liang, C.-C. Wang, C.-C. Kei, Y.-C. Hsueh, W.-H. Cho and T.-P. Perng, *J. Phys. Chem. C*, 2011, **115**, 9498.
86. A. L. Linsebigler, G. Lu and J. T. Yates, Jr., *Chem. Rev.*, 1995, **95**, 735.
87. M. Kemell, V. Pore, M. Ritala and M. Leskelä, *Chem. Vap. Deposition*, 2006, **12**, 419.
88. Y. Zhou, D. M. King, X. Liang, J. Li and A. W. Weimer, *Appl. Catal., B*, 2010, **101**, 54.
89. Y. Zhou, D. M. King, J. Li, K. S. Barrett, R. B. Goldfarb and A. W. Weimer, *Ind. Eng. Chem. Res.*, 2010, **49**, 6964.
90. P. C. Stair, C. Marshall, G. Xiong, H. Feng, M. J. Pellin, J. W. Elam, L. Curtiss, L. Iton, H. Kung, M. Kung and H. H. Wang, *Top. Catal.*, 2006, **39**, 181.
91. H. Feng, J. W. Elam, J. A. Libera, M. J. Pellin and P. C. Stair, *Chem. Eng. Sci.*, 2009, **64**, 560.
92. H. Feng, J. W. Elam, J. A. Libera, M. J. Pellin and P. C. Stair, *J. Catal.*, 2010, **269**, 421.
93. W. Deng, S. Lee, J. A. Libera, J. W. Elam, S. Vajda and C. L. Marshall, *Appl. Catal., A*, 2011, **393**, 29.
94. Y. J. Pagán-Torres, J. M. R. Gallo, D. Wang, H. N. Pham, J. A. Libera, C. L. Marshall, J. W. Elam, A. K. Datye and J. A. Dumesic, *ACS Catal.*, 2011, **1**, 1234.
95. I. Muylaert, J. Musschoot, K. Leus, J. Dendooven, C. Detavernier and P. Van Der Voort, *Eur. J. Inorg. Chem.*, 2012, **2012**, 251.
96. S. P. Sree, J. Dendooven, T. I. Korányi, G. Vanbutsele, K. Houthoofd, D. Deduytsche, C. Detavernier and J. A. Martens, *Catal. Sci. Technol.*, 2011, **1**, 218.
97. *Studies in surface science and catalysis*, ed. H. van Bekkum, E. M. Flanigen, P. A. Jacobs and J. C. Jansen, Elsevier, Amsterdam, vol. 137, 2001.
98. A. Corma, *Chem. Rev.*, 1997, **97**, 2373.
99. D. Trong On, D. Desplantier-Giscard, C. Danumah and S. Kaliaguine, *Appl. Catal., A*, 2001, **222**, 299.
100. L. Y. Chen, Z. Ping, G. K. Chuah, S. Jaenicke and G. Simon, *Microporous Mesoporous Mater.*, 1999, **27**, 231.
101. R. Mokoya and W. Jones, *J. Mater. Chem.*, 1999, **9**, 555.
102. S. Kawi and S. C. Chen, *Stud. Surf. Sci. Catal.*, 2000, **129**, 219.
103. S. Kawi and S. C. Chen, *Stud. Surf. Sci. Catal.*, 2000, **129**, 227.
104. V. R. Choudhary and K. Mantri, *Microporous Mesoporous Mater.*, 2002, **56**, 317.

105. S. Zeng, J. Blanchard, M. Breysse, Y. Shi, X. Shu, H. Nie and D. Li, *Microporous Mesoporous Mater.*, 2005, **85**, 297.
106. M. J. Remy, D. Stanica, G. Poncelet, E. J. P. Feijen, P. J. Grobet, J. A. Martens and P. A. Jacobs, *J. Phys. Chem.*, 1996, **100**, 12440.
107. M. R. Hoffmann, S. T. Martin, W. Choi and D. W. Bahnemann, *Chem. Rev.*, 1995, **95**, 69–96.
108. H. Zhou, Y. Qu, T. Zeid and X. Duan, *Energy Environ. Sci.*, 2012, **5**, 6732.
109. M. Anpo and M. Takeuchi, *J. Catal.*, 2003, **216**, 505.
110. Y. Shioya, K. Ikeue, M. Ogawa and M. Anpo, *Appl. Catal., A*, 2003, **254**, 251.
111. V. Zelenak, V. Hornebecq, S. Mornet, O. Schaf and P. Llewellyn, *Chem. Mater.*, 2006, **18**, 3184.
112. W. Dong, Y. Sun, C. W. Lee, W. Hua, X. Lu, Y. Shi, S. Zhang, J. Chen and D. Zhao, *J. Am. Chem. Soc.*, 2007, **129**, 13894.
113. W. Dong, C. W. Lee, X. Lu, Y. Sun, W. Hua, G. Zhuang, S. Zhang, J. Chen, H. Hou and D. Zhao, *Appl. Catal., B*, 2010, **95**, 197.
114. F. Bosc, A. Ayral and C. Guizard, *Thin Solid Films*, 2006, **495**, 252.
115. D. Fattakhova-Rohlfing, J. M. Szeifert, Q. Yu, V. Kalousek, J. Rathousky and T. Bein, *Chem. Mater.*, 2009, **21**, 2410.
116. M. Gohin, I. Maurin, T. Gacoin and J. Boilot, *J. Mater. Chem.*, 2010, **20**, 8070.
117. Z. L. Hua, J. L. Shi, L. X. Zhang, M. L. Ruan and J. N. Yan, *Adv. Mater.*, 2002, **14**, 830.
118. A. Kohno, T. Gondo, K. Koga and T. Tajiri, *IOP Conf. Ser.: Mater. Sci. Eng.*, 2011, **24**, 012019.
119. S. P. Sree, J. Dendooven, D. Smeets, D. Deduytsche, A. Aerts, K. Vanstreels, M. Baklanov, J. W. Seo, K. Temst, A. Vantomme, C. Detavernier and J. A. Martens, *J. Mater. Chem.*, 2011, **21**, 7692.
120. P. Gilbert, *J. Theor. Biol.*, 1972, **36**, 105.
121. D. Longrie, D. Deduytsche and C. Detavernier, *J. Vac. Sci. Technol., A*, 2014, **32**, 10802.

CHAPTER 8

Preparation and Characterization of Model Catalysts for the HCl Oxidation Reaction

CHRISTIAN KANZLER, HERBERT OVER,*
BERND M. SMARSLY* AND CLAAS WESSEL

Physikalisch-Chemisches Institut, Justus-Liebig-Universität Gießen, Heinrich-Buff-Ring 58, D-35392, Gießen, Germany
*Email: Herbert.Over@phys.chemie.uni-giessen.de;
Bernd.Smarsly@phys.chemie.uni-giessen.de

8.1 The Deacon Process: Oxidation of HCl

In this chapter we concentrate on the heterogeneously catalyzed HCl oxidation (the so-called Deacon process), which is an industrially important reaction for recycling Cl_2 from the toxic and unavoidable waste by-product HCl, which amounts worldwide to some 10 million tons annually.[1] Heterogeneously catalyzed HCl oxidation is known to be an extremely corrosive reaction which had prevented industrial application for some 140 years after the invention by Henry Deacon.[2] The original Deacon process was patented in 1868 using a CuO catalyst and consists of consecutive solid-state reactions. First, CuO is chlorinated to $CuCl_2$ and in a subsequent step the $CuCl_2$ is re-oxidized to CuO, thereby releasing the desired product of molecular Cl_2. The required reaction temperature of 720 K is dictated by the high activation for the re-oxidation of $CuCl_2$ so that the conversion of HCl is limited to 75% for thermodynamic reasons. Even worse the catalyst is not stable since the $CuCl_2$ formed is volatile at such high temperatures leading to rapid and uncontrolled loss of the active component during the Deacon reaction.

Alternative catalysts for the Deacon process have been proposed but most of them have not been industrialized. Instead electrolysis or simple neutralization to salts have been the standard ways to get around the waste disposal problem of the toxic HCl.[3,4] Recently (1999), Sumitomo Chemical introduced and commercialized the first stable and active catalyst for the Deacon process which is based on RuO_2. The reaction temperature can be kept as low as 600 K so that the equilibrium conversion is 90–95%.[5,6] In addition, RuO_2 is stable, partly due to the low reaction temperature and partly due to the surface chlorination of RuO_2. Under reaction conditions a fraction of the surface lattice O atoms is replaced by chlorine[7]; bridging chlorine is further stabilized by Ti doping.[8]

The scarcity and the high cost of Ru have initiated an intensive search for alternative materials as potential catalysts for the HCl oxidation.[4,9,10] Recently, it was demonstrated that CeO_2 represents an interesting option for replacing RuO_2.[11,12] The activity of CeO_2-based catalysts in the HCl oxidation reaction was proposed to be governed by the oxygen storage capacity (OSC). Unfortunately, pure CeO_2 is chemically not stable under HCl-rich reaction conditions,[12,13] and currently the activity of CeO_2 is too low to be economically competitive with RuO_2-based catalysts.[10,12] However, it is known that doping isovalent elements like Zr^{4+} into the CeO_2 lattice strongly affects the redox properties of ceria[14–17] and improves the thermal stability of CeO_2 against sintering.[18] Therefore, Zr doping may be a promising strategy to stabilize CeO_2 against in-depth chlorination in the Deacon process *via* formation of a solid solution of CeO_2 and ZrO_2 which may have higher activity as well.

8.2 Why Model Catalysis?

In heterogeneous catalysis the reaction mixture comes in contact with the surface of the catalyst, which consists of supported particles of the active component dispersed and immobilized over high surface carriers, often oxides such as MgO, SiO_2, or Al_2O_3. The particles expose various facets with defects such as edges and kinks. Therefore, from a structural point of view heterogeneous catalysts represent a quite intricate situation. In general, the reaction mechanism consists of the adsorption of the reactants and the subsequent formation of the desired product in a surface reaction sometimes including lattice atoms such as oxygen in oxides.[19] The interaction of the active component with the support may improve the activity and selectivity of the catalyst.[20,21] But also particle size and particle morphology may affect the performance of a catalytic reaction. This effect is called structure sensitivity in catalysis research.[22]

In order to gain atomic scale insights into a heterogeneously catalyzed reaction system one has to resort on model catalysis.[23,24] The chosen model system should be related as closely as possible to the considered effect of the real reaction system.[25] Besides model catalysts also simple model reactions are in the focus of model catalysis. In surface science the most frequently studied reaction system is the CO oxidation reaction;[26] simply in the sense that only activity counts and stability of catalyst is not of major concern. For the case of $RuO_2(110)$, the CO oxidation has been thoroughly discussed in previous papers[27,28] and found to be structure insensitive.[29]

Here we will focus on the heterogeneously catalyzed HCl oxidation. Various model catalysts can be envisioned for the Sumitomo process: single crystals of RuO_2, single crystalline $RuO_2(110)$ films, supported $RuO_2(110)$ films of various thicknesses on a rutile $TiO_2(110)$ substrate, micro-scale RuO_2 powder, and RuO_2 nanofibers (cf. Figure 8.1).[13] The proper choice of the model system is dictated by the scientific question in mind when taking off with the scientific project. If one is interested in a molecular level understanding of the reaction mechanism, then one should retain to single crystal $RuO_2(110)$ or $RuO_2(100)$ films with well-defined surface structure of the catalytically active sites provided by one-fold under-coordinated surface atoms (cf. Figure 8.1a). If steps and kinks are important in the catalytic

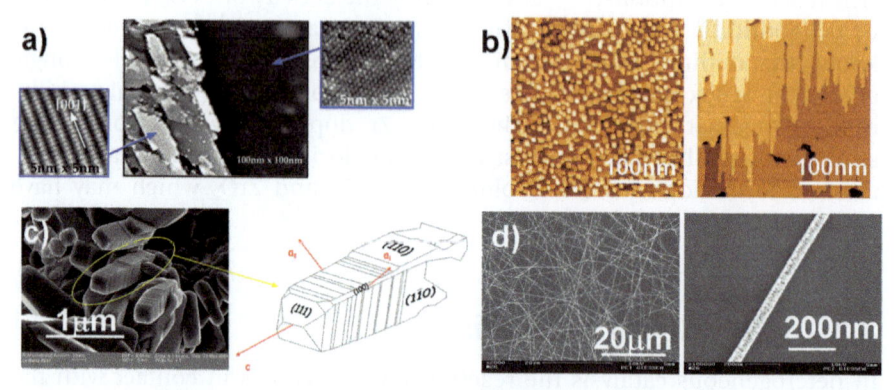

Figure 8.1 Various types of model catalysts forming a hierarchy of increasing complexity exemplified by a RuO_2-based catalyst. (a) Single crystalline $RuO_2(110)$ films grown on a single crystalline Ru(0001) surface. On the atomic scale, the oxide film is well ordered exposing under-coordinated Ru and O sites. (b) Separated single crystalline $RuO_2(110)$ islands (height: 2 nm, lateral size 10 nm) are evenly distributed across the $TiO_2(110)$ surface and dense RuO_2 films of 10 nm thickness on $TiO_2(110)$. (c) Micro-scale RuO_2 powder exposing various facets in various orientations, predominantly along the (110) orientation. (d) Polycrystalline RuO_2-based nanofibers revealing a well-defined cylinder morphology. Morphological alteration under harsh reaction conditions should easily be visible in scanning electron microscopy (SEM).

conversion of HCl, one has to study stepped surface films with a periodic array of steps.[30] If the interaction of the support with the catalytically active RuO$_2$ component is dominating the catalytic performance, then one needs to look at a layered system consisting of the active component RuO$_2$(110) coated for instance on TiO$_2$(110) (*cf.* Figure 8.1b).

Single crystalline model catalysts allow for employing the whole battery of surface science techniques to comprehend on the surface reaction on the molecular level.[31,32] This encompasses the application of electron spectroscopy and electron diffraction for determining the electronic and atomic structure of the surface. With well-prepared co-adsorption experiments of the reactants one can gain deep insights into the elementary reaction steps. With *in situ* techniques such as infrared spectroscopy (RAIRS), reaction intermediates can be identified. All this kind of information can directly be substantiated by *ab initio* calculations. The great drawback of this kind of single crystalline model system is their small active surface area of about 1 cm^2 which results only in a small overall conversion. Therefore, the kinetics of the actual reaction can hardly be studied under flow conditions (neither in ultrahigh vacuum (UHV) nor in the mbar range) if the observed turn-over number is not as high as with the CO oxidation reaction. Rather, kinetic studies on single crystalline model catalysts need the use of a dedicated corrosion resistant batch reactor with infinite residence time.

To overcome the problem with the small active surface area of single crystalline model catalysts, one can use RuO$_2$ powder catalysts as shown in Figure 8.1c. The active surface area of these model catalysts is high enough (20–50 m^2 g^{-1}) for reliable kinetic studies in flow reactor experiments.[32] Extensive sets of kinetic data are available for the HCl oxidation.[33] However, the structure of RuO$_2$ powder catalysts is less well-defined than that of single crystalline surfaces. Various surface orientations with low surface energies are exposed, and the concentration of defects (steps, vacancies) on these particles are much higher than on single crystalline RuO$_2$(110) films. Assuming that the HCl oxidation is structure sensitive, density functional theory (DFT) calculations for the catalyzed reaction over RuO$_2$ powder catalysts have to be performed for various low-surface-energy RuO$_2$ facets, which in total is quite computer time consuming.[33] In addition, the low surface energy (100) and (101) orientations undergo severe reconstructions with so far unknown atomic structure[29,34] so that DFT simulations of bulk-truncated surfaces are not meaningful. Reaction intermediates seen with *in situ* experiments can hardly be identified with theoretical studies.

Besides activity, the stability of the catalyst is of major concern under such harsh reaction conditions as encountered with the HCl oxidation. The terminology 'stability' implies several aspects and needs careful consideration. Regarding practical catalysis 'stability' addresses not only chemical stability, but also morphological changes, *e.g.*, transformation of the catalyst nanostructure, grain structure, surface area, *etc.* Such morphological transformations can be unfavorable for catalysis, even if the chemical composition of the catalyst is unaffected.

For stability issues special model catalysts have to be designed with distinct morphologies in order to be able to visualize and follow structural degradation of the model catalyst after reaction with scanning electron microscopy (SEM). In stability studies we applied RuO_2-based and CeO_2-based nanofibers as model catalysts (cf. Figure 8.1d). These model systems bear the additional advantage of having large active surface areas of about 20–30 m^2 g^{-1} to conduct kinetic studies in a typical flow reactor set-up.

8.3 Synthesis of Single Crystalline RuO_2 Films for Gaining Molecular Information on Stability and Activity

In order to produce single crystalline RuO_2 films with specific surface orientations one can start from metallic Ru single crystals which are cut along particular directions. A well-ordered crystalline RuO_2 film is grown by exposing the single crystalline Ru surfaces to large amounts of molecular oxygen (say 10^6 L, at 10^{-5} mbar; 1 L (where L means Langmuir) corresponds to an exposure of 1.33×10^{-6} mbars) at temperatures in the range from 600 to 750 K. This procedure leads to the growth of a 1–2 nm thick RuO_2 film. On the Ru(0001) surface a single crystalline RuO_2 film in (110) orientation is preferentially formed (cf. Figure 8.1a),[35,36] while on the Ru(10$\bar{1}$0) a high quality RuO_2 film in (100) orientation can be grown.[37,38] Occasionally other orientations are observed on the Ru(0001) and Ru(10$\bar{1}$0) single crystal surfaces, such as (101).[39,40] Unfortunately RuO_2 films thicker than 3–5 nm can only be formed at higher surface temperatures, where the oxide surface roughens considerably, forming facets with various orientations.[40]

Thicker (and flat) single crystalline RuO_2(110) films with variable thickness of several 10 nm can be prepared on TiO_2(110) single crystals serving as templates.[41–43] Ru-carbonyls are frequently used as Ru-containing precursors and the actual oxidation process is assisted by an oxygen plasma. The adsorption and decomposition of $(Ru)_3(CO)_{12}$ over TiO_2(110) has been studied in great detail by XPS and RAIRS.[44,45] However, carbon contamination is unavoidable. At growth temperatures above 700 K Ti and Ru interdiffuse to form mixed $Ru_xTi_{2-x}O_2$ epitaxial films. This mixed oxide buffer-film can be utilized to minimize the lattice mismatch between RuO_2(110) and TiO_2(110), thereby producing flat single crystalline unstrained RuO_2(110) films of more than 20 nm thickness.[43] Very recently, RuO_2 islands and closed films epitaxially grown on TiO_2(110) could be prepared (cf. Figure 8.1b) using an electron beam induced Ru evaporation as a carbon-free Ru source.[46]

Thin films of RuO_2 can be produced either by deposition of Ru and subsequent oxidation of Ru films, or directly by deposition of RuO_2. For instance, ultra-thin highly textured Ru films in (0001) orientation can be deposited on Si(001) by magnetron sputtering[47] and subsequently be oxidized at various temperatures,[48] exposing the substrate to molecular oxygen, ozone, NO$_2$ or plasma-activated O$_2$.

Sputtering[49,50] is the most commonly used technique to deposit thin RuO$_2$ films on dissimilar substrates. O$_2$-containing plasmas are frequently utilized for Ru etching.[51] The Ru etching rate can be increased by Cl$_2$ addition to an O$_2$ plasma.[52] Magnetron plasma sputtering[53–55] or even better reactive sputtering in O$_2$ atmosphere produces RuO$_2$ films with a Ru:O film stoichiometry of 1:2.[56,57] These oxide films are in general poly-crystalline and therefore less suited as model catalysts.

Metal organic chemical vapor deposition (MOCVD) provides another technology to form thin RuO$_2$ films.[58–61] However, these films suffer sometimes from carbon contamination. With this method conductive RuO$_2$ films could be prepared either with (110)- or with (101)-textured orientations on SiO$_2$/Si(001).[62] The structural texture of the RuO$_2$ films can be controlled by both temperature and growth rate. The roughness of MOCVD-grown RuO$_2$ films can be reduced by co-deposition of iodine containing molecules.[63] Very clean and thermally stable RuO$_2$ films can be produced by chemical vapor deposition (CVD) using RuO$_4$[64–67] or hydrous-RuO$_2$[68] as the metallic precursor.

Pulsed laser deposition (PLD) is a powerful method to deposit RuO$_2$ films on dissimilar substrates such as LaAlO$_3$[69] with an epitaxial RuO$_2$ film in (100) orientation normal to the surface. RuO$_2$ films were also grown on a MgO(100) substrate using PLD.[70] High quality films on MgO(100) could be produced at sample temperatures above 870 K resulting in RuO$_2$ films that are expitaxial and (110) oriented.[71] Under oxygen deficient conditions the RuO$_2$ film grows in (101) orientation on Si.[72] Epitaxial single crystalline films of Ru(0001) and other platinum group metals can be produced on YSZ-buffered Si(111) wafers by PLD.[73] Subsequent oxidation leads to the growth of ultra thin RuO$_2$(110) films which serve as perfect single crystalline model catalysts.

Thin films of metallic Ru have been grown on thin Al$_2$O$_3$ and TiO$_2$ films by atomic layer deposition (ALD), employing bis(cyclopentadienyl)ruthenium (RuCp$_2$) and oxygen as precursors.[74] ALD film growth is self-limited and based on surface reactions, which offers the possibility to control the deposition process on the atomic scale. By keeping the precursors separate throughout the coating process, the grown film layer can be controlled as precise as ∼0.1 angstroms per monolayer. Oxygen has been used in Ru-ALD as a reactant gas for several Ru precursors. The supplied Ru precursor is adsorbed on the surface in a first pulse and reacts with the oxygen in the second pulse. The ligands of the Ru precursor are completely or partly oxidized to volatile by-products, mainly H$_2$O and CO$_2$. In the next pulse the Ru precursor is adsorbed onto the surface and again oxidized. The cyclization of this two-step process leads in a conformal layer-by-layer growth of the Ru or RuO$_2$[75,76] on a substrate with arbitrary topography.

8.4 Atomic-Scale Properties of RuO$_2$(110)

RuO$_2$ crystallizes in the rutile structure where the O atoms adopt sp^2 hybridization, while the Ru atoms are coordinated to six O atoms forming a slightly distorted octahedron (d^{2}sp^3 hybridization of Ru). The bulk-truncated RuO$_2$(110) surface exposes two kinds of under-coordinated surface atoms. These are the bridging oxygen atoms (O$_{br}$), which are coordinated to two (instead to three) Ru atoms underneath. These are the 1f-cus Ru atoms, *i.e.*, one-fold under-coordinated Ru atoms, which are coordinated to five (instead of six) O atoms (*cf.* Figure 8.2a). We may recall that RuO$_2$ is a metallic oxide with electronic conductivity half of that of Ru itself,[77] thereby facilitating both the experimental and computational work.[28]

Counting the formal charges of the Ru and O with +4 and −2, respectively, the bulk-truncated RuO$_2$(110) surface turns out to be auto compensated in that the number of electrons missing at the surface O atoms are compensated by the surplus electrons at the 1f-cus Ru sites (electron counting rule[78]). The actual charge on the Ru and O atoms in RuO$_2$ is however much smaller as estimated by a Bader analysis based on DFT calculations,[79] namely +1.74 for bulk-Ru and −0.87 for bulk-O, while 1f-cus Ru carries a charge of +1.60 and O$_{br}$ −0.80. For comparison: Assuming that each Ru–O bond polarizes 2/3 of a unit charge the formal charge of O$_{br}$ is −4/3 and that of the 1f-cus Ru is +10/3.

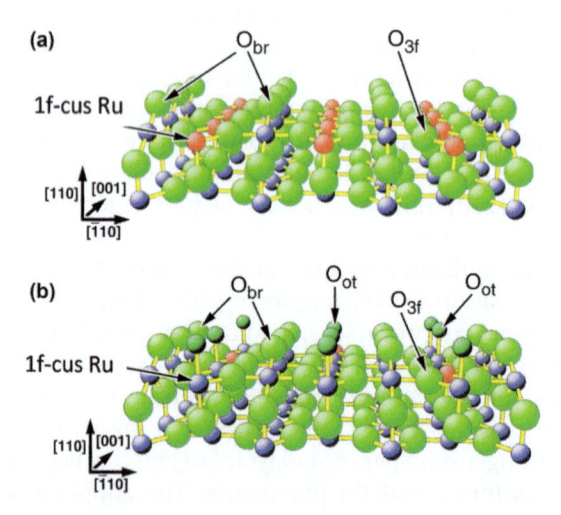

Figure 8.2 Ball and stick model of the stoichiometric RuO$_2$(110) surface (a) and the oxygen exposed RuO$_2$(110) surface (b), where most of the under-coordinated 1f-cus Ru site are occupied by on-top O (O$_{ot}$). The big green balls are the oxygen atoms, the small blue and red balls are the Ru atoms. At the stoichiometric RuO$_2$(110) surface there are two types of under-coordinated atoms, the bridging O atoms (O$_{br}$) and the 1f-cus Ru site; 1f-cus stands for one-fold coordinatively unsaturated site. Copyright © 2012 American Chemical Society.

Most of the molecules studied so far on the RuO$_2$(110) surface (CO,[80] H$_2$O,[81,82] O$_2$[83,84] N$_2$,[80] methanol,[85] CO$_2$,[86] NO,[87] ethylene,[88,89] ethane, methane[90] and NH$_3$[91,92]) adsorb from the gas phase directly above the 1f-cus Ru atoms. Therefore, the 1f-cus Ru atoms are considered to be the active sites of RuO$_2$(110), governing the interaction with the surrounding gas atmosphere.[93]

Dissociative adsorption of molecular oxygen from the gas phase leads to the formation of atomic O$_{ot}$ species in terminal position above the 1f-cus Ru atoms (cf. Figure 8.2b). O$_{ot}$ has been shown to be prone to readily pick up hydrogen from bridging O$_{br}$H groups.[82] O$_{br}$ forms two σ bonds with Ru, whereas O$_{ot}$ forms only one single σ bond. This difference in bond order is also reflected by total energy calculations: The O$_{br}$ species is by 130–150 kJ mol^{-1} stronger bound than O$_{ot}$.[83,94,95]

Another intriguing feature of the RuO$_2$(110) surface is the high diffusion barriers for the lateral migration of adsorbed species. With DFT calculations the diffusion barrier for on-top O along the 1f-cus rows is determined to be 120–160 kJ mol^{-1}.[94,96]

8.5 What can be Learnt from Single Crystalline RuO$_2$ Model Catalyst?

The first question we want to address is the interaction of the reactants HCl and O$_2$ with the stoichiometric model catalyst RuO$_2$(110). Oxygen exposure at low temperatures (100 K and below) leads to the population of molecular oxygen on the RuO$_2$(110) surface which desorbs at 140 K.[83] Using plane wave DFT calculations Wang et al.[84] found that all molecular O$_2$ species are prone to dissociate on the RuO$_2$(110) as long as there are two neighboring vacant 1f-cus Ru sites available. Only when isolated free 1f-cus Ru sites are generated by dissociative oxygen adsorption, molecular O$_2$ can be stabilized in this vacancy by 40 kJ mol^{-1}.

Oxygen exposure at room temperature to the stoichiometric RuO$_2$(110) surface leads to the population of the atomic oxygen adsorbed on-top of the 1f-cus Ru atoms (cf. O$_{ot}$ in Figure 8.2b). Exposure of 5 L of O$_2$ at room temperature saturates 86% of the 1f-cus Ru atoms by on-top O atoms (O$_{ot}$) if diffusion is ignored.[79,83] Exposing the RuO$_2$(110) surface to molecular oxygen at 500K and cooling the sample to room temperature in an oxygen atmosphere of 10^{-7} mbar leads however to a surface where practically all 1f-cus Ru sites are occupied by on-top O.[97] The initial dissociative sticking coefficient of O$_2$ on the stroichiometric RuO$_2$(110) is 0.7, i.e., 70% of the impinging O$_2$ molecules adsorb dissociatively on the surface, requiring two neighboring vacant 1f-cus Ru sites.[98]

On the basis of high resolution core level shift spectroscopy (HRCLS) experiments the stability of the model catalyst RuO$_2$(110) in the Sumitomo process has been attributed to the selective replacement of bridging oxygen (O$_{br}$, cf. Figure 8.3a) by chlorine, a process which is confined to the top-most

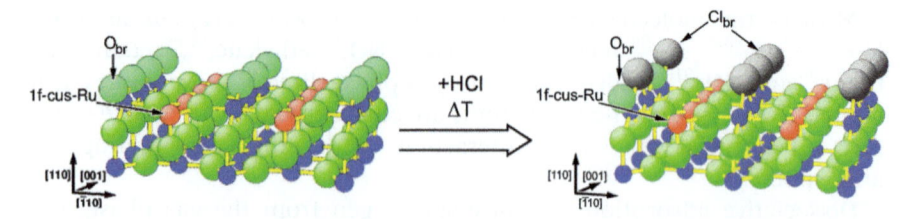

Figure 8.3 Schematic representation of the chlorination of RuO$_2$(110). The oxygen atoms are indicated as green balls and the Ru atoms as blue (bulk-coordinated) and red (under-coordinated) balls. Ball-and-stick model of bulk truncated RuO$_2$(110) (left) revealing under-coordinated surface atoms: bridging O atoms (O$_{br}$) and one-fold coordinatively unsaturated Ru sites (Ru 1f-cus: red balls). Upon HCl exposure at higher temperatures the stoichiometric surface transforms into a chlorinated surface (right) where the bridging O$_{br}$ atoms are partly replaced by bridging chlorine (Cl$_{br}$) atoms (shown in gray color).[7,101,102]
Reprinted with permission from ref. 102. Copyright © 2010 American Chemical Society.

layer of RuO$_2$(110). A deeper reduction/chlorination of RuO$_2$(110) has not been observed under ultra-high vacuum (UHV)-typical conditions[7] nor at higher pressures in the mbar range.[99,100] The maximum surface chlorination of RuO$_2$ attained has been estimated to be 70–80%.[101]

Surface chlorination of RuO$_2$(110) proceeds *via* a multi-step process. The adsorption energy of molecular HCl on RuO$_2$(110) is only 40–60 kJ mol^{-1}. Therefore, above $T = 200$ K HCl can only be stabilized on the catalyst's surface by dissociative adsorption in that the Cl binds to 1f-cus Ru forming Cl$_{ot}$ and hydrogen is transferred to under-coordinated bridging O atoms forming bridging (O$_{br}$H) hydroxyl groups. The dissociative adsorption of HCl takes place on RuO$_2$(110) without activation barrier and is exothermic by 130 kJ mol^{-1}. *Via* a second hydrogen transfer by dissociative HCl adsorption O$_{br}$H is transformed to bridging water. Finally, bridging water is replaced (either concerted or sequentially) by chlorine, a process which is activated by 140 kJ mol^{-1} as determined by DFT calculations.[101,102]

In situ surface X-ray diffraction (SXRD)[100] revealed that chlorinated RuO$_2$(110) and RuO$_2$(100) model catalysts are long-term stable under reaction conditions where the gas feed p(HCl):p(O$_2$) was varied from 1 : 4 to 4 : 1 for pressures in the mbar range and temperatures as high as 850 K. Even pure HCl exposure in the mbar range is not able to chemically reduce RuO$_2$ below 600 K, since the bridging oxygen positions are mainly populated by chlorine, and on-top adsorbed chlorine blocks part of the under coordinated Ru sites. Without the presence of under-coordinated surface oxygen further HCl uptake is suppressed due to missing partners for the H-transfer. Therefore, the RuO$_2$(110) surface is resistant against pure HCl exposure in the mbar range as long as the temperature is below 600 K. Above 650 K chemical reduction of RuO$_2$(110) to metallic Ru sets in. Substitution of a bridging oxygen atom by a bridging chlorine (Cl$_{br}$) is associated with a

Preparation and Characterization of Model Catalysts for the HCl Oxidation Reaction 207

change of the formal oxidation state of the underlying 2f-cus Ru atoms from +IV 2/3 to +III 2/3. This change in the oxidation state upon chlorination (being closer to +IV) suggests that Cl exerts a stabilizing influence on the RuO$_2$(110) surface.

The surface reaction of HCl oxidation over chlorinated RuO$_2$(110) is summarized in Figure 8.4 in the form of a catalytic cycle. Both O$_2$ and HCl adsorb dissociatively on the partly chlorinated RuO$_2$(110) surface, where part of the bridging O atoms are replaced by bridging Cl atoms (Cl$_{br}$).[7] The adsorption energy of oxygen is 200 kJ mol^{-1} with respect to O$_2$ molecules in the gas phase forming O$_{ot}$ species.[94] The adsorption energy of the first HCl is 125 kJ mol^{-1}, where hydrogen from HCl is directly transferred to the

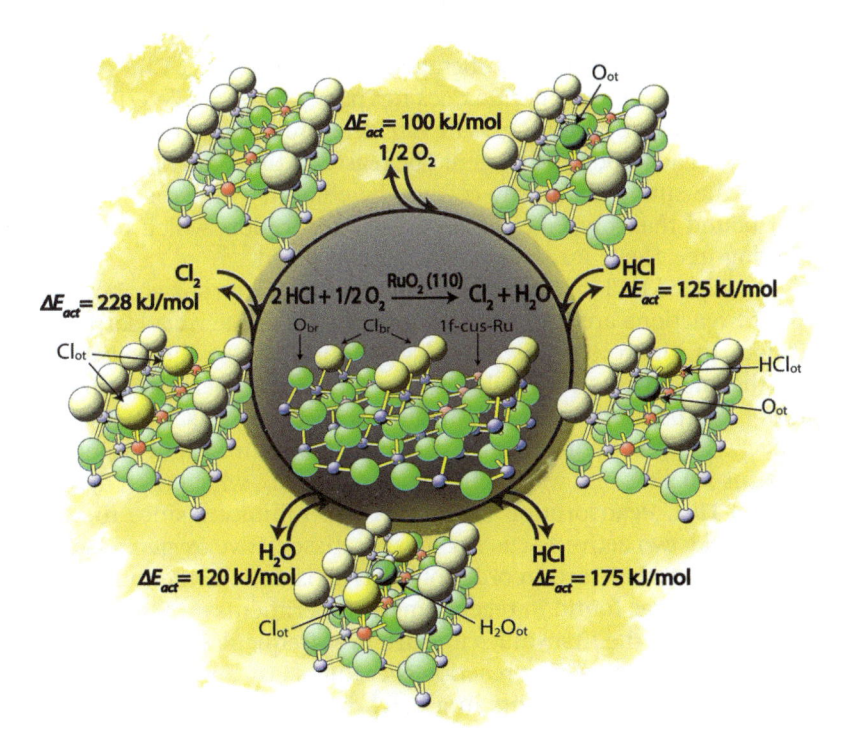

Figure 8.4 In the HCl oxidation reaction (the Deacon process) over RuO$_2$(110) both reactants, O$_2$ and HCl, adsorb dissociatively. Subsequently, surface oxygen is reduced to the by-product water by H stemming from dissociative HCl adsorption. Water desorbs around 420 K, and the remaining adsorbed chlorine atoms can recombine to form the desired product Cl$_2$ at temperatures around 600 K. Besides surface reactions, the Deacon process is governed by the adsorption/desorption equilibria of Cl$_2$, O$_2$, H$_2$O, and HCl gases with the catalyst's surface, leading to reaction inhibitions for the cases of H$_2$O, Cl$_2$ and to promotion for the case of O$_2$ and HCl. All energies given are calculated by DFT calculations.[8,103] ΔE defines the adsorption energies which determine the dynamic adsorption/desorption equilibrium between the surface and the gas phase. Copyright © 2012 American Chemical Society.

adjacent O_{ot} during the dissociative adsorption process, while Cl_{ot} is bound on top to 1f-cus Ru. The second HCl molecule dissociates, transferring its H atoms to the $O_{ot}H$ species, thereby forming water $O_{ot}H_2$. The adsorption energy of the second HCl molecule is even 175 kJ mol^{-1}. Alternatively, HCl adsorption can form a second $O_{ot}H$ species so that a hydrogen transfer between neighboring $O_{ot}H$ species may form water. This process is slightly activated by 30 kJ mol^{-1}.[82] In any case, the formed water molecule $O_{ot}H_2$ is bound by 120 kJ mol^{-1} to the surface so that it desorbs around 420 K from the surface.[81]

At elevated temperatures (about 600 K) the remaining surface Cl_{ot} species can recombine to form Cl_2. This process is activated by 228 kJ mol^{-1} and constitutes the elementary step with the highest activation barrier encountered in this catalytic cycle. Accordingly the catalyst temperature must be chosen higher than 600 K to be able to liberate the desired product Cl_2.[33,100,103] Yet, the association of surface chlorine does not present the rate determining step under typical reaction conditions. The rate determining step represents rather the dissociative adsorption of oxygen! The reason is that under typical reaction conditions the surface is mostly covered by chlorine (which is substantially stronger bound than oxygen), so that oxygen adsorption, which requires two neighboring free 1f-cus Ru sites, is efficiently blocked.[83,97]

In the catalyzed HCl oxidation over $RuO_2(110)$, which runs at about 600 K, the surface reactions are intimately coupled to gas phase *via* adsorption and desorption. In Figure 8.4 this interplay is illustrated by the yellow clouds around the model catalyst/catalytic cycle. The experimentally determined desorption temperatures of water, oxygen, or HCl are 420 K, 400 K or 550 K, respectively. Readsorption of water inhibits the reaction since water blocks active 1f-cus Ru sites and water can transfer H to Cl_{ot} which then desorbs in the form of HCl. Readsorption of Cl_2 acts also inhibiting since the Cl_2 adsorption blocks two active 1f-cus Ru sites for dissociative oxygen adsorption. Quite in contrast, the reaction order of the HCl oxidation reaction is positive with respect to O_2 as found in recent kinetic experiments on RuO_2 powder.[32]

Only recently,[100] the first values for turn-over frequencies (TOF) have been reported for the HCl oxidation reaction over $RuO_2(110)$ and $RuO_2(100)$ model catalysts. Reactivity experiments in a batch reactor indicate that independent of the used surface orientation 0.6 Cl_2 molecules are produced per second and active site at 650 K (*i.e.*, TOF = 0.6 s^{-1}), when starting with a reaction mixture of P(HCl) = 2 mbar and P(O_2) = 0.5 mbar. Therefore, the HCl oxidation maybe considered as being not structure sensitive. A similar result was reported for the CO oxidation.[29]

8.6 Synthesis of Metal Oxide Fibers *via* Electrospinning

Electrospinning is a versatile method to prepare fibers with diameters in the range of typically several hundred nanometers. Fundamental requirements

Preparation and Characterization of Model Catalysts for the HCl Oxidation Reaction 209

for electrospinning are the solubility of the desired material (or its precursor(s)) and a sufficiently high viscosity of the solution to be electrospun (typically in the range of 0.1–2 Pa s).[104,105] The first electrospun materials were polymers due to the fact that already moderate polymer concentrations give solutions of high viscosity (Figure 8.5a). The electrospinning process itself and the underlying physical processes were studied quite extensively on simple polymer solutions. For a general introduction to the electrospinning technique the reader is referred to the literature.[106–109]

Anisotropic oxidic nanostructures such as nanotubes, nanorods, or nanofibers are very promising for diverse applications including model

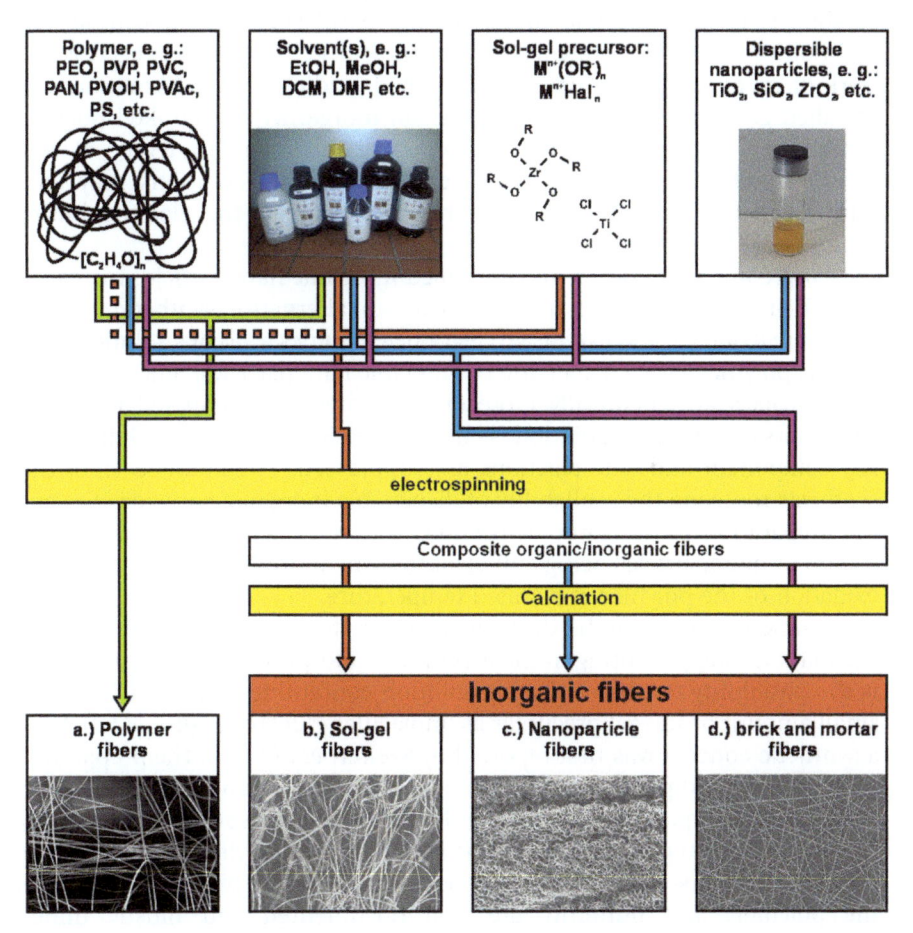

Figure 8.5 Synthetic concepts for the preparation of electrospun fibers: a.) and b.): "conventional" synthetic concepts; c.) and d.): "new" synthetic concepts. a.) preparation of polymeric fibers; b.) preparation of inorganic fibers *via* the sol–gel route (with or without the addition of organic polymers); c.) preparation of inorganic fibers *via* the nanoparticle approach; d.) preparation of inorganic fibers *via* the 'bricks and mortar' approach.

catalysis.[110–113] Inorganic nanofibers can be synthesized by electrospinning *via* different synthetic concepts (see Figure 8.5). The first possibility is to add some kind of inorganic precursor, typically a metal halide or a metal alkoxide, to a spinable polymer solution (*cf.* Figure 8.5b). *Via* sol–gel reaction these precursors can be chemically transformed to the corresponding metal oxide. This sol–gel reaction may already start during the spinning process and will be completed after thermal treatment of the electrospun composite fibers at elevated temperature (depending on the desired metal oxide material). The thermal treatment is necessary in any case to remove the spinning polymer to obtain 'pure' metal oxide fibers.

As sol–gel solutions may become very viscous during gelation of the precursors, it is also possible to electrospin a sol–gel solution without the addition of an organic polymer. When the sol–gel precursors react, a network of metal–oxygen bonds is established leading to the formation of a so-called gel that can be considered as an inorganic, cross-linked polymer exhibiting a high viscosity in solution. Depending on the degree of condensation of the precursors the viscosity of sol–gel solutions may alter during the reaction which in turn negatively affects the electrospinning process and finally prevents it.

To overcome these problems preformed inorganic nanoparticles instead of sol–gel precursors can be introduced as an alternative synthetic concept (*cf.* Figure 8.5c).[114,115] With this strategy one combines a viscous solution of an organic polymer with a dispersion of inorganic nanoparticles. Since inorganic nanoparticles are chemically more or less unreactive compared to sol–gel precursors the spinning solution is chemical stable as long as precipitation of any component and side reactions are suppressed. The bottlenecks of this synthesis route are the availability of dispersible inorganic nanoparticles and the preparation of inorganic dispersions with a certain 'threshold' concentration of nanoparticles being sufficient to form an inorganic fiber after total oxidation of the polymer at elevated temperature.

In some cases it may be difficult to sinter the nanoparticles during the calcinations step, resulting in the loss of fiber morphology. In this case there is the possibility to use a spinning solution containing nanoparticles and a small amount of sol–gel precursor as 'molecular glue' (*cf.* Figure. 8.5d). Such a synthetic concept was first reported by Szeifert *et al.*[116] for the preparation of thin films *via* dip-coating and is known as the 'bricks and mortar' concept.

While the preparation and dependencies of electrospun polymer nanofibers on physical parameters such as solution conductivity, surface tension, viscosity, solvents vapor pressure, solvent composition, molecular weight of the polymer, and humidity are well documented and mostly understood,[106,108,109,117,118] the preparation process of inorganic fibers suffers from a much lower degree of mechanistic understanding. In addition to the given parameters influencing the electrospinning of polymer fibers there are further spinning parameters to be considered for chemically more complex solutions containing inorganic materials or precursors (*i.e.*, sol–gel precursor(s) and/or inorganic nanoparticles) and polymers. The impact of the aforementioned parameters on the spinning process of a sol–gel solution may differ from that of a polymer solution as revealed for the case of dip-coating: Humidity exerts a strong impact on the hydrolysis and condensation of a sol–gel solution, leading to varying viscosity that influences the structure formation during the so-called evaporation-induced self-assembly process (EISA-Process) in dip-coating.[119,120] As evaporation of the solvent during electrospinning takes place on a much shorter timescale, i.e., in sub-seconds,[117,121] the EISA process of dip-coating may not be directly transferable to electrospinning so that a structure-directing model for electrospinning of sol–gel solutions or solutions containing inorganic nanoparticles has not been firmly established so far. If one aims to develop a model describing the structural evolution of electrospun inorganic fibers, the following parameters have to be carefully considered:

- The influence of temperature and humidity on the sol–gel reaction during electrospinning;
- The evolution of the ionic conductivity of the spinning solution during hydrolysis and condensation;
- The phase behavior of the spinning system consisting of polymer, (reacting) sol–gel precursor(s) and evaporating solvent(s) that may lead to phase separation.

All these parameters probably have a strong influence on the emerging fiber morphologies, e.g., the evolution of porous vs. dense fibers or the occurrence of so-called bead structures. To investigate the influences of the above mentioned parameters in detail, it is necessary to simplify the spinning system as far as possible so that the dependence on a single parameter can be examined. In this sense the use of nanoparticles can be beneficial, because the fiber morphology is controlled only by *physical* parameters during the electrospinning such as the solution viscosity, the phase behavior, or the solvent evaporation, while the structure forming process of the sol–gel approach is dominated by sol–gel chemistry (*cf.* Figure 8.5b and c). The particle concentration in the spinning solution should be of the same order of magnitude as the polymer concentration but at least one third of the polymer concentration.

8.7 Electrospun RuO$_2$-based Fibers as Model Catalysts in the HCl Oxidation Reaction

RuO$_2$ fibers can be prepared by electrospinning a polymeric solution of Poly(vinylacetate) (PVAc) or Poly(vinylpyrrolidon) (PVP) containing sol–gel precursors like Ru(acac)$_3$ or RuCl$_3 \cdot$ H$_2$O followed by calcination at elevated temperatures sufficient to remove the polymer completely.[122,123]

For our own experiments we prepared RuO$_2$ or Ru$_x$Ti$_{1-x}$O$_2$ mixed fibers by the following procedure:[124] A solution of 50–90 mg (0.22–0.4 mmol) RuCl$_3 \cdot$ H$_2$O was refluxed in 200 mg dimethylformamide (DMF) at 125 °C to produce a Ru–DMF complex and HCl gas was expelled. By adding 20 mg acetic acid before refluxing, it was possible to increase the release of HCl. Then 300 mg ethanol and 200 mg of 12 wt% polyvinylpyrrolidone in methanol were added and mixed vigorously. For the solution to be used for electrospinning, ethanol was used possessing analysis grade. After calcination at 475 °C for 30 min. (ramp of 5 °C min^{-1}) hollow nanotubes were observed, especially if low amounts of precursor were used. Adding 25 mg H$_2$O or using polyvinylbutyral (PVB, Mowital® B60 H) as spinning polymer resulted in compact nanofibers instead of nanotubes. The electrospinning parameters were as follows: voltage: 5 kV; distance: 6 cm; feed rate: 0.3 mL h^{-1}; relative humidity: <25%

The Ru$_x$Ti$_{1-x}$O$_2$ nanofibers could be prepared by adding the appropriate amount of Ti(OiPr)$_4$ after refluxing. In case of $x<0.2$ it was also possible to use RuCl$_3 \cdot$ H$_2$O as precursor without refluxing.

In order to visualize reaction-induced morphological changes of the catalysts after the HCl oxidation reaction we used RuO$_2$-based nanofibers which were synthesized by electrospinning as described above. The crystallite size of the RuO$_2$ nanofibers is about 9 nm according to Rietveld analysis with a BET surface area of 30 m^2 g^{-1}. Reaction-induced alteration of the fiber morphology becomes quite easily visible in scanning electron microscopy (SEM). In Figure 8.6a and b we show how the morphology of pure RuO$_2$ nanofibers degrades after HCl oxidation reaction in a flow reactor under oxidizing feed gas composition (p(HCl) = p(O$_2$) = 200 mbar using a buffer gas p(Ar) = 600 mbar; the total flow rate of the reaction mixture was 50 mL min^{-1} (STP)) keeping the catalyst bed at 650 K for 2 h on stream. Clearly, pure RuO$_2$ is structurally not stable under typical Deacon conditions which may lead also to catalyst loss under industrial conditions. Temperature treatment alone in Ar + O$_2$ flow does not lead to sintering at temperatures up to 750 K. Consequently, the observed transport of Ru even at temperatures as low as 573 K suggests the participation of a volatile Ru species formed during the HCl oxidation reaction. We presume that a chlorine-containing Ru species serves as chemical transporter. The BET surface area remained nearly constant which is in line with the almost constant reactivity of the fibers during the atmospheric pressure HCl oxidation at 573 K. Obviously, activity experiments alone are not able to identify morphological instabilities of a catalyst.

This result conflicts apparently with the observed stability of RuO$_2$(110) and RuO$_2$(100) model catalysts. Therefore, the morphology changes of RuO$_2$ fibers suggest that only RuO$_2$ facets with orientations other than (110) and (100) are corroded by HCl. Quite in contrast, mixed RuO$_2$–TiO$_2$ fibers are morphologically stable under the same HCl oxidation reaction conditions (*cf.* Figure 8.6c and d).

The Ru-doped TiO$_2$ nanofibers were prepared with 15 at.% Ru, resulting in phase-pure rutile mixed RuO$_2$–TiO$_2$ fibers (Rietveld analysis). The normalized catalytic activity of mixed RuO$_2$–TiO$_2$ nanofibers is as high as that of pure RuO$_2$ nanofibers.[13] As shown in Figure 8.6b and d, mixed RuO$_2$–TiO$_2$

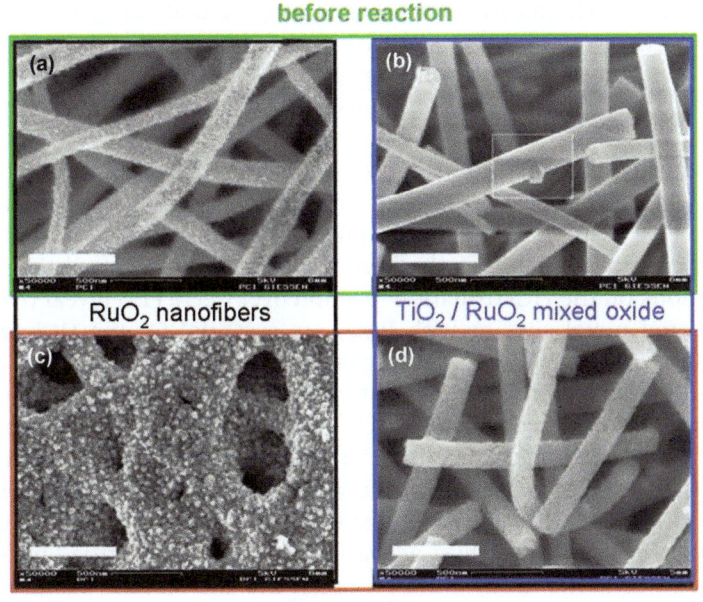

Preparation and Characterization of Model Catalysts for the HCl Oxidation Reaction 213

before reaction

(a) RuO$_2$ nanofibers

(b) TiO$_2$ / RuO$_2$ mixed oxide

(c)

(d)

after reaction

Figure 8.6 High-resolution SEM images (the white bars correspond to 500 nm): RuO$_2$ and TiO$_2$–RuO$_2$ mixed nanofibers before and after HCl oxidation reaction in a flow reactor under oxidizing feed gas composition (p(HCl) = p(O$_2$) = 200 mbar; buffer gas p(Ar) = 600 mbar; total flow rate 50 mL min^{-1} (STP)) at 650 K for 2 h. Clearly the RuO$_2$ nanofibers disintegrate while the TiO$_2$–RuO$_2$ mixed fibers are stable under such harsh reaction conditions.[13]

nanofibers are much more stable than pure RuO$_2$ fibers. Stability improvement of RuO$_2$ by titanium has been observed on single crystal surfaces of RuO$_2$ as well, namely on RuO$_2$(110) and RuO$_2$(100).[125] It turned out that Ti deposition substantially improved the thermal stability of both orientations of RuO$_2$ in UHV which is fully reconciled with the found stabilization of RuO$_2$–TiO$_2$ nanofibers. To understand this stabilization effect of Ti we have to recall the atomic scale reduction process on RuO$_2$(110). Chemical reduction and also the disintegration of RuO$_2$(110) starts off with the removal of bridging O by water formation.[126] In order to stabilize the RuO$_2$(110) surface it helps to replace the Ru atoms coordinated to O$_{br}$ by Ti. DFT calculations[8] indeed have shown that that O$_{br}$ and Cl$_{br}$ bind by 20 kJ mol^{-1} stronger to Ti than to Ru.

8.8 CeO$_2$-based Catalysts: Alternatives to RuO$_2$-based Deacon Catalysts

While industrial and model-like RuO$_2$-based catalysts will certainly keep playing a key role in industrial and model catalysis, the scarcity of Ru pushes the search for alternative metal oxides as potential catalysts for the HCl oxidation. CeO_2 represents a promising alternative to RuO_2 for several reasons.[127] In addition to its abundance, CeO_2 exhibits reversible oxidation/reduction between the oxidation states Ce^{3+} and Ce^{4+}, thus being of interest for any catalytic redox reaction.

The mechanism of HCl oxidation over CeO_2 has been studied by DFT calculations over $CeO_2(111)$, the most stable surface termination of $CeO_2(111)$.[127] Experimental studies on single crystalline CeO_2 films or CeO_2 crystals are not available in the literature. Density functional theory (DFT) calculations on the HCl oxidation reaction over $CeO_2(111)$ reveal that under-coordinated metal sites as with RuO_2 are essential for the HCl oxidation. While those metal sites are openly available on $RuO_2(110)$, these sites must be created first on $CeO_2(111)$ by the formation of oxygen vacancies. Here, the high oxygen storage capacity of CeO_2 may come beneficially into play. Into these vacancies HCl adsorbs dissociatively transferring the hydrogen to adjacent surface oxygen sites. A second HCl molecule adsorbs above a surface Ce site forming water and a second surface chlorine species. The elementary reaction step with the highest activation energy was identified with the activation of Cl atoms from lattice vacancies to the surface positions where Cl–Cl association occurs.

In the following we will focus on the morphological stability of CeO_2-based catalysts. Nanofibers of CeO_2 can be of particular interest in this context, because the stability or morphological changes upon catalytic reactions can be easily visualized by microscopic methods.

The procedures for electrospinning of CeO_2, $Zr_{1-x}Ce_xO_2$, and ZrO_2 nanofibers had been reported recently.[128–132] CeO_2 fibers can be prepared using $CeCl_3 \cdot 7H_2O$ as the molecular precursor. Electrospinning was conducted at the electric field of 1.2 kV cm^{-1}. The electrospun fibers were placed in a preheated oven (400 K) and dried for 20 min. Subsequently, this material was heated to 973 K at the rate of 3.8 K min^{-1} and kept for 1 h at 973 K, yielding grayish-colored mats of CeO_2 fibers.

The Deacon reaction was performed in a fixed bed flow reactor with 25 mg of CeO_2 nanofibers, a reaction mixture of $HCl/O_2/Ar = 2/2/3$ at a flow rate of 15 sccm at $T = 430$ °C for 3 h on stream. Only for the strongly reducing reaction conditions, the XRD data in Figure 8.7 indicate the partial transformation of CeO_2 to $CeCl_3$. This result itself is quite interesting, demonstrating that a relevant portion of CeO_2 is finally, not only temporarily, transformed into chlorides. However, such data do not provide any microscopic insights into the transformation process itself, for instance the location of the chloride formation and the concomitant morphological changes. By contrast, the study of CeO_2 nanofibers before and after HCl oxidation reaction (*cf.* Figure 8.7) clearly indicates significant corrosion of CeO_2 fibers, as seen by the largely increased roughness of the fibers under identical conditions of the electron microscopy experiment. Even more interestingly, additional small crystallites can be observed to grow orthogonally from the fiber surface, which may be correlated to the formation of cerium chlorides and cerium

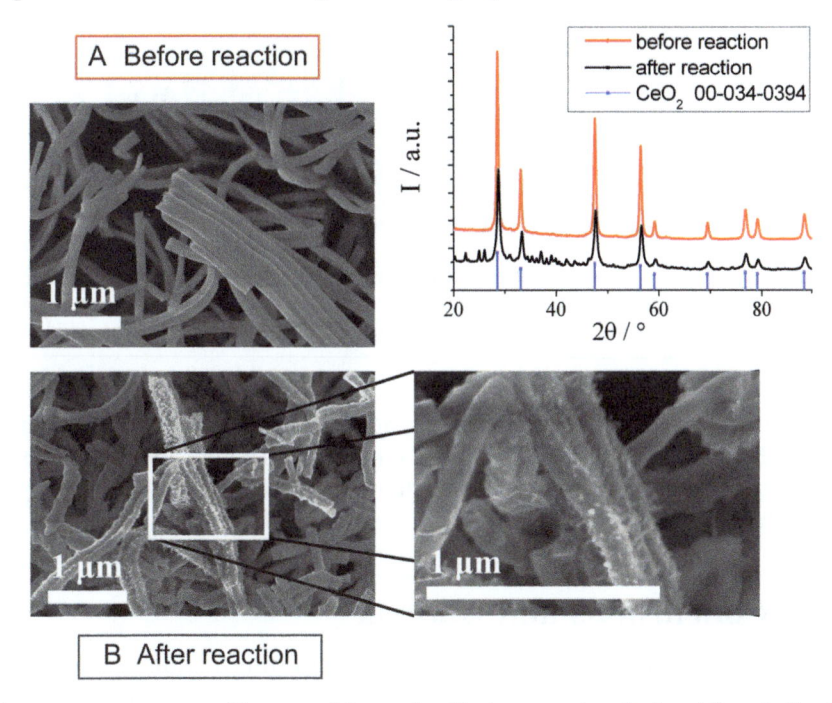

Figure 8.7 CeO$_2$ nanofibers used for HCl oxidation reaction, before (A) and after (B) reaction.[135] A reaction mixture of HCl/O$_2$/Ar = 2/2/3 at a flow rate of 15 mL min^{-1} (STP) at T = 703 K for 3 h on stream was used.

oxychlorides. The degradation of CeO$_2$ does not change the BET surface area, as determined by nitrogen physisorption measurements (T = 77 K) performed on the material before and after the reaction (S_{BET} = 40 m^2 g^{-1}). These results are consistent with recently reported ones for the HCl oxidation over CeO$_2$ powder catalyst.[127]

In addition to HCl oxidation experiments with pure CeO$_2$ nanofiber catalysts, mixed CeO$_2$–ZrO$_2$ fibers were also investigated (cf. Figure 8.8). For the synthesis of mixed CeO$_2$–ZrO$_2$ fibers, CeCl$_3$·7H$_2$O and ZrOCl$_2$·8H$_2$O were mixed and then processed analogously as for the pure CeO$_2$ material. The final mat of Zr$_{0.2}$Ce$_{0.8}$O$_2$ fibers exhibits pale yellow color. The mixed oxide fibers consist of nanograins (typically 5–28 nm in diameter as determined by Debye–Scherrer equation).

Detailed experiments show that Zr$_{1-x}$Ce$_x$O$_{2-\delta}$ fibers are more robust against chemical corrosion than CeO$_2$ fibers.[135] From Figure 8.8 it can be seen that the morphology of the Zr$_{0.20}$Ce$_{0.80}$O$_2$ nanofibers does not change upon HCl oxidation reaction. This conclusion is supported by XRD as no CeCl$_3$, hydrous or otherwise, shows up as reflections in the XRD data. The Zr$_{0.20}$Ce$_{0.80}$O$_2$ nanofibers form a single cubic/tetragonal solid solution and this structure is not affected by the HCl oxidation reaction. In comparison with the pure CeO$_2$ nanofibers, the nanograins of Zr$_{0.20}$Ce$_{0.80}$O$_2$ are

Figure 8.8 $Zr_{0.20}Ce_{0.80}O_2$ nanofibers used for a HCl oxidation reaction, before (A) and after (B) reaction. A reaction mixture: $HCl/O_2/Ar = 2/2/3$ at a flow of 15 mL min^{-1} (STP) at $T = 703$ K for 60 hours on stream was used.[135]

substantially smaller as revealed by broader reflections in the XRD data.[135] The BET surface area of the $Zr_{0.20}Ce_{0.80}O_2$ nanofibers before and after reaction is about 30 m^2 g^{-1}. Obviously, no substantial sintering took place during the reaction.

8.9 Conclusions and Outlook

Model catalysis is the approach to gain in-depth information of a catalyzed reaction such as the harsh HCl oxidation reaction (Deacon process). With surface sensitive experiments on single crystalline $RuO_2(110)$ and $RuO_2(100)$ films we have gained molecular level understanding of the reaction mechanism and on the surface chlorination process. Analogous experiments on single crystalline CeO_2 surfaces (such as CeO(111) are completely missing.

With the use of nanofibers, *i.e.*, model catalyst with well-defined cylindrical morphology, morphological changes caused by the HCl oxidation reaction can be studied and followed. We could demonstrate that the addition of Ti is able to stabilize RuO_2-based catalysts, while the addition of Zr is able to stabilize CeO_2-based catalysts.

However, whilst nanofibrous CeO_2-based materials are quite ideal model systems to study stability issues, they are not that suitable as model to study strategies to combine enhanced stability with optimized activity of nanostructured CeO_2 or $Zr_{1-x}Ce_xO_{2-\delta}$, as the surface area is only of the order of *ca.* 20–30 m^2 g^{-1}. Here, CeO_2[133] and $Zr_{1-x}Ce_xO_{2-\delta}$[134] in the form of mesoporous architectures appears as a promising alternative, featuring surface areas of up to 250 m^2 g^{-1} and possessing a well-defined porosity in terms of the pore shape and pore size distribution. These materials have been reported recently in the form of both, thin films and powders, with pore sizes of *ca.* 10–20 nm and quite regular mesoscopic order.[133,134,136]

Aside from crystalline nanoparticles,[133] CeO_2 was synthesized in the form of mesoporous powders and films.[134,136–139] Only very few other crystalline oxides (most notably TiO_2) have been prepared in such a diversity of nanostructures and morphologies. This flexibility in morphological control using sol–gel chemistry thus enables the synthesis of basically any nanostructure and texture. Moreover, also solid solutions with Zr ($Zr_{1-x}Ce_xO_2$), possessing higher stability than pure CeO_2, can be structured on the nanometer scale using sol–gel methods, providing similar morphological diversity.[133,136]

The use of electrospun nanofibers in stability studies is versatile and not restricted to HCl oxidation as it can easily be extended to other types of corrosive reactions such as the NH_3 oxidation reaction (Ostwald process). In order to gain deeper insights into the underlying microscopic processes of reaction induced corrosion molecular modeling such as kinetic Monte Carlo methods or molecular dynamics is required. Transport phenomena have extensively been studied for the Ostwald process over Rh-stabilized Pt gauzes.[140–142] These semi-empirical models may serve as a starting point in future studies to comprehend on the morphological changes of Pt and Rh-doped Pt nanowires prepared by electrospinning.

One of the major problems faced in photocatalysis is photostability. Also, here, photocatalysis research may benefit from the fibrous morphology of electrospun model photocatalysts to learn more about the photodegradation.[143,144] Stability is also a major concern in electrocatalysis such as the oxygen evolution reaction (OER),[145] oxygen reduction reaction (ORR)[146,147] or the chlorine evolution reaction (CER).[148] Novel model electrodes can be prepared from active nanofibers in the form of mats which can then be beneficially applied for stability investigations.

Electrospun nanofibers can also be used as templates (carrier) which are further functionalized by impregnation with an active component or other cofactors. Not only corrosion but also other types of degradation could be studied by nanofibrous model catalysts such as coking and carbide formation of the catalysts in *e.g.*, the Fischer–Tropsch synthesis.

Therefore we are confident that the use of electrospun nanofibers as model catalysts is a quite general and versatile approach to study the stability of catalysts in heterogeneous catalysis, electrocatalysis and photocatalysis under typical reaction conditions.

Acknowledgements

This Deacon project was supported by the German Science council (DFG) and the federal ministry of science and education (BMBF_Deacon: 033R018C). We acknowledge financial support within the LOEWE program of excellence of the Federal State of Hassia (project initiative STORE-E). This project was supported by the Laboratory of Materials Research (LaMa) of the Justus-Liebig University. All our co-workers in this project are acknowledged who have contributed to this exciting project: Dr Daniela Crihan, Dr Marcus Knapp, Dr Stefan Rohrlack, Dr Jan-Philipp Hofmann, Dr Nikolay Tarabanko, Franziska Hess, Dr Attila Farkas, Sven Urban, Katarzyna Zalewska, Dr Ari Seitsonen, and Dr Rainer Ostermann.

References

1. www.eurochlor.org (retrieved 20. Okt. 2013).
2. H. Deacon, *U.S. patent*, 1868, **85**, 370.
3. H. Y. Pan, R. G. Minet, S. W. Benson and T. T. Tsotsis, *Ind. Eng. Chem. Res.*, 1994, **33**, 2996.
4. J. Perez-Ramirez, C. Mondelli, T. Schmidt, O. F.-K. Schlüter, A. Wolf, L. Mleczko and T. Dreier, *Energy Environ. Sci.*, 2011, **4**, 4786.
5. K. Iwanaga, K. Seki, T. Hibi, K. Issoh, T. Suzuta, M. Nakada, Y. Mori and T. Abe, *Sumitomo Kagaku*, 2004, **I**, 1–11.
6. K. Seki, *Catal. Surv. Asia*, 2010, **14**, 168.
7. D. Crihan, M. Knapp, S. Zweidinger, E. Lundgren, C. J. Weststrade, J. N. Andersen, A. P. Seitsonen and H. Over, *Angew. Chem., Int. Ed.*, 2008, **47**, 2131.
8. H. Over, *J. Phys. Chem. C*, 2012, **116**, 6779.
9. M. Hammes, M. Valtchev, M. B. Roth, K. Stöwe and W. F. Maier, *Appl. Catal., B*, 2013, **132–133**, 389.
10. A. P. Amrute, C. Mondelli, M. Moser, G. Novell-Leruth, N. Lopez, D. Rosenthal, R. Farra, M. E. Schuster, D. Schuster, T. Schmidt and J. Perez-Ramirez, *J. Catal.*, 2012, **286**, 287.
11. R. Farra, M. Eichelbaum, R. Schlögl, L. Szentmiklosi, T. Schmidt, A. P. Amrute, C. Mondelli, J. Perez-Ramirez and D. Teschner, *J. Catal.*, 2013, **297**, 119.
12. M. Moser, C. Mondelli, T. Schmidt, F. Girgsdies, M. E. Schuster, R. Farra, L. Szentmikolosi, D. Teschner and J. Perez-Ramirez, *Appl. Catal., B*, 2013, **132–133**, 123.
13. Ch. Kanzler, S. Urban, K. Zalewska-Wierzbicka, F. Hess, S. F. Rohrlack, C. Wessel, R. Ostermann, J. P. Hofmann, B. M. Smarsly and H. Over, *ChemCatChem*, 2013, **5**, 2621.
14. A. Trovarelli, *Catal. Rev.: Sci. Eng.*, 1996, **38**, 439.
15. M. Ozawa, M. Kimura, H. Sobukawa and K. Yokota, *J. Alloys Compd.*, 1993, **193**, 43.
16. R. di Monte and J. Kasper, *Catal. Today*, 2005, **100**, 27.
17. H.-F. Wang, H. Y. Li, X. Q. Ging, Y. L. Guo, G.-Z. Lu and P. Hu, *Phys. Chem. Chem. Phys.*, 2012, **14**, 16521.
18. D. Duprez and C. Descorme, in *Catalysis by Ceria and Related Materials*, ed. A. Trovarelli, Imperial College, vol. 2, 2002, p. 243.
19. *Handbook of Heterogeneous Catalysis*, ed. G. Ertl, H. Knözinger, F. Schüth and J. Weitkamp, VCH, Weinheim, 2008.
20. I. E. Wachs, *Catal. Today*, 2005, **100**, 79.
21. H. J. Freund, *Chem. – Eur. J.*, 2010, **16**, 9384.
22. F. Gao and D. W. Goodman, *Annu. Rev. Phys. Chem.*, 2012, **63**, 265.
23. G. Ertl, *Angew. Chem., Int. Ed.*, 2008, **47**, 3524.
24. G. A. Somorjai and Y. Li, *Proc. Natl. Acad. Sci. U. S. A.*, 2011, **108**, 918.
25. R. Imbihl, R. J. Behmn and R. Schlögl, *Phys. Chem. Chem. Phys.*, 2007, **9**, 3459.

26. H. J. Freund, G. Meijer, M. Scheffler, R. Schlogl and M. Wolf, *Angew. Chem., Int. Ed.*, 2011, **50**, 10064.
27. J. Assmann, V. Narkhede, N. A. Breuer, M. Muhler, A. P. Seitsonen, M. Knapp, D. Crihan, A. Farkas, G. Mellau and H. Over, *J. Phys.: Condens. Matter*, 2008, **20**, 184017.
28. H. Over, *Chem. Rev.*, 2012, **112**, 3356.
29. Y. D. Kim, H. Over, G. Krabbes and G. Ertl, *Top. Catal.*, 2001, **14**, 95.
30. A. Scheibe, S. Gutnher and R. Imbihl, *Catal. Lett.*, 2003, **86**, 33.
31. *Frontiers in Surface and Interface Science*, ed. C. B. Duke and E. W. Plummer, Elsevier, Amsterdam, 2002.
32. (a) ed. K. Wandelt, *Special issue of Surface Science dedicated to Prof. Dr. Gerhard Ertl Nobel-Laureate in Chemistry 2007, Surf. Sci.*, 2009, 603, 1; (b) N. Lopez, J. Gomez-Segura, R. P. Marin and J. Perez-Ramirez, *J. Catal.*, 2008, **255**, 29.
33. D. Teschner, R. Farra, L. Yao, R. Schlogl, H. Soerijanto, R. Schomäcker, T. Schmidt, L. Szentmiklosi, A. P. Amrute, C. Modelli, J. Perez-Ramirez, G. Novell-Leruth and N. Lopez, *J. Catal.*, 2012, **285**, 273.
34. M. Knapp, A. P. Seitsonen, Y. D. Kim and H. Over, *J. Phys. Chem.*, 2004, **108**, 14392–14397.
35. H. Over, Y. D. Kim, A. P. Seitsonen, S. Wendt, E. Lundgren, M. Schmid, P. Varga, A. Morgante and G. Ertl, *Science*, 2000, **287**, 1474–1476.
36. Y. B. He, M. Knapp, E. Lundgren and H. Over, *J. Phys. Chem.*, 2005, **109**, 21825.
37. Y. D. Kim, S. Schwegmann, A. P. Seitsonen and H. Over, *J. Phys. Chem. B*, 2001, **105**, 2205.
38. H. J. Zhang, B. Lu, Y. H. Lu, Y. F. Xu, H. Y. Li, S. N. Bao and P. He, *Surf. Sci.*, 2007, **601**, 2297.
39. H. J. Zhang, B. Lu, Y. H. Lu, Y. F. Xu, H. Y. Li, S. N. Bao and P. He, *Surf. Sci.*, 2007, **601**, 2297.
40. H. Over, A. P. Seitsonen, M. Knapp, E. Lundgren, M. Schmid and P. Varga, *ChemPhysChem*, 2004, **5**, 167.
41. Y. J. Kim, Y. Gao and S. A. Chambers, *Appl. Surf. Sci.*, 1997, **120**, 250.
42. G. A. Rizzi, A. Magrin and G. Granozzi, *Surf. Sci.*, 1997, **443**, 277.
43. S. A. Chambers, *Adv. Mater.*, 2010, **22**, 219.
44. D. C. Meier, G. A. Rizzi, G. Granozzi, X. Lai and D. W. Goodman, *Langmuir*, 2002, **18**, 698.
45. X. Y. Zhao, J. Hrbek and J. A. Rodriguez, *Surf. Sci.*, 2005, **575**, 115.
46. Y. B. He, D. Langsdorf and H. Over, unpublished STM investigations (Febr. 2014).
47. S. Bajt, J. B. Alameda, T. W. Barbee Jr., W. M. Clift, J. A. Folta, B. Kaufman and E. A. Spiller, *Opt. Eng.*, 2002, **41**, 1797.
48. Y. B. He, A. Goriachko, C. Korte, A. Farkas, G. Mellau, P. Dudin, L. Gregoriatti, A. Barinov, M. Kiskinova, A. Stierle, N. Kasper, S. Bajt and H. Over, *J. Phys. Chem. C*, 2007, **111**, 10988.
49. P. J. Kelly and R. D. Arnell, *Vacuum*, 2000, **56**, 159.
50. C. C. Hsu, J. W. Coburn and D. B. Graves, *J. Vac. Sci. Technol., A*, 2006, **24**, 1.

51. I. Yunogami and K. Nojiri, *J. Vac. Sci. Technol., B*, 2000, **18**, 1911.
52. J. H. Kim, J. H. Lee, H. Kato, N. Horimoto, M. Kawai and Y. S. Lee, *J. Korean Phys. Soc.*, 2003, **42**, 408.
53. Y. Abe, M. Kawamura and K. Sasaki, *Jpn. J. Appl. Phys.*, 2002, **41**, 6857.
54. S. H. Oh, C. G. Park and C. Park, *Thin Solid Films*, 2000, **359**, 118.
55. G. Battaglin, V. Rigato, S. Zndolin, A. Benedetti, S. Ferro, L. Nanni and A. De Battisti, *Chem. Mater.*, 2004, **16**, 946.
56. E. Kolawa, F. C. T. So, W. Flick, X.-A. Zhao, E. T.-S. Pan and M.-A. Nicolet, *Thin Solid Films*, 1989, **173**, 217.
57. S. Y. Mar, C. S. Chen, Y. S. Huang and K. K. Tiong, *Appl. Surf. Sci.*, 1995, **90**, 497.
58. M. L. Green, M. E. Gross, L. E. Papa, K. J. Schones and D. Brasen, *J. Electrochem. Soc.*, 1985, **132**, 2677.
59. J. Si and S. B. Desu, *J. Mater. Res.*, 1993, **8**, 2644.
60. T. Takagi, I. Oizuki, I. Kobuyashi and M. Okada, *Jpn. J. Appl. Phys.*, 1995, **34**, 4104.
61. P. C. Liao, S. Y. Mar, W. S. Ho, Y. S. Huang and K. K. Tiong, *Thin Solid Films*, 1996, **287**, 74.
62. G.-R. Bai, A. Wang, C. M. Foster and J. Vetrone, *Thin Solid Films*, 1997, **310**, 75.
63. J. J. Kim, D. H. Jung, M. S. Kim, S. H. Kim and D. Y. Yoon, *Thin Solid Films*, 2002, **409**, 28.
64. Z. Yuan, R. J. Puddlephatt and M. Sayer, *Chem. Mater*, 1993, **5**, 908.
65. J. H. Han, S. W. Lee, G.-J. Choi, S. Y. Lee, C. S. Hwang, C. Dussarat and J. Gatineau, *Chem. Mater.*, 2009, **21**, 207.
66. J. H. Han, S. W. Lee, S. K. Kim, S. Han, C. S. Hwang, C. Dussarat and J. Gatineau, *Chem. Mater.*, 2010, **22**, 5700.
67. A. Kleinmann-Shwarsctein, A. B. Laursen, F. Cavalcy, W. Tang, S. Dahl and I. Chorckendorff, *Chem. Commun.*, 2012, **48**, 967.
68. S. Bhaskar, S. B. Majumder, P. S. Dobal, R. S. Katiyar, A. L. Morales Cruz and E. R. Fachini, *Mater. Res. Soc. Symp. Proc.*, 2000, **606**, 69.
69. Q. X. Jia, Z. Q. Shi, K. L. Jiao, W. A. Anderson and F. M. Colins, *Thin Solid Films*, 1991, **196**, 29.
70. H. Over, Y. B. He, A. Farkas, G. Mellau, C. Korte, M. Knapp, M. Chandhok and M. Fang, *J. Vac. Sci. Technol., B*, 2007, **25**, 1123.
71. X. Fang, M. Tachiki and T. Kobayashi, *Jpn. J. Appl. Phys.*, 1997, **36**, L511.
72. M. Hiratani, Y. Matsui, K. Imagawa and S. Kimura, *Thin Solid Films*, 2000, **366**, 102.
73. S. Gsell, M. Fischer, M. Schreck and B. Stritzker, *J. Cryst. Growth*, 2009, **311**, 3731.
74. T. Aaltonen, P. Alen, M. Ritala and M. Leselä, *Chem. Vap. Deposition*, 2003, **9**, 45.
75. J.-H. Kim, D.-S. Kil, A.-J. Yeom, J.-S. Roh, N.-J. Kwak and J.-W. Kim, *Appl. Phys. Lett.*, 2007, **91**, 052908.

76. H. Wang, R. Alvis and R. M. Ulfrig, *Chem. Vap. Deposition*, 2009, **15**, 312.
77. W. D. Ryden, A. W. Lawson and C. C. Sartain, *Phys. Rev. B: Condens. Matter Mater. Phys.*, 1970, **1**, 1494.
78. C. Noguera, *J. Phys.: Condens. Matter*, 2000, **12**, R367.
79. H. Wang and W. F. Schneider, *J. Chem. Phys.*, 2007, **127**, 064706.
80. Y. D. Kim, A. P. Seitsonen and H. Over, *Phys. Rev. B: Condens. Matter Mater. Phys.*, 2001, **63**, 5419.
81. A. Lobo and H. Conrad, *Surf. Sci*, 2003, **523**, 279.
82. M. Knapp, D. Crihan, A. P. Seitsonen and H. Over, *J. Am. Chem. Soc.*, 2005, **127**, 3236.
83. Y. D. Kim, A. P. Seitsonen, S. Wendt, J. Wang, C. Fan, K. Jacobi, H. Over and G. Ertl, *J. Phys. Chem. B*, 2001, **105**, 3752.
84. H. Wang, W. F. Schneider and D. Schmidt, *J. Phys. Chem. C*, 2009, **113**, 15266.
85. N. Lopez and G. Novell-Leruth, *Phys. Chem. Chem. Phys.*, 2010, **12**, 12217.
86. Y. Wang, A. Lafosse and K. Jacobi, *J. Phys. Chem. B*, 2002, **106**, 5476.
87. Y. Wang, K. Jacobi and G. Ertl, *J. Phys. Chem. B*, 2003, **107**, 13918.
88. U. A. Paulus, Y. Wang, H. P. Bonzel, K. Jacobi and G. Ertl, *Surf. Sci.*, 2004, **566**, 989.
89. U. A. Paulus, Y. Wang, H. P. Bonzel, K. Jacobi and G. Ertl, *J. Phys. Chem. B*, 2005, **109**, 2139.
90. U. Erlekam, U. A. Paulus, Y. Wang, H. P. Bonzel, K. Jacobi and G. Ertl., *Z. Phys. Chem.*, 2005, **219**, 891.
91. A. P. Seitsonen, D. Crihan, M. Knapp, A. Resta, E. Lundgren, J. N. Andersen and H. Over, *Surf. Sci.*, 2009, **603**, L113.
92. Y. Wang, K. Jacobi, W. D. Schoene and G. Ertl, *J. Phys. Chem. B*, 2005, **109**, 7883.
93. H. Over, *Appl. Phys. A: Mater. Sci. Process.*, 2002, **75**, 37.
94. A. P. Seitsonen and H. Over, *Surf. Sci*, 2009, **603**, 1717.
95. K. Reuter and M. Scheffler, *Phys. Rev. B: Condens. Matter Mater. Phys.*, 2003, **68**, 045407.
96. K. Reuter and M. Scheffler, *Phys. Rev. B: Condens. Matter Mater. Phys.*, 2001, **65**, 035406.
97. D. Crihan, M. Knapp, A. P. Seitsonen and H. Over, *J. Phys. Chem. B*, 2006, **110**, 22947.
98. A. Böttcher and H. Niehus, *Phys. Rev. B: Condens. Matter Mater. Phys.*, 1999, **60**, 14396.
99. M. A. Hevia, A. P. Amrute, T. Schmidt and J. Perez-Ramirez, *J. Catal.*, 2010, **276**, 141.
100. S. Zweidinger, J. P. Hofmann, O. Balmes, E. Lundgren and H. Over, *J. Catal.*, 2010, **272**, 169.
101. J. P. Hofmann, S. Zweidinger, A. P. Seitsonen, A. Farkas, M. Knapp, O. Balmes, E. Lundgren, J. N. Andersen and H. Over, *Phys. Chem. Chem. Phys.*, 2010, **12**, 15358.

102. J. P. Hofmann, S. Zweidinger, M. Knapp, A. P. Seitsonen, K. Schulte, J. N. Andersen, E. Lundgren and H. Over, *J. Phys. Chem. C*, 2010, **114**, 10901.
103. S. Zweidinger, D. Crihan, M. Knapp, J. P. Hofmann, A. P. Seitsonen, E. Lundgren, C. J. Weststrate, J. N. Andersen and H. Over, *J. Phys. Chem. C*, 2008, **112**, 9966.
104. H. Fong, I. Chun and D. H. Reneker, *Polymer*, 1999, **40**, 4585.
105. Z.-M. Huang, Y.-Z. Zhang, M. Kotaki and S. Ramakrishna, *Compos. Sci. Technol.*, 2003, **63**, 2223.
106. D. Li and Y. Xia, *Adv. Mater.*, 2004, **16**, 1151.
107. A. Greiner and J. H. Wendorff, *Angew. Chem., Int. Ed.*, 2007, **46**, 5670.
108. S. Ramakrishna, K. Fujihara, W.-E. Teo, T.-C. Lim and Z. Ma, *An Introduction to Electrospinning and Nanofibers*, World Scientific Publishing Co. Pte. Ltd., Singapore, 2005.
109. J.-H. Wendorff, S. Agarwal and A. Greiner, *Electrospinning Materials, Processing and Applications*, Wiley-VCH, Weinheim, 2012.
110. G. R. Patzke, F. Krumeich and R. Nesper, *Angew. Chem., Int. Ed.*, 2002, **114**, 2554.
111. D. Li, J. T. McCann, Y. Xia and M. Marquez, *J. Am. Ceram. Soc.*, 2006, **89**, 1861.
112. S. Cavaliere, S. Subianto, J. Savych, D. J. Jones and J. Rozière, *Energy Environ. Sci.*, 2011, **4**, 4761.
113. V. Thavasi, G. Singh and S. Ramakrishna, *Energy Environ. Sci.*, 2008, **1**, 205.
114. M. Kanehata, B. Ding and S. Shiratori, *Nanotechnology*, 2007, **18**, 315602.
115. C. Wessel, R. Ostermann, R. Dersch and B. M. Smarsly, *J. Phys. Chem C*, 2011, **115**, 362.
116. J.-M. Szeifert, D. Fattakhova-Rohlfing, D. Georgiadou, V. Kalousek, J. Rathouský, D. Kuang, S. Wenger, S. M. Zakeeruddin, M. Grätzel and T. Bein, *Chem. Mater.*, 2009, **21**, 1260.
117. D. H. Reneker, A. L. Yarin, H. Fong and S. Koombhongse, *J. Appl. Phys.*, 2000, **87**, 4531.
118. D. H. Reneker and A. L. Yarin, *Polymer*, 2008, **49**, 2387.
119. J. Brinker, Y. Lu, A. Sellinger and H. Fan, *Adv. Mater.*, 1999, **11**, 579.
120. D. Grosso, F. Cagnol, G. J. Soler-Illia, E. L. de Crepaldi, H. Amenitsch, A. Brunet-Bruneau, A. Bourgeois and C. Sanchez, *Adv. Funct. Mater.*, 2004, **14**, 309.
121. P. K. Baumgarten, *J. Colloid Interface Sci.*, 1971, **36**, 71.
122. T.-S. Hyun, H. L. Tuller, D.-Y. Youn, H.-G. Kim and I.-D. Kim, *J. Mater. Chem.*, 2010, **20**, 9172.
123. Y. Wu, R. Balakrishna, M. V. Reddy, A. Sreekumaran Nair, B. V. R. Chowdari and S. Ramakrishna, *J. Alloys Compd.*, 2012, **69**, 517.
124. R. Ostermann, PhD Thesis *Design and physica-chemical characterization of nanowires and multicomponent metal oxide films with tailored*

mesostructure and crystallinity, JLU, Gießen, 2011. http://geb.uni-giessen.de/geb/volltext/2011/8086.
125. Lj. Atanasoska, R. T. Atanansoski, F. H. Pollak and W. E. O'Grady, *Surf. Sci.*, 1990, **230**, 95.
126. A. Farkas, G. Mellau and H. Over, *J. Phys. Chem. C*, 2009, **113**, 14341.
127. C. Mondelli, A. P. Amrute, M. Moser, G. Novell-Leruth, N. Lopez, D. Rosenthal, R. Farra, M. E. Schuster, D. Teschner, T. Schmidt and Y. Perez-Ramirez, *J. Catal.*, 2012, **286**, 287.
128. Y. F. Zhang, J. Y. Li, Q. Li, L. Zhu, X. D. Liu, X. H. Xinghua Zhong, J. Meng and X. Q. Cao, *J. Colloid Interface Sci.*, 2007, **307**, 567.
129. C. L. Shao, H. Y. Guan, Y. C. Liu, J. Gong, N. Yu and X. H. Yang, *J. Cryst. Growth*, 2004, **267**, 380.
130. T. Tatte, M. Hussainov, M. Paalo, M. Part, R. Talviste, V. Kiisk, H. Mandar, K. Pohako, T. Pehk, K. Reivelt, M. Natali, J. Gurauskis, A. Lohmus and U. Maeorg, *Sci. Technol. Adv. Mater.*, 2011, **12**, 034412.
131. G. C. Pontelli, R. P. Reolon, A. P. Alves, F. A. Berutti and C. P. Bergmann, *Appl. Catal., A*, 2011, **405**, 79.
132. S. Xu, D. F. Sun, H. Liu, X. L. Wang and X. B. Yan, *Catal. Commun.*, 2011, **12**, 516.
133. A. S. Deshpande, N. Pinna, P. Beato, M. Antonietti and M. Niederberger, *Chem. Mater.*, 2004, **16**, 2599–2604.
134. A. S. Deshpande, N. Pinna, B. Smarsly, M. Antonietti and M. Niederberger, *Small*, 2005, **1**, 313.
135. S. Urban, C. H. Kanzler, N. Tarabanko, K. Zalewska-Wierzbicka, R. Ellinghaus, S. F. Rohrlack, L. Chen, P. Klar, B. M. Smarsly and H. Over, *Catal. Lett.*, 2013, **143**, 1362.
136. T. Brezesinski, C. Erpen, K. I. Iimura and B. Smarsly, *Chem. Mater.*, 2005, **17**, 1683.
137. T. Brezesinski, M. Groenewolt, N. Pinna, M. Antonietti and B. Smarsly, *New J. Chem.*, 2005, **29**, 237.
138. T. Brezesinski, B. Smarsly, M. Groenewolt, M. Antonietti, D. Grosso, C. Boissière and C. Sanchez, *Stud. Surf. Sci. Catal.*, 2005, **156**, 243.
139. T. W. Wang, O. Sel, I. Djerdj and B. Smarsly, *Colloid Polym. Sci.*, 2006, **285**, 1.
140. M. Baerns, R. Imbihl, V. A. Kondratenko, R. Kraehnert, W. K. Offermans, R. A. van Santen and A. Scheibe, *J. Catal.*, 2015, **232**, 226.
141. R. Imbihl, A. Scheibe, Y. F. Zeng, S. Gunther, R. Kraehnert, V. A. Kondratenko, M. Baerns, W. K. Offermans, A. P. J. Jansen and R. A. van Santen, *Phys. Chem. Chem. Phys.*, 2007, **9**, 3522.
142. L. Hannevold, O. Nilsen, A. Kjekshus and H. Fjellvag, *Appl. Catal., A*, 2005, **284**, 163.
143. H. Tributsch, *Int. J. Hydrogen Energy*, 2008, **33**, 5911.
144. H. J. Lewerenz, *Chem. Soc. Rev.*, 1997, **26**, 239.

145. S. Trasatti, in *Electrochemical Hydrogen Technologies*, ed. H. Wendt, Elsevier, Amsterdam, 1990, p. 104.
146. N. M. Markovic and P. N. Ross Jr., *Surf. Sci. Rep.*, 2002, **45**, 117.
147. G. Zehl, G. Schmithals, A. Hoell, S. Haas, C. Hartnig, I. Dorbandt, P. Bogdanoff and S. Fiechter, *Angew. Chem., Int. Ed.*, 2007, **46**, 7311.
148. S. Trasatti, *Electrochim. Acta*, 2000, **45**, 2377.

CHAPTER 9

Controllable Synthesis of Metal Nanoparticles for Electrocatalytic Activity Enhancement

QING LI, WENLEI ZHU AND SHOUHENG SUN*

Department of Chemistry, Brown University, Providence, RI 02912, USA
*Email: ssun@brown.edu

9.1 Introduction

The accelerated consumption of fossil fuels for energy use, coupled with the limited supply of these natural resources, has motivated the serious search for alternative sources of energy that are renewable and sustainable. Electrochemical systems, such as fuel cells, batteries, and water splitting devices, represent the most efficient and environmentally friendly technologies for energy conversion and storage to date.[1] They are operated in a similar electrochemical principle and require oxidation and reduction reactions occurring at two separated electrodes to generate power or to produce fuels. Some typical electrochemical reactions explored extensively in recent years are hydrogen oxidation reaction (HOR) and oxygen reduction reaction (ORR) in fuel cells, ORR and oxygen evolution reaction (OER) in metal–air batteries, and hydrogen evolution reaction (HER) and OER in water splitting. So far, the major obstacles hampering the commercialization of these technologies lie in the significant reaction over-potentials observed on the catalysts used for such conversions and the high costs associated with the use of some of the most expensive noble metals as these catalysts.

Electrocatalytic reactions at surfaces are very sensitive to the nanoscale and atomic-level structure of the heterogeneous interface at which they take place.[2] Due to the large surface to volume ratio and unique active binding sites present on the surface, metal nanoparticles (NPs) differ significantly from their bulk counterparts and have been considered to be excellent alternatives for advanced electrocatalysis. Currently, NPs based on Pt, Ru, Ir and their alloys are the universal choice as catalysts to improve reaction kinetics and to reduce reaction over-potentials. However, the-state-of-the-art NP electrocatalysts developed for advanced electrochemical devices such as fuel cells and metal–air batteries are still far from their predicted catalytic potentials, leaving much room for rational design of NP catalysts towards electrocatalytic activity enhancement.[3] In particular, it has been well documented that the size, composition, and structure of metal NPs are important factors in determining their catalysis. In this chapter, we summarize our recent efforts in controllable synthesis of monodisperse metal NPs for electrocatalytic reactions, including ORR, formic acid oxidation (FAOR), and selective CO_2 reduction. We intend to use these high quality NPs as models to illustrate the important controls achieved in NP catalysis and to understand structure–activity correlations. We believe such studies are important for rational design and optimization of metallic NPs as commercially viable catalysts for alternative energy applications.

9.2 Synthesis of Monodisperse Metal NPs

Being capable of producing monodisperse NPs (standard deviation in diameter <10%) is the first step to study NP properties inherent to nanoscale dimensions and to distinguish them from those associated with structural heterogeneities observed in large particles or polydisperse NPs.[4] For example, the color sharpness of an optical device based on semiconductive NPs is strongly NP-uniformity dependent,[5] and the magnetic orientation and narrow magnetization transitions from NP to NP in a magnetic NP array is critical for the next-generation multi-terabit magnetic recording applications.[6] This section outlines the general synthetic strategies applied in solution phase chemistry to prepare monodisperse metal NPs.

9.2.1 General Concept on NP Formation

Currently, NPs are often prepared from the high temperature organic phase reaction. This synthetic method has become highly efficient for the fabrication of metallic NPs with well-defined dimensions.[4,7,8] Specifically, it allows for the fine control over a number of reaction variables including reactant, solvent, surfactant, reaction temperature. Compared to the reaction performed in an aqueous phase solution, organic phase synthesis can be carried out in a much wider temperature range (from below 0 °C to over

300 °C depending on the solvent chosen) and variety of surfactant choice (from bipolar lipid molecules to multifunctional polymers), making it especially important for controlling NP sizes, shapes and compositions.

The mechanism proposed by LaMer et al. shows that the production of monodisperse colloids requires a temporally discrete nucleation event followed by slower controlled growth on the existing nuclei (Figure 9.1).[9] The nucleation process is thermodynamically unfavorable due to the decrease in entropy. With the injection of more energy into the reaction system, nuclei can still be formed and, more importantly, this process is associated tightly with the enthalpy change upon the formation of chemical bonds in the nuclei. At a critical size, the enthalpy factor (ΔH) overcomes the entropy one ($T\Delta S$) and the free energy is decreased, favoring the growth of nuclei into NPs. Kinetically, the NP growth process, and therefore the NP sizes and shapes, can be controlled by reaction parameters such as precursor/surfactant types, concentrations and reaction temperatures. More often than not, the NP shapes are determined by the surfactant binding difference on crystal facets with weaker binding surface facilitating the faster growth. Since the growth of any one NP is similar to all others, the size distribution of the NP product is largely decided by the time over which the nuclei are formed and begin to grow. Therefore, the rapid addition of reagents to the reaction solution is a key to preparing monodisperse NPs. Besides growth *via* atomic addition, Ostwald ripening is widely considered to be a second growth phase in many systems.[10,11] In this process the high surface energy of

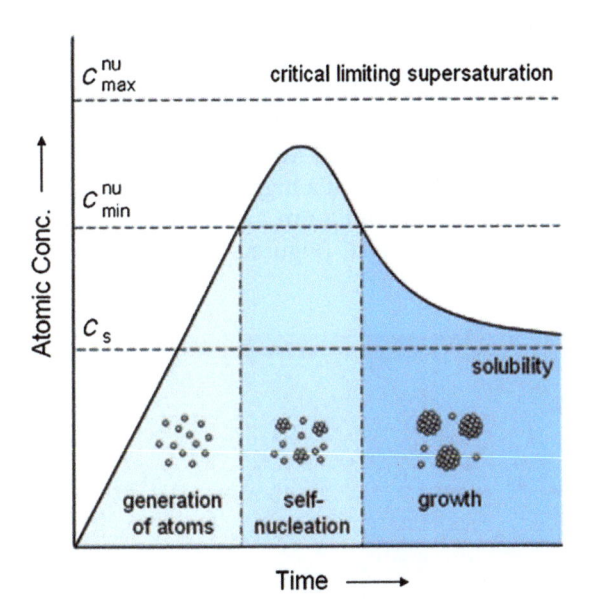

Figure 9.1 Plot of atomic concentration against time, illustrating the generation of atoms, nucleation, and subsequent growth.
Reprinted from ref. 9 with permission by Wiley-VCH.

the small NPs leads to their disintegration and the smaller species of atoms to clusters are redeposited onto the larger NPs. Finally, the average NP size increases over time with a concurrent decrease in NP number.

9.2.2 NP Size Control

Controlling the size and size distribution of NPs is essential for understanding physical and chemical properties of the NPs and this can be achieved by controlling reaction parameters during the solution phase synthesis. Metal precursor decomposition and metal salt reduction are two common chemical methods used to generate metal atoms, the building blocks for metal NPs.[9] The nucleation and growth process can be controlled by organic solvent and surfactant used for the synthesis and NP stabilization. In the synthesis, a non-polar solvent is often used and surfactant co-exists with a metal precursor. This, plus high temperature reaction condition, ensures fast nucleation and growth of metal atoms into well-structured NPs. If NPs are made from metal salt reduction, then the strength of reducing agent is an important parameter to control the nucleation and growth process. In an organic phase reaction condition, a stronger reducing agent often leads to a fast nucleation process and the formation of large amount of nuclei, which consumes more metal salt precursor and, as a result, yields relatively smaller NPs.

Temperature serves as another key role in determining the size of NPs and is demonstrated in the synthesis of Au NPs.[12,13] These Au NPs are prepared by reducing $HAuCl_4$ with borane *tert*-butylamine (BBA) in tetralin (1,2,3,4-tetrahydronaphalene) in the presence of oleylamine (OAm). In this reaction system, OAm is used to help the $HAuCl_4$ dissolution in tetralin by forming an amine–Au complex, and BBA acts a reducing agent to convert Au(III) to Au(0) for the formation of OAm-stabilized Au NPs. The NP sizes are controlled by reaction temperatures: reduction at a higher temperature leads to smaller Au NPs. As a result, 3 nm, 6 nm and 8 nm Au NPs have been synthesized from reaction at 40 °C, 20 °C, and 3 °C (Figure 9.2).

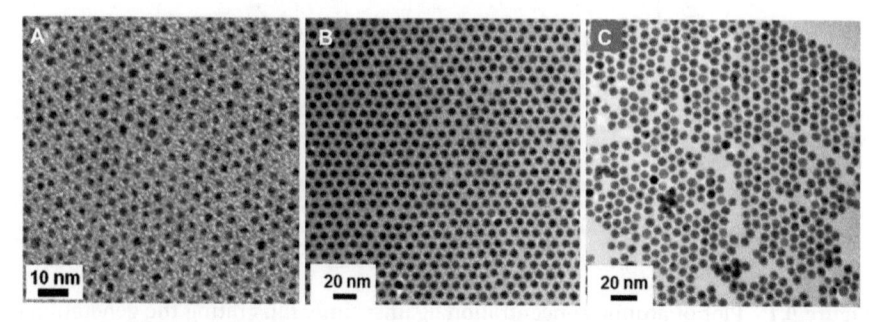

Figure 9.2 TEM images of the as-synthesized (A) 3, (B) 6, and (C) 8 nm Au NPs. Reprinted from ref. 12 with permission by American Chemical Society.

9.2.3 NP Shape Control

The shape of metal NPs has a significant effect on their catalytic behaviors. In an electrocatalytic reaction, the reactants are usually adsorbed on the NP catalyst surface to facilitate electron transfer and product formation. The energy barriers of reactant adsorption and product desorption play the key role in determining the reaction activity and reaction selectivity. Catalytic NPs with a uniform size and shape are often formed under a thermodynamic growth condition, which means that the NP growth is relatively slow and adatoms migrate into low energy sites following each step of deposition, resulting in NPs covered by low energy facets. In order to obtain a desired NP shape, surfactant-facet interaction or NP growth environment need to be well-controlled so that the adatoms can be added on a specific crystal direction. Currently, studies on reactions in organic phase have led to a variety of methods to control NP shapes, including: (i) kinetically controlled growth,[14] (ii) growth in reverse micelle,[15] (iii) aggregation directed growth,[16] (iv) seed-mediated growth,[17] and (v) oxidative etching.[18] These techniques induce the NP growth away from thermodynamic equilibrium state, leading to the formation of high energy surfaces that contain a large amount of low-coordinated atoms for active electrocatalysis. In this section, we focus on the shape control of metal NPs by kinetic control and by growth in reverse micelle.

9.2.3.1 Kinetic Control

Growing NPs kinetically is an effective way to achieve shape control. Such a control growth is realized *via* either surface energy differentiation or selectively bonding of surfactants on NP surface. In the first case, the diffusion time of an adatom from solution to the NP surface is typically shorter than the time it takes for an adatom to migrate to an energetically preferable site. Therefore, adatoms become kinetically locked into high energy positions, resulting in diffusion controlled growth. This can lead to a number of high energy surface including steps, kinks, terraces, islands and vacancies.[19] In the presence of surfactants, the bonding strength of the surfactant on different crystallographic planes dominates the relative growth rates with the weakly bound surface having a preferred growth.

Pt NPs with tetrahedral, cubic, and truncated octahedral shapes have been prepared by controlling the concentration of surfactant.[20,21] One example is in the synthesis of Pt NPs with shapes controlled to be polyhedral, truncated cubic and cubic.[22] In the reaction, Pt(acac)$_2$ (acac = acetylacetonate) was mixed with OAm and oleic acid (OA) in octadecene (ODE) and the mixture was heated to 200 °C while a small amount (less than 10%) of iron pentacarbonyl (Fe(CO)$_5$) was injected to initiate nucleation of Pt. Injection of Fe(CO)$_5$ at 180 °C led to fast nucleation/growth process, producing 3 nm Pt NPs with thermodynamically more stable polyhedral shape (Figure 9.3a). Injecting Fe(CO)$_5$ at lower temperatures (for example at 160 °C or 120 °C)

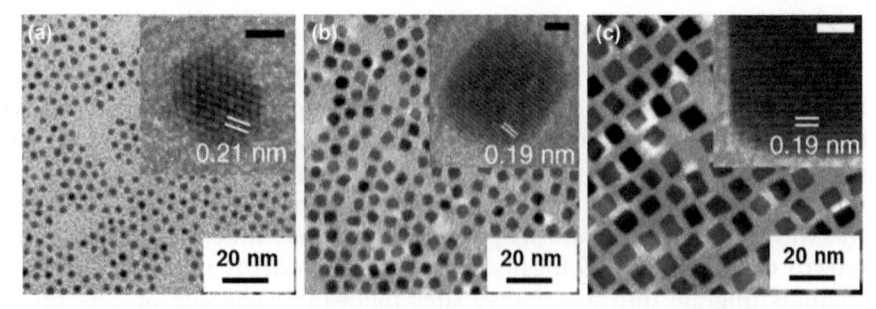

Figure 9.3 Representative TEM images of (a) the 3 nm , (b) the 5 nm truncated cubic, and (c) the 7 nm cubic Pt NPs. The insets are the representative HRTEM images of corresponding single particles, showing (a) Pt(111), (b) Pt(100), and (c) Pt(100) lattice fringes. All scale bars in the insets correspond to 1 nm.
Reprinted from ref. 22 with permission by Wiley-VCH.

Figure 9.4 TEM images of (a) FePt NPs and (b) FePt nanocubes.
Part (a) reprinted from ref. 6 with permission by the American Association for the Advancement of Science. Part (b) reprinted from ref. 23 with permission by American Chemical Society.

followed by a slow rate of heating slowed down the rate of nucleation and growth. As a result, more Pt-atoms were allowed to grow on (111), forming 5 nm truncated cubes (Figure 9.3b) or 7 nm cubes (Figure 9.3c).

Using surfactant mediated growth, we have succeeded in preparing FePt alloy NPs *via* thermal decomposition of Fe(CO)$_5$ and reduction of Pt(acac)$_2$ with size and shape controls. Monodisperse 6 nm FePt NPs were synthesized by the simultaneous reduction of Pt(acac)$_2$ and decomposition of Fe(CO)$_5$ in the presence of OA and OAm (Figure 9.4a).[6] By controlling the addition sequence of OAm and OA as well as Fe(CO)$_5$/Pt(acac)$_2$ ratio, 6.9 nm FePt nanocubes were prepared (Figure 9.4b).[23]

9.2.3.2 Anisotropic Growth in Reverse Micelle

Despite the examples demonstrated in shape-controlled synthesis of Pt-based NPs, the growth along either ⟨100⟩ or ⟨111⟩ tends to yield NPs with

Controllable Synthesis of Metal Nanoparticles 231

isotropic shape. To achieve an anisotropic growth condition, the reaction may be performed in a micelle structure. This micelle is formed when amphiphilic molecules are dispersed in water. The hydrophilic groups are exposed to water, leaving the hydrophobic tail regions buried in the micelle center. The reversed case, *i.e.*, reverse micelle, is amphiphilic molecules in a nonpolar solvent, where the hydrophobic tails are in contact with solvent with the hydrophilic end embedded in the center of the structure. Depending on the dimension of hydrophobic and hydrophilic units, anisotropic environment can be created and reaction in such a structure at a controlled temperature often leads to the formation of one-dimensional structure. This synthetic approach has been highlighted in the synthesis of Au nanorods,[15] where amphiphilic cetyltrimethylammonium bromide (CTAB) molecules are used to form double-layer micelle structures in the aqueous solution, and the Au precursors (usually $HAuCl_4$) are encapsulated at the micelle center. Due to the slightly different bonding strength of CTAB with Au crystal facets, preferential growth is introduced by forming the elongated micelles. Another example is in the synthesis of FePt nanowires (NWs) *via* decomposition of $Fe(CO)_5$ and reduction of $Pt(acac)_2$ in the mixture of OAm and ODE.[24] The lengths of the resulting FePt NWs can be tuned from 20 to 200 nm by changing the volume ratio of OAm to ODE (Figure 9.5a–c). It is believed that OAm self-organizes into an elongated reverse micelle-like structure within which the FePt nuclei are formed. The elongated nuclei result in the different OAm packing densities on different surfaces, as indicated by (1), (2), and (3) in Figure 9.5d. In area (1), the

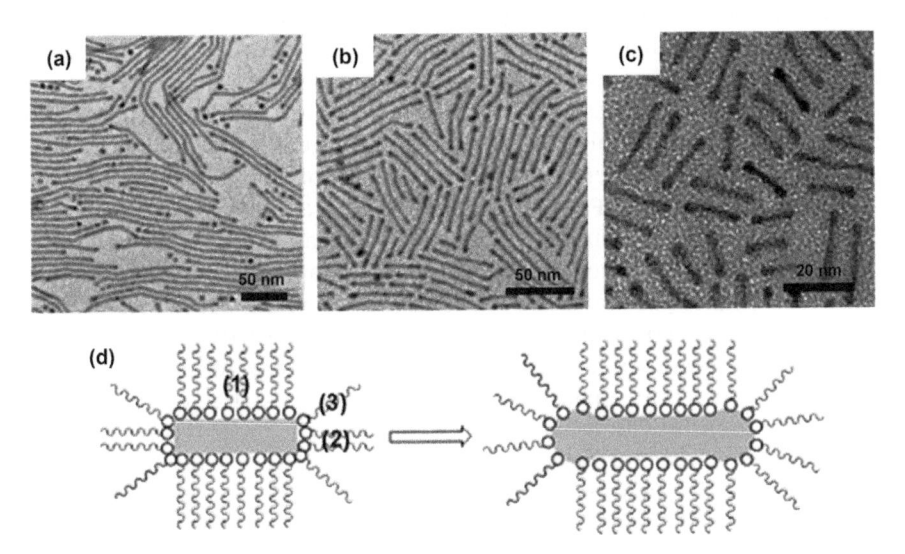

Figure 9.5 (a)–(c) TEM images of $Fe_{55}Pt_{45}$ NWs with a length of 200 nm (a), 50 nm (b), and 20 nm (c). (d) Schematic illustration of the growth of FePt NWs in reverse micelle.
Reprinted from ref. 24 with permission by Wiley-VCH.

molecules are well-organized and addition of FePt in this direction is more difficult owing to the presence of the hydrophobic barrier. Area (2) has less densely packed OAm and facilitates the growth of FePt along this direction and the formation of NWs. Area (3) is the most readily accessible place for the addition of FePt, leading to the fast growth of FePt and the rounded end of the NWs.

9.3 NP Activation for Catalysis

Metal NPs prepared from solution phase reactions are coated with a layer of surfactant, which creates an insulating barrier around each NP, making them inactive for electrocatalysis. Therefore, to function as electrocatalysts, these metal NPs must first be activated by removing the surface coating. To activate the NP catalysts, the NPs are first deposited onto a high-surface-area and conductive support (*e.g.*, carbon black) to ensure the NP dispersion over the surface of the support and stabilization against aggregation upon the surfactant removal. Such deposition is often done by mixing the NP dispersion with the support under a sonication condition.

The most straightforward way of removing surfactant is to anneal NPs at high temperature in air to burn off or vaporize the organic surfactants.[13] For example, OAm coated Au NPs deposited on a solid support can be treated at 300 °C in 8% O_2/He mixture for 1 h to remove the OAm surfactant, after which the supported Au NPs are highly active for CO oxidation, showing 100% CO conversion at −45 °C. An alternative approach is to neutralize the surfactants so that their binding to the metal NP surface is weakened and the NPs can be washed clean.[25] For example, on activating the OAm-coated Pd NPs, the NPs deposited on the carbon support are immersed in acetic acid. This acid neutralizes –NH_2 into –NH_3^+, facilitating OAm removal from the NP surface by simple washing. After this treatment, the Pd NPs are catalytically active for electrochemical oxidation of formic acid.

If NPs, for example, FePt, CoPt, NiPt NPs, are subject to changes (NP oxidation, NP aggregation or NP sintering) in air annealing or acid washing conditions, surfactant exchange reaction may be applied to replace the long chain surfactant with small molecules that bind weakly on the NP surface. For instance, to activate OAm-coated $NiPt_3$ NPs, the NPs are dissolved in butylamine at room temperature. Large amount of butylamine facilitates its replacement of OAm, giving butylamine-coated NPs that can be further cleaned by washing with other common solvents. The process preserves both NP size and shape for electrocatalytic studies.[26] This strategy has been extended successfully to activate Co and Co/CoO NPs for oxygen reduction reaction in alkaline media.[27] Similarly, tetramethylammonium hydroxide (TMAOH)[28] and tetrabutylammonium hydroxide (TBAOH)[29] can be used to replace long-chain surfactant to facilitate NP cleaning. Recently, it is reported that the organic surfactant present around NP can be even replaced by small inorganic metal chalcogenide complexes.[30]

Controllable Synthesis of Metal Nanoparticles

9.4 Applications of Metal NPs in Electrocatalysis

Metal NPs prepared from solution phase reactions and activated by a proper treatment can be used as ideal models for studying electrocatalytic reactions, such as oxygen reduction reaction (ORR), formic acid oxidation reaction (FAOR), and CO_2 reduction, that are all important for energy applications. The fine-tuning of NP parameters achieved in the synthesis allows the detailed studies and optimizations of NP catalytic performance at the atomic scale.

9.4.1 NP Catalysis for ORR

Electrochemical reduction of oxygen, commonly referred to oxygen reduction reaction (ORR), is an important half reaction used to couple with fuel oxidation for the desired energy conversion. Electrochemical devices using ORR for efficient energy conversion include low-temperature fuel cells and metal–air batteries. For ORR to proceed at lowered electrochemical overpotentials, NP catalysts, especially Pt-based NP catalysts, must be present at the cathode. ORR involves a 4-electron reduction process to convert O_2 to H_2O. Although it is difficult to fully understand the ORR processes on catalyst surface, three possible mechanisms for ORR have been proposed, as summarized in Figure 9.6.[31] Oxygen may be selectively reduced to H_2O via the dissociative mechanism if it is not adsorbed in its molecular form. The reduction process may also undergo via the association mechanism in which OOH_{ads} is formed on the catalyst surface and further cleaved into O_{ads} and OH_{ads}. Alternatively, the ORR may proceed through the peroxo mechanism when two electron-transfer steps lead successively to OOH_{ads} and to HOO-H_{ads} before the O–O bond is cleaved and OH_{ads} is formed. The 4-electron

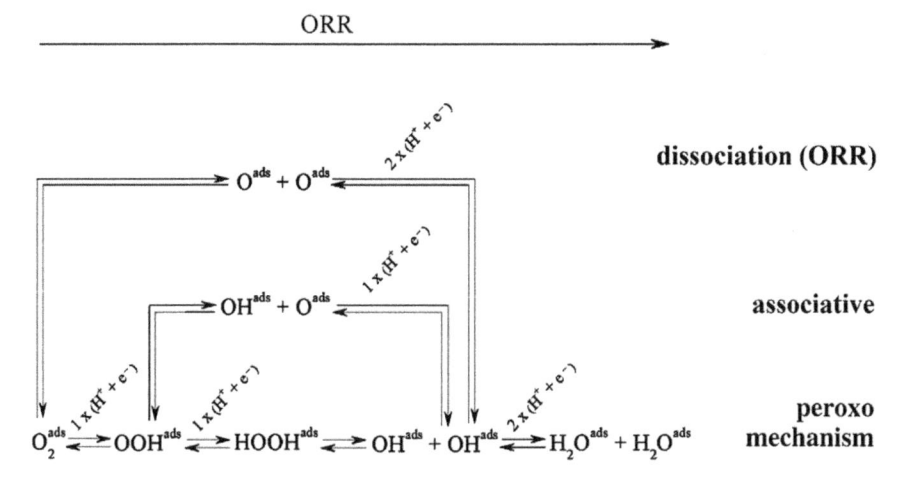

Figure 9.6 Proposed mechanisms for ORR.
Reprinted from ref. 31 with permission by Wiley-VCH.

reduction of O_2 to H_2O is much preferred in a fuel cell device and the 2-electron reduction pathway leading to the formation of H_2O_2, a strong oxidizer, should be avoided.

Currently, the state-of-the-art electrocatalysts for ORR in polymer electrolyte fuel cells (PEFCs) are carbon supported Pt NPs. Despite their acceptable activity, these catalysts do have the stability issues that need to be resolved. Furthermore, Pt is limited in natural reserve and is costly to use. Therefore, to minimize the Pt usage or ideally to develop more advanced catalysts without the use of any noble metals for ORR has attracted tremendous interests.[32,33]

9.4.1.1 Pt-based Alloy NP Catalysts

In spite of the great progress achieved in last decade,[34] the ORR activity of the pure Pt catalysts is still limited. During the ORR, the formation of the adsorbed oxygenated species (e.g., O_{ads}, OH_{ads}, OOH_{ads}), originating from the aqueous electrolyte and/or the reaction itself, is considered as the decisive process for determining the catalyst performance. As a result, a balance between adsorption and desorption of O_{ads}, OH_{ads} and OOH_{ads} is believed to be the key to successful development of a viable ORR catalyst. Recent density functional theory (DFT) calculations indicate the Pt–O binding energy of Pt(111) surface is 0.2 eV away from optimal energy level.[35–38] Theoretical studies further show that the addition of an early transition metal (M) to Pt can downshift the d-band center of the Pt catalyst, leading to a lower degree of adsorption of oxygenated species and increases the number of active sites accessible to oxygen.[39,40] Importantly, volcano-type relationships between the oxygen adsorption energy (or the d-band center) and the ORR activity for various PtM alloys have been also established.[39,41,42] In parallel to the theoretical studies, extensive efforts have been directed towards the preparation of multi-metallic PtM electrocatalysts with enhanced activities for the ORR. Among the catalysts studied, Pt alloyed with Fe, Co, Ni, and Cu show much improved ORR activity relative to that of other transition metals and Pt itself, which is in good agreement with DFT calculations.[40,43,44]

One of the most promising bimetallic systems developed in our group is the FePt alloy. FePt NPs are considered an important material for a range of magnetic (data storage, high performance magnets) and electrocatalytic applications.[6,45] Although preliminary studies demonstrate that FePt NPs are indeed ORR active,[46] there exists a serious concern when using these alloy NPs to catalyze ORR in acidic solution: the alloy structure is chemically unstable and the Fe components of the FePt NPs are subject to fast dissolution. The easy dissolution of Fe and the formation of the Pt skeleton structure make FePt NPs even less stable than Pt NPs in acidic environments. As synthesized, these FePt NPs have the chemically disordered solid solution structure with Fe and Pt arranged randomly in the face centered cubic (fcc) fashion (Figure 9.7a), the FePt NPs are often denoted as fcc-FePt.

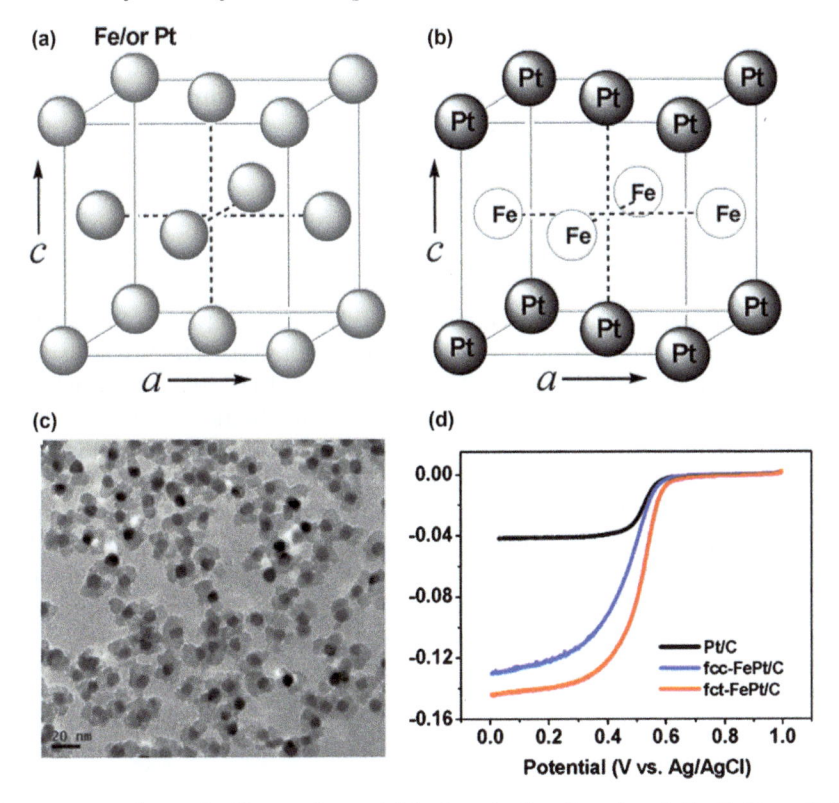

Controllable Synthesis of Metal Nanoparticles 235

Figure 9.7 Schematic illustration of (a) chemically disordered fcc-FePt and (b) chemically ordered fct-FePt. (c) TEM image of the 8 nm fct-FePt/MgO NPs annealed at 750 °C for 6 h under Ar + 5% H$_2$. (d) ORR activity of the commercial Pt, fcc-FePt, and fct-FePt catalysts in oxygen saturated 0.5 M H$_2$SO$_4$ after stability tests.
Reprinted from ref. 47 with permission by American Chemical Society.

One strategy to stabilize such FePt NPs is to convert their disordered structure into a chemically ordered tetragonal intermetallic structure (often denoted as face-centered tetragonal (fct) structure) with Fe and Pt being in alternating layers (Figure 9.7b). Recently, this structure transformation from fcc to fct FePt was achieved by high temperature annealing (6 h in forming gas at 750 °C) of the FePt NPs embedded in MgO matrix (to prevent FePt NPs from aggregation/sintering) (Figure 9.7c).[47] The monodisperse fct-FePt NPs were found to be stable in 0.5 M H$_2$SO$_4$ solution and showed superior durability and higher activity than the fcc-FePt NPs and commercial Pt catalyst (Figure 9.7d). Theoretical calculations indicate that the Pt–O binding energy value of the fct-FePt is closer to the optimal value than fcc-FePt and correspondingly, the fct-FePt NPs are more active than the fcc-FePt NPs for ORR.[48]

ORR catalysis can be further enhanced when Pt-based catalysts are prepared in a controlled shape, especially in a one-dimensional (1D) NWs,[49] where interactions between a crystal facet and oxygenated species as well as the carbon support can be optimized to achieve superior catalytic activity and durability. These PtM NWs combine both alloy and shape effects to enhance ORR catalysis and provide a new approach to highly efficient Pt-based catalysts for the ORR. MPt NWs (M = Fe or Co) with controlled diameters (2.5–6.3 nm) are synthesized through decomposition of Fe(CO)$_5$ (or Co$_2$(CO)$_8$) and reduction of Pt(acac)$_2$ in sodium oleate solution of ODE and OAm at 240 °C.[50] By controlling the molar ratio of Fe(CO)$_5$ or Co$_2$(CO)$_8$ to Pt(acac)$_2$, the FePt (or CoPt) NW composition can be readily tuned. When treated with acetic acid, these FePt (or CoPt) NWs are converted to FePt/Pt (or CoPt/Pt) and become active and stable for ORR catalysis. The surface specific and mass activities of the FePt NWs at 0.90 V vs. RHE reach 1.53 mA cm^{-2} and 844 mA mg$^{-1}_{Pt}$, which are 4.7 and 5.5 times higher than those of the commercial Pt catalyst. By adding a third metal salt precursor in the synthesis of FePt NWs, we have successfully produced ternary FePtM NWs (M = Cu or Ni) for ORR activity enhancement.[51] When treated with acetic acid and etched electrochemically in 0.1 M HClO$_4$, core–shell FePtM/Pt NWs are formed, as illustrated in Figure 9.8. In electrochemical measurements, the ORR activity of the FePtCu NWs with relatively high initial Cu content (Fe$_{29}$Pt$_{41}$Cu$_{30}$) are superior to that of other FePtCu compositions, binary FePt NWs and the commercial Pt catalyst, demonstrating the promotional role of

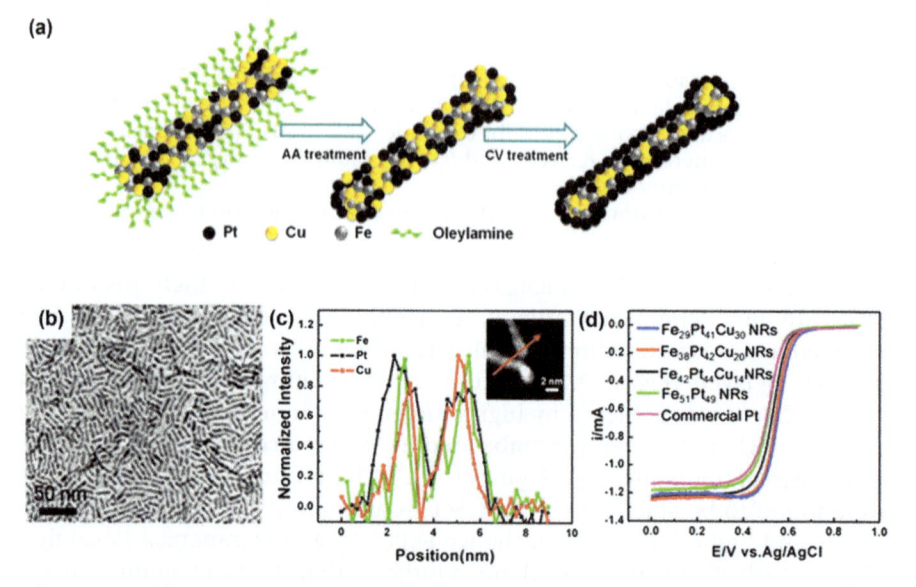

Figure 9.8 (a) Schematic illustration of the formation and activation of the core/shell-type FePtCu/Pt NW catalyst. (b) TEM image of the 20×2 nm Fe$_{29}$Pt$_{41}$Cu$_{30}$ NWs. (c) HAADF–STEM image (inset) and STEM–EELS line scan of the representative Fe$_{10}$Pt$_{75}$Cu$_{15}$ NWs obtained after acetic acid washing and potential cycling of Fe$_{29}$Pt$_{41}$Cu$_{30}$ NRs.
Reprinted from ref. 51 with permission by American Chemical Society.

Controllable Synthesis of Metal Nanoparticles 237

Cu in FePt system and the significance of 1D nanostructures toward ORR enhancement.

9.4.1.2 Pt-based Core–Shell NPs

Pt-based core–shell NPs are another class of electrocatalysts studied for ORR. In particular, a thin shell of Pt or PtM alloy is deposited on the non-Pt NP core and thus the usage of Pt can be greatly reduced. Furthermore, the rational design of core–shell nanostructures with Pt or PtM shell could introduce (i) a 'ligand effect', which is caused by the atomic vicinity of two dissimilar surface metal atoms that induces electronic charge transfer between the atoms, and thus affects their electronic band structure, and (ii) a 'strain effect', which originates from the lattice mismatch at the core–shell interface that may include compressed or expanded arrangements of surface atoms (surface strain).[43] These two effects are of great importance for ORR activity enhancement. Therefore, design and synthesis of core–shell structures allows the tuning of the Pt electronic structure and surface strain for an optimal ORR activity. So far several approaches have been applied to synthesize Pt-based core–shell NPs for electrochemical applications and there are already quite a few excellent reviews in this area.[52–54]

Multi-metallic core–shell NPs with a layer of uniform FePt over Au,[55] Pd,[56] and even core–shell Pd/Au NPs[57] can be prepared by seed-mediated growth method in an organic solution at a controlled temperature. For instance, Pd/FePt NPs with 5 nm Pd core and uniform FePt shell were synthesized by the nucleation and growth of Fe(CO)$_5$ in the presence of Pt(acac)$_2$ over the 5 nm Pd NPs at 180 °C (Figure 9.9a–c).[56] The FePt thickness (1–3 nm) was controlled by the amount of Fe(CO)$_5$, Pt(acac)$_2$, and the Pd seeds used in the

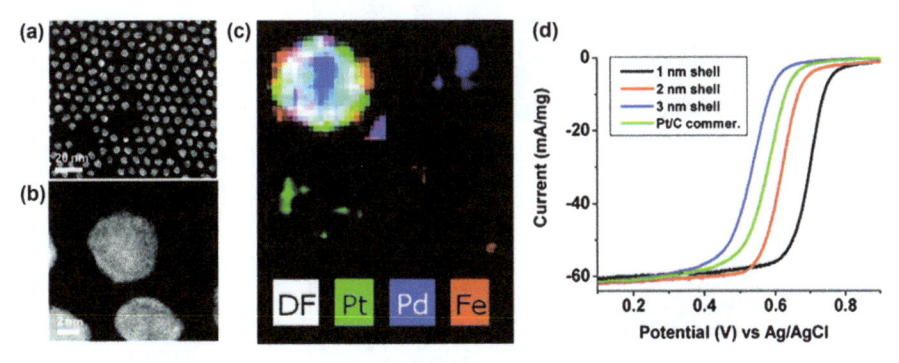

Figure 9.9 (a) High-angle annular dark-field scanning TEM (HAADF–STEM), (b) high-resolution HAADF–STEM, and (c) elemental mapping images of the 5 nm/1 nm Pd/FePt NPs. (d) ORR polarization curves for three types of Pd/FePt NPs and the commercial Pt/C catalyst in 0.1 M HClO$_4$. The current was normalized against the total mass of NPs used, and the electrode rotation rate was kept at 1600 rpm.
Reprinted from ref. 56 with permission by American Chemical Society.

reaction. The ORR activity measured on different Pd/FePt NPs in 0.1 M HClO$_4$ are highly dependent on the FePt shell thickness, with the thinner FePt shell (1 nm) having higher activity (Figure 9.9d).

The seed-mediated approach to Pt-based core–shell NPs can be extended to the preparation of core–shell NWs, as demonstrated in FePtM/FePt where M = Pd or Au.[58] Through controlled decomposition of Fe(CO)$_5$ and reduction of Pt(acac)$_2$ in the presence of 2.5 nm wide FePtM NW seeds, 0.3–1.3 nm thick FePt shells are deposited around the FePtM core (Figure 9.10a–d). These FePtM/FePt NWs have shell thickness and core composition-dependent activity for ORR in 0.1 M HClO$_4$ solution and the FePtPd/FePt NWs with 0.8 shell coating have the highest activity compared to the similar FePtAu/FePt NWs and the commercial Pt catalyst (Figure 9.10e and f). The higher ORR activity observed from the FePtPd/FePt NWs than from the FePtAu/FePt NWs is attributed to the electronic effect of Pd to FePt, which further downshifts the d-band center of Pt, facilitating O$_2$ adsorption, activation, and desorption. The obtained FePtM/FePt-0.8 NWs are also stable after 5000 potential cycling between 0.4 and 0.8 V (vs. Ag/AgCl). This stability enhancement comes likely from both strong NW/carbon support interaction and the controlled Pt–Pt strain in the core–shell structure. In the study, the 0.8 nm coating is critical to form uniform coating and for the shell to 'feel' the core effect on its d-band shift (electronic effect) and Pt–Pt distance (strain effect).

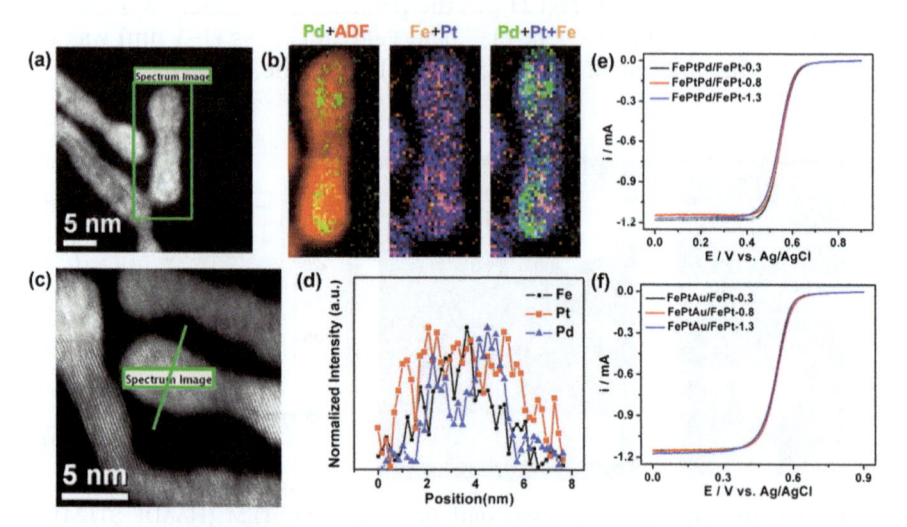

Figure 9.10 HAADF–STEM (a and c), and STEM–EELS mapping (29×66 pixels, spatial resolution of 3 Å); (b) images of FePtPd/FePt-0.8. (c and d) High-resolution (1.7 Å) line-scan EELS analysis across one NW. ORR polarization curves of C–FePtPd/FePt (e) and C–FePtAu/FePt (f) in O$_2$-saturated 0.1 M HClO$_4$ solution at 293 K.
Reprinted from ref. 58 with permission by American Chemical Society.

9.4.2 NP Catalysts for FAOR

Recently, the rapid development of portable devices, such as mobile phones, notebooks and digital cameras has significantly increased the demand for high-output power sources. The H_2-based PEFCs are inappropriate for portable power applications due to the inconvenience and potential danger in the transport and use of hydrogen. Direct methanol fuel cells (DMFCs), a type of direct-feed fuel cell, hold a promise for these portable power applications due to the high energy density and easy transportation of methanol. However, the performance of DMFC is limited by many factors, including the slow kinetics of methanol oxidation, cross-over of methanol from the anode to the cathode side of the cell, and safety concerns about the methanol itself. Alternatively, formic acid has been investigated as an alternative fuel for the direct-feed fuel cells thanks to its lower fuel cross-over than methanol and non-toxicity.[59] In the presence of the Pt catalyst, direct oxidation of formic acid releases two electrons per molecule and the most commonly accepted mechanism is the so-called parallel or dual pathway mechanism.[59] Pathway 1 is a direct oxidation mechanism where a dehydrogenation reaction occurs, without forming CO as a reaction intermediate:

$$HCOOH \rightarrow CO_2 + 2H^+ + 2e^- \tag{9.1}$$

The second reaction pathway (pathway 2) forms adsorbed carbon monoxide (CO) as a reaction intermediate by dehydration followed by a 2e oxidation:

$$HCOOH \rightarrow CO_{abs} + H_2O \rightarrow CO_2 + 2H^+ + 2e^- \tag{9.2}$$

The reaction mechanism described by pathway 2 would poison the Pt catalyst due to the formation of strong Pt–CO bond. Thus, dehydrogenation is the desired reaction pathway, to enhance overall cell efficiency and to avoid poisoning of the NP catalysts.

9.4.2.1 Pd-based Alloy NP Catalyst for FAOR

Recent studies reveal that Pd catalysts are much less prone to CO poisoning/deactivation than Pt catalysts, making Pd an extremely promising Pt-alternative catalyst for direct formic acid fuel cell applications.[60] Using high temperature organic phase reaction, we prepared monodisperse 4.5-nm Pd NPs by the reduction of Pd(acac)$_2$ with OAm and borane tributylamine (BBA) for FAOR.[25] Here OAm acted as the solvent, surfactant, and reductant, and BBA served as a co-reductant. The Pd NPs obtained, supported on Ketjen carbon, are catalytically active and stable for FAOR in 0.1 M HClO$_4$ solution.

It is well known that the commercial PtRu catalysts for methanol oxidation reaction (MOR) could efficiently remove the chemisorbed CO intermediate on the Pt surface by a bi-functional mechanism, in which Ru activates water molecules to form OH species at lower potentials than Pt, favoring CO

240   *Chapter 9*

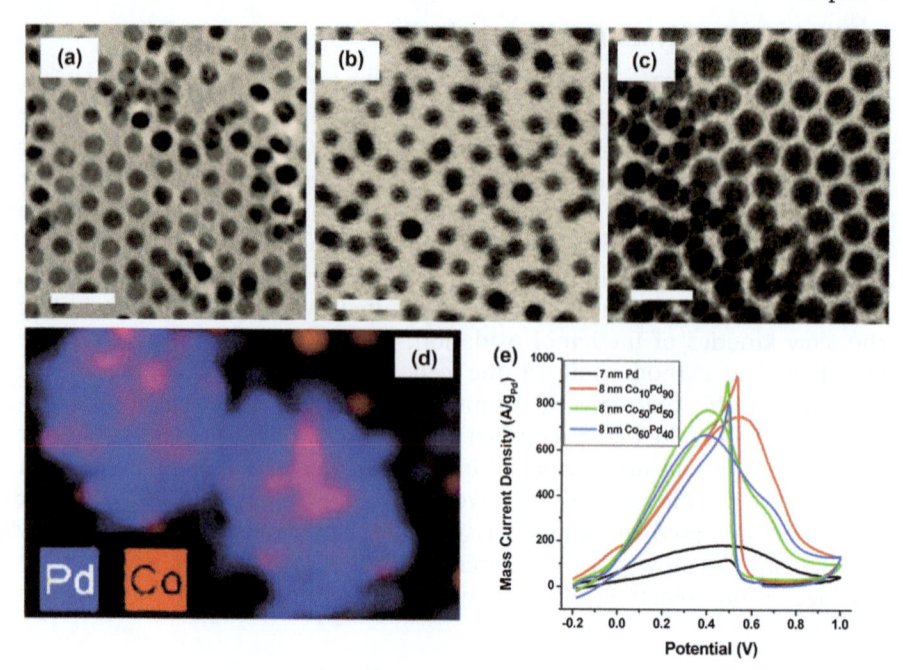

Figure 9.11 TEM images of (a) 8 nm, (b) 5 nm, and (c) 12 nm Co$_{60}$Pd$_{40}$ NPs (scale bar = 25 nm). (d) EDS elemental maps for Co (red) and Pd (blue) within two individual 8 nm Co$_{60}$Pd$_{40}$ NPs indicating the distribution of the two elements within each NP. (e) NP mass current densities *vs.* applied potential in 0.1 M HClO$_4$ and 2 M HCOOH at 25 °C (scan rate: 50 mV s^{-1}). Reprinted from ref. 62 with permission by American Chemical Society.

oxidation.[61] Inspired by the MOR, one strategy to achieve FAOR activity enhancement is to incorporate an early transition metal (M) into the Pd structure to make MPd alloys, because the oxophilic property of M in the alloy structure which should benefit the CO removal. Recently, the synthesis of monodisperse CoPd NPs by co-reduction of Co(acac)$_2$ and PdBr$_2$ at 260 °C in OAm and trioctylphosphine (TOP) was reported.[62] NP sizes were controlled to be 5 to 12 nm by either the heating ramp rate or metal salt concentration (Figure 9.11a–d), and NP compositions ranging from Co$_{10}$Pd$_{90}$ to Co$_{60}$Pd$_{40}$ were tuned by metal molar ratios. It represents a general approach to Pd-based bimetallic NPs and could also be extended to make CuPd NPs when Co(acac)$_2$ was replaced by Cu(ac)$_2$. The 8 nm CoPd NPs show composition dependent FAOR activity with Co$_{50}$Pd$_{50}$ > Co$_{60}$Pd$_{40}$ > Co$_{10}$Pd$_{90}$ > Pd (Figure 9.11e).

9.4.2.2 Intermetallic NP Catalyst for FAOR

As discussed in Section 9.4.1.1, the chemical structure of the alloy NPs is of great importance towards the activity and stability enhancement in electrochemical reactions. Recently, the tetragonal FePt/PtAu NPs were

Controllable Synthesis of Metal Nanoparticles 241

developed as robust catalysts for FAOR.[63] In this study, the cubic solid solution FePtAu NPs (denoted as fcc-FePtAu) were synthesized by co-reduction of Pt(acac)$_2$ and HAuCl$_4$A·H$_2$O and thermal decomposition of Fe(CO)$_5$. The composition of the FePtAu NPs was controlled by varying metal precursor ratios. Thermal annealing in Ar + 5% H$_2$ at 600 °C for 1 h led to the formation of tetragonal FePtAu (denoted as fct-FePtAu) while annealing at 400 °C only led to the removal of the surfactant without causing the structure change, as displayed in Figure 9.12a. To facilitate fcc–fct phase transformation, the Fe/Pt ratio was kept near 1 : 1, while Au was tuned with three different percentages to demonstrate Au composition-dependent fcc–fct transition and FAOR catalysis. Figure 9.12b shows the representative TEM image of the monodisperse FePtAu NPs on Ketjen carbon. Analyses from STEM–EELS and STEM–EDS indicated that in the fct-FePtAu, Au diffuses out and distribute as clusters around the shell. Such a unique core–shell fct-FePt/PtAu structure pattern is essential to demonstrate the enhanced activity and durability for FAOR. Results on FAOR (Figure 9.12c and d) demonstrate that the presence of surface Au alleviates the CO poisoning effect, as evidenced by the disappearance of the small peak at ~0.7 V. The much less degree of CO poison observed on fct-FePt/PtAu than on Pt or FePt indicates that Au on the NP surface is able to promote the dehydrogenation reaction of

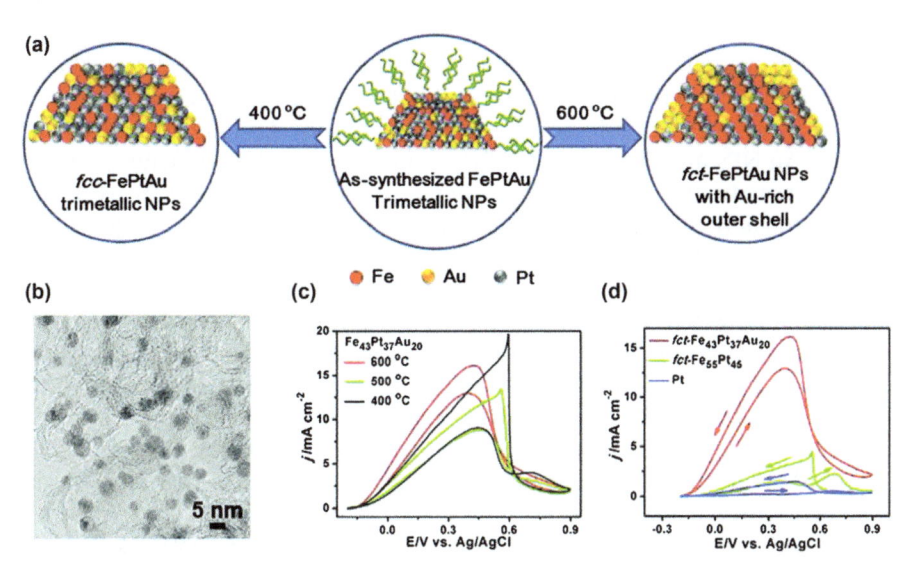

Figure 9.12 (a) Schematic illustration of the structural change of the FePtAu NPs upon annealing. When annealed at 400 °C, the FePtAu NPs have fcc structure, but at 600 °C, the fct-FePtAu structure is formed, with Au segregating on the NP surface. (b) TEM image of the fct-Fe$_{43}$Pt$_{37}$Au$_{20}$/C NPs. J–V curves of (c) the Fe$_{43}$Pt$_{37}$Au$_{20}$ NPs annealed at different temperatures and (d) the specific activity of the fct-Fe$_{43}$Pt$_{37}$Au$_{20}$, fct-Fe$_{55}$Pt$_{45}$ and commercial Pt catalysts in N$_2$-saturated 0.5 M H$_2$SO$_4$ + 0.5 M HCOOH solutions.
Reprinted from ref. 63 with permission by American Chemical Society.

HCOOH and to inhibit the dehydration reaction that leads to the formation of CO. In durability test, the fct-$Fe_{43}Pt_{37}Au_{20}$ NPs are found to be more stable than the fcc counterparts, suggesting that the fct structure does favor the durability enhancement of FePtAu NPs for FAOR.

9.4.3 Metal NPs as Catalysts for Electrochemical Reduction of CO_2

The monodisperse metal NPs prepared from the organic solution phase reaction allows for detailed studies of NP catalysts not only for fuel cell reactions, but many other reactions as well. One such reaction that has stimulated great interest recently is the catalytic reduction of CO_2 into other active carbon forms to reduce the problems caused by the accelerated consumption of fossil fuels and the resultant over-production of CO_2.[64] Among many different approaches developed thus far for CO_2 reactivation, electrochemical reduction of CO_2 is considered a potentially 'clean' method as the reduction proceeds at the expense of a sustainable supply of electric energy.[65] Theoretically, CO_2 can be reduced to different kinds of hydrocarbons such as carbon monoxide, formic acid, methane or other hydrocarbons at potentials around +0.2 to −0.2 V (*vs.* RHE). Experimentally, however, very negative potentials must be applied to initiate CO_2 reduction.[66]

Recent studies have shown that nanocatalyst can be used to enhance electrochemical reduction of CO_2, as demonstrated in the reduction of CO_2 to CO by Ag NPs in ionic liquid solution,[67] and by Au nanostructured surface.[68] Au NPs are also active for CO_2 reduction and their activity is NP size dependent. In the study, monodisperse polycrystalline Au NPs were first prepared *via* the organic phase reduction of $HAuCl_4$. 4 nm and 6 nm Au NPs (Figure 9.13a) were formed by fast injection of borane *tert*-butylamine complex. Au NPs of sizes 8 nm and 10 nm were synthesized by seed-mediated growth of Au over the 6 nm Au seeding NPs.[69] These NPs were first deposited on the carbon support and activated by thermal annealing at 180 °C in air. Electrochemical analyses show that CO_2 is reduced CO on the Au NP surface and the 8 nm Au NPs are the most selective catalyst with the reduction faraday efficiency (FE) reaching 90% at −0.67 V (Figure 9.13b), while the 4 nm NPs have the highest mass activity due to their smaller size (Figure 9.13c). The high selectivity of the 8 nm Au NPs arises from the dominant edge sites present on their surface due to the polycrystalline nature (with an average crystal domain diameter of 4 nm) of the NPs. This is supported by the DFT calculations that the edge sites on the Au NP surface facilitate CO_2 reduction to CO while the corner sites are active for proton reduction to hydrogen. When the Au crystallite have the diameter greater than 4 nm, the edge sites become dominant over corner sites on the NP surface (Figure 9.13d). These studies suggest that metal NPs, when prepared with the size and structure control, can be optimized for electrochemical reduction of CO_2.

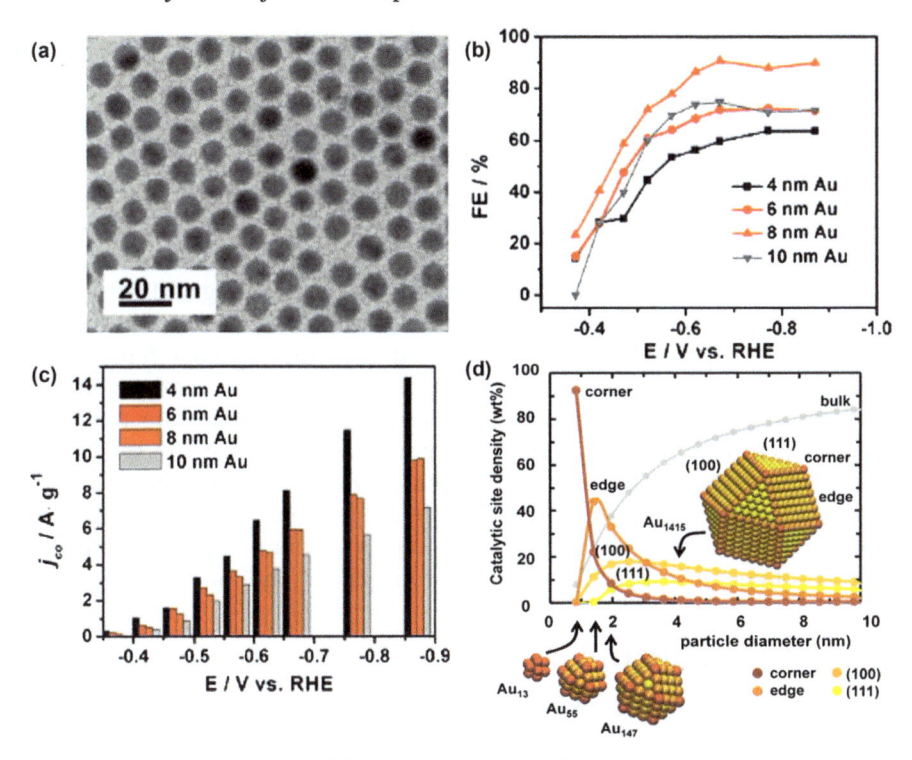

Controllable Synthesis of Metal Nanoparticles 243

Figure 9.13 TEM images of (a) the 8 nm Au NPs. (b) Potential-dependent FEs of the C–Au on electrocatalytic reduction of CO_2 to CO. (c) Current densities for CO formation (mass activities) on the C–Au at various potentials. (d) Density of adsorption sites (yellow, light orange, dark orange, or red symbols for (111), (001), edge, or corner on-top sites, respectively) on closed-shell cuboctahedral Au clusters vs. the cluster diameter. The weight fraction of Au bulk atoms is marked with gray dots.
Reprinted from ref. 69 with permission by American Chemical Society.

9.5 Summary and Perspectives

The worldwide increase in energy demand has driven the intensive research on next-generation electrochemical energy conversion and storage technologies, including fuel cells, batteries, and water electrolyzers. Metal NPs prepared *via* solution phase reactions with controlled sizes, shapes, and compositions are considered essential in developing commercially viable catalysts for applications in energy conversion devices.

This chapter summarizes the general concept in preparing monodisperse NPs. It provides some typical examples, mostly from the authors' laboratory, on using solution phase reactions to control NP nucleation and growth. Specifically, the chapter highlights the syntheses of noble metal (Pt or Au) based multi-metallic NPs with controlled sizes, shapes, compositions, and structures. Such synthetic controls allow the fine tuning of both surface and electronic features, as well as chemical stability, of the NPs, making it possible to control NP catalysis in a more rational way. Several reactions, including ORR, FAOR, and selective CO_2 reduction, are discussed to demonstrate the promising solutions offered by the well-controlled solution phase reaction to rational design of NP catalysts for enhancing or even optimizing their catalysis.

Based on the current understanding of the existing catalysts and on the synthetic fine-tuning on NP parameters, it is now possible to: (i) fundamentally understand the reaction mechanisms using both theoretical calculations and experimental verifications; (ii) rationally design NP structures leveraging ligand and strain effects to maximize the activity and to reduce the usage of noble metals; (iii) develop efficient methodology for the synthesis of fully-ordered intermetallic NPs to achieve further catalytic optimizations; (iv) combine the metal NPs with advanced supporting materials such as graphene and nitrogen-doped graphene to harvest the synergistic effect between NPs and the support to maximize NP catalytic output; and (v) streamline the synthesis strategy for scale-up production of metal NPs. The optimized NP catalysts will have a deep impact on our search for alternative sources of energy, especially on energy conversions and fuel re-generation.

Acknowledgements

Work at Brown University was supported by the U.S. Department of Energy, Office of Energy Efficiency and Renewable Energy, the Fuel Cell Technologies Program, by the U.S. Army Research Laboratory and the U.S. Army Research Office under the Multi University Research Initiative MURI (W911NF-11-1-0353) on 'Stress-Controlled Catalysis *via* Engineered Nanostructures', and by the National Science Foundation under the Center for Chemical Innovation 'CO_2 as a Sustainable Feedstock for Chemical Commodities', CHE-1240020.

References

1. H. L. Wang and H. J. Dai, *Chem. Soc. Rev.*, 2013, **42**, 3088.
2. M. T. M. Koper, *Nanoscale*, 2011, **3**, 2054.
3. J. B. Wu and H. Yang, *Acc. Chem. Res.*, 2013, **46**, 1848.
4. C. B. Murray, C. R. Kagan and M. G. Bawendi, *Annu. Rev. Mater. Sci.*, 2000, **30**, 545.
5. A. P. Alivisatos, *Science*, 1996, **271**, 933.
6. S. H. Sun, C. B. Murray, D. Weller, L. Folks and A. Moser, *Science*, 2000, **287**, 1989.
7. Y. N. Xia, Y. J. Xiong, B. Lim and S. E. Skrabalak, *Angew. Chem., Int. Ed.*, 2009, **48**, 60.
8. M. Niederberger, *Acc. Chem. Res.*, 2007, **40**, 793.
9. V. K. LaMer and R. H. Dinegar, *J. Am. Chem. Soc.*, 1950, **72**, 4847.
10. Y. De Smet, L. Deriemaeker and R. Finsy, *Langmuir*, 1997, **13**, 6884.
11. H. Gratz, *Scr. Mater.*, 1997, **37**, 9.
12. Y. Lee, A. Loew and S. H. Sun, *Chem. Mater.*, 2010, **22**, 755.

13. S. Peng, Y. M. Lee, C. Wang, H. F. Yin, S. Dai and S. H. Sun, *Nano Res.*, 2008, **1**, 229.
14. Y. W. Jun, J. S. Choi and J. Cheon, *Angew. Chem., Int. Ed.*, 2006, **45**, 3414.
15. L. F. Gou and C. J. Murphy, *Chem. Mater.*, 2005, **17**, 3668.
16. R. L. Penn and J. F. Banfield, *Geochim. Cosmochim. Acta*, 1999, **63**, 1549.
17. S. Hofmann, R. Sharma, C. T. Wirth, F. Cervantes-Sodi, C. Ducati, T. Kasama, R. E. Dunin-Borkowski, J. Drucker, P. Bennett and J. Robertson, *Nat. Mater.*, 2008, 7, 372.
18. B. Lim and Y. N. Xia, *Angew. Chem., Int. Ed.*, 2011, **50**, 76.
19. Z. Y. Zhou, N. Tian, J. T. Li, I. Broadwell and S. G. Sun, *Chem. Soc. Rev.*, 2011, **40**, 4167.
20. T. S. Ahmadi, Z. L. Wang, T. C. Green, A. Henglein and M. A. ElSayed, *Science*, 1996, **272**, 1924.
21. J. M. Petroski, Z. L. Wang, T. C. Green and M. A. El-Sayed, *J. Phys. Chem. B*, 1998, **102**, 3316.
22. C. Wang, H. Daimon, T. Onodera, T. Koda and S. H. Sun, *Angew. Chem., Int. Ed.*, 2008, **47**, 3588.
23. M. Chen, J. Kim, J. P. Liu, H. Y. Fan and S. H. Sun, *J. Am. Chem. Soc.*, 2006, **128**, 7132.
24. C. Wang, Y. L. Hou, J. M. Kim and S. H. Sun, *Angew. Chem., Int. Ed.*, 2007, **46**, 6333.
25. V. Mazumder and S. H. Sun, *J. Am. Chem. Soc.*, 2009, **131**, 4588.
26. J. B. Wu, J. L. Zhang, Z. M. Peng, S. C. Yang, F. T. Wagner and H. Yang, *J. Am. Chem. Soc.*, 2010, **132**, 4984.
27. S. J. Guo, S. Zhang, L. H. Wu and S. H. Sun, *Angew. Chem., Int. Ed.*, 2012, **51**, 11770.
28. V. Salgueirino-Maceira, L. M. Liz-Marzan and M. Farle, *Langmuir*, 2004, **20**, 6946.
29. Y. Hou, S. Sun, C. Rong and J. P. Liu, *Appl. Phys. Lett.*, 2007, 91.
30. M. V. Kovalenko, M. Scheele and D. V. Talapin, *Science*, 2009, **324**, 1417.
31. I. Katsounaros, S. Cherevko, A. R. Zeradjanin and K. J. J. Mayrhofer, *Angew. Chem., Int. Ed.*, 2014, **53**, 102.
32. S. J. Guo, S. Zhang and S. H. Sun, *Angew. Chem., Int. Ed.*, 2013, **52**, 8526.
33. F. Jaouen, E. Proietti, M. Lefevre, R. Chenitz, J. P. Dodelet, G. Wu, H. T. Chung, C. M. Johnston and P. Zelenay, *Energy Environ. Sci.*, 2011, **4**, 114.
34. Z. W. Chen, M. Waje, W. Z. Li and Y. S. Yan, *Angew. Chem., Int. Ed.*, 2007, **46**, 4060.
35. J. K. Norskov, J. Rossmeisl, A. Logadottir, L. Lindqvist, J. R. Kitchin, T. Bligaard and H. Jonsson, *J. Phys. Chem. B*, 2004, **108**, 17886.
36. V. Stamenkovic, B. S. Mun, K. J. J. Mayrhofer, P. N. Ross, N. M. Markovic, J. Rossmeisl, J. Greeley and J. K. Norskov, *Angew. Chem., Int. Ed.*, 2006, **45**, 2897.
37. L. Li, A. H. Larsen, N. A. Romero, V. A. Morozov, C. Glinsvad, F. Abild-Pedersen, J. Greeley, K. W. Jacobsen and J. K. Norskov, *J. Phys. Chem. Lett.*, 2013, **4**, 222.

38. L. Qi and J. Li, *J. Catal.*, 2012, **295**, 59.
39. V. R. Stamenkovic, B. S. Mun, M. Arenz, K. J. J. Mayrhofer, C. A. Lucas, G. F. Wang, P. N. Ross and N. M. Markovic, *Nat. Mater.*, 2007, **6**, 241.
40. V. R. Stamenkovic, B. Fowler, B. S. Mun, G. F. Wang, P. N. Ross, C. A. Lucas and N. M. Markovic, *Science*, 2007, **315**, 493.
41. J. L. Zhang, M. B. Vukmirovic, Y. Xu, M. Mavrikakis and R. R. Adzic, *Angew. Chem., Int. Ed.*, 2005, **44**, 2132.
42. I. E. L. Stephens, A. S. Bondarenko, F. J. Perez-Alonso, F. Calle-Vallejo, L. Bech, T. P. Johansson, A. K. Jepsen, R. Frydendal, B. P. Knudsen, J. Rossmeisl and I. Chorkendorff, *J. Am. Chem. Soc.*, 2011, **133**, 5485.
43. P. Strasser, S. Koh, T. Anniyev, J. Greeley, K. More, C. F. Yu, Z. C. Liu, S. Kaya, D. Nordlund, H. Ogasawara, M. F. Toney and A. Nilsson, *Nat. Chem.*, 2010, **2**, 454.
44. D. L. Wang, H. L. L. Xin, R. Hovden, H. S. Wang, Y. C. Yu, D. A. Muller, F. J. DiSalvo and H. D. Abruna, *Nat. Mater.*, 2013, **12**, 81.
45. S. H. Sun, *Adv. Mater.*, 2006, **18**, 393.
46. W. Chen, J. M. Kim, S. H. Sun and S. W. Chen, *J. Phys. Chem. C*, 2008, **112**, 3891.
47. J. Kim, Y. Lee and S. H. Sun, *J. Am. Chem. Soc.*, 2010, **132**, 4996.
48. S. Zhang, X. Zhang, G. M. Jiang, H. Y. Zhu, S. J. Guo, D. Su, G. Lu and S. H. Sun, *J. Am. Chem. Soc.*, 2014, **136**, 7734.
49. C. H. Cui and S. H. Yu, *Acc. Chem. Res.*, 2013, **46**, 1427.
50. S. J. Guo, D. G. Li, H. Y. Zhu, S. Zhang, N. M. Markovic, V. R. Stamenkovic and S. H. Sun, *Angew. Chem., Int. Ed.*, 2013, **52**, 3465.
51. H. Y. Zhu, S. Zhang, S. J. Guo, D. Su and S. H. Sun, *J. Am. Chem. Soc.*, 2013, **135**, 7130.
52. Q. Zhang, I. Lee, J. B. Joo, F. Zaera and Y. D. Yin, *Acc. Chem. Res.*, 2013, **46**, 1816.
53. R. G. Chaudhuri and S. Paria, *Chem. Rev.*, 2012, **112**, 2373.
54. H. Yang, *Angew. Chem., Int. Ed.*, 2011, **50**, 2674.
55. C. Wang, D. van der Vliet, K. L. More, N. J. Zaluzec, S. Peng, S. H. Sun, H. Daimon, G. F. Wang, J. Greeley, J. Pearson, A. P. Paulikas, G. Karapetrov, D. Strmcnik, N. M. Markovic and V. R. Stamenkovic, *Nano Lett.*, 2011, **11**, 919.
56. V. Mazumder, M. F. Chi, K. L. More and S. H. Sun, *J. Am. Chem. Soc.*, 2010, **132**, 7848.
57. V. Mazumder, M. F. Chi, K. L. More and S. H. Sun, *Angew. Chem., Int. Ed.*, 2010, **49**, 9368.
58. S. J. Guo, S. Zhang, D. Su and S. H. Sun, *J. Am. Chem. Soc.*, 2013, **135**, 13879.
59. X. W. Yu and P. G. Pickup, *J. Power Sources*, 2008, **182**, 124.
60. R. Larsen, S. Ha, J. Zakzeski and R. I. Masel, *J. Power Sources*, 2006, **157**, 78.
61. J. Rossmeisl, P. Ferrin, G. A. Tritsaris, A. U. Nilekar, S. Koh, S. E. Bae, S. R. Brankovic, P. Strasser and M. Mavrikakis, *Energy Environ. Sci.*, 2012, **5**, 8335.

62. V. Mazumder, M. F. Chi, M. N. Mankin, Y. Liu, O. Metin, D. H. Sun, K. L. More and S. H. Sun, *Nano Lett.*, 2012, **12**, 1102.
63. S. Zhang, S. J. Guo, H. Y. Zhu, D. Su and S. H. Sun, *J. Am. Chem. Soc.*, 2012, **134**, 5060.
64. M. Gattrell, N. Gupta and A. Co, *J. Electroanal. Chem.*, 2006, **594**, 1.
65. D. T. Whipple and P. J. A. Kenis, *J. Phys. Chem. Lett.*, 2010, **1**, 3451.
66. Y. Hori, *Modern Aspects of Electrochemistry*, ed. C. G. Vayenas, *et al.*, Springer-Verlag, London, 2008, vol. 42, pp. 89–189.
67. B. A. Rosen, A. Salehi-Khojin, M. R. Thorson, W. Zhu, D. T. Whipple, P. J. A. Kenis and R. I. Masel, *Science*, 2011, **334**, 643.
68. Y. H. Chen, C. W. Li and M. W. Kanan, *J. Am. Chem. Soc.*, 2012, **134**, 19969.
69. W. L. Zhu, R. Michalsky, O. Metin, H. F. Lv, S. J. Guo, C. J. Wright, X. L. Sun, A. A. Peterson and S. H. Sun, *J. Am. Chem. Soc.*, 2013, **135**, 16833.

CHAPTER 10

Investigating Nano-structured Catalysts at the Atomic scale by Field Ion Microscopy and Atom Probe Tomography

CÉDRIC BARROO,[a] PAUL A. J. BAGOT,*[b] GEORGE D. W. SMITH[b] AND THIERRY VISART DE BOCARMÉ*[a]

[a] Université Libre de Bruxelles, Faculty of Sciences, Chemical Physics of Materials, CP243, B-1050, Brussels, Belgium; [b] University of Oxford, Department of Materials, Parks Road, Oxford, OX1 3PH, UK
*Email: paul.bagot@materials.ox.ac.uk; tvisart@ulb.ac.be

10.1 Introduction

Since the early developments of surface science, methods for the study of catalysis have raised questions regarding the transferability of the results obtained with high resolution microscopies and spectroscopies of very well-defined chemical systems to the macroscopic scale of a catalytic reaction occurring in an industrial reactor. These issues, known as the materials gap and the pressure gap, still provoke intense discussions between researchers of both communities.[1] As a catalyst is often a dispersion of nanoparticles on a support of high specific area, the ultimate goal for a researcher is to get access to the structures and the instantaneous surface composition of a particle in its catalytically working state. Many reactions are such that only selected surface atoms are to be considered as active sites. Surface

---

RSC Catalysis Series No. 22
Atomically-Precise Methods for Synthesis of Solid Catalysts
Edited by Sophie Hermans and Thierry Visart de Bocarmé
© The Royal Society of Chemistry 2015
Published by the Royal Society of Chemistry, www.rsc.org

reconstructions can occur and modify the number of active sites, either prior to the establishment of a steady state regime, or during the lifetime of the catalyst as it ages. These reconstructions are driven by the minimization of the total free energy of a system[2-4] and can be triggered by thermal and/or chemical treatments. In addition to that complexity, alloys are very often used in applied catalysis. At the microscopic scale, the surface composition of the very first atomic layers often differs from that of the bulk leading in some cases to extreme situations where a core–shell structure develops. This surface enrichment is also affected by external thermal treatments and exposures to reactive species, sometimes with opposing trends, making it possible to fine-tune the surface composition with a high degree of accuracy. In systems where expensive noble metals are employed, surface segregation can deliberately be engineered to reduce the total overall load of these metals. The variations of the relative concentrations of metals usually affect mainly the very first atomic layers of the surface. To control and to adjust these concentrations to desired values using atomically precise methods, experimental techniques are necessary that are able to map the three-dimensional structure of nanocrystals before and after physico-chemical treatments.

Field emission (FEM), field ion (FIM) and field desorption microscopies (FDM) are analytical techniques based respectively on field electron, field ion, and field desorption emission in a projection-type microscope. A sharp tip sample points towards a screen, sometimes in the presence of a gas. When the voltage on the tip is raised to adequate values, the electric field allows for gas ionization or electron emission.[5] The charged particles are accelerated towards the screen where their kinetic energy is converted into light. These methods can thus be used to image the arrangement of surfaces or adsorbed species at the surface of the extremity of a nanosized metal tip. Well before scanning tunnelling microscopy (STM), FIM was the first technique able to spatially resolve individual atoms in direct space by ionization of gaseous species,[6,7] with a typical resolution of ~ 2 Å. It was developed from FEM which is based on the emission of electrons in the presence of an electric field, with a lower resolution of some 20 Å. Both FEM and FIM are imaging techniques that—by contrast to STM—provide a complete picture of the instant electron/ion emission distribution from the entire imaged surface area. This allows the study of spatially correlated phenomena and investigations of local surface reaction kinetics on an atomic-scale.[8-13] The studied sample, the apex of a nanosized needle, exhibits a heterogeneous surface formed by differently oriented nanofacets and can thus serve as a suitable model for a catalytic particle of similar sizes. Figure 10.1 illustrates this general idea. Figure 10.1a shows a transmission electron micrograph of a supported catalyst. One single catalyst particle is modeled by the extremity of a sharp tip (Figure 10.1b). A typical FIM picture of a metal tip (rhodium in this case) is presented in Figure 10.1c, together with a comparison ball model having the same crystal lattice (Figure 10.1d). Each ball of the model represents one single atom. Specimens of the type shown in Figure 10.1c usually serve as the starting point for experiments under conditions where

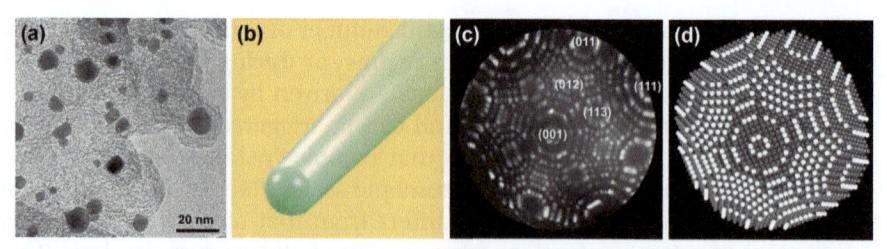

Figure 10.1 (a) Transmission electron micrograph of a carbon supported platinum catalyst (Adapted with permission from ref. 14. Copyright (2014) American Chemical Society). (b) Typical tip shape used for field emission/ion microscopies. (c) Field ion micrograph of a clean rhodium tip sample imaged in the presence of neon gas at ~ 50 K. (d) Ball model depicting the structure of a curved face-centered cubic metal crystal, built with an end radius similar to the sample imaged in (c).

surface reconstruction, catalysis or surface enrichment occurs. The tip surfaces can be prepared to high reproducibility while characterizing them with atomic resolution by FIM imaging.

Studies using FIM and FEM on model systems mimicking the catalytic behavior of individual metal nanoparticles provide data that are not accessible to conventional studies of supported catalysts. The temporal and spatial resolution of field emission techniques are such that the local and transient features of the individual nanoparticles can be imaged and even—as will be described in detail—chemically probed. This makes possible *e.g.*, the detection of fluctuation-induced effects.[15] The fluctuations are confined to one single particle (a few nanometers in size) and can cause noise-induced kinetic transitions in the nanosized reaction systems. These lead to severe deviations from the kinetics of the same reaction studied on the mesoscopic or macroscopic scale.

The application of FIM and FEM can provide a number of kinetic features over a single nanosized catalyst particle; however, the determination of the local chemical composition remains an issue that is still the subject of current developments.[16,17] This identification is achieved by removing surface material from the tip extremity as ions by field desorption (if ionization is restricted to the adsorbed species), or by field evaporation (if the atoms of the tip material are ionized). The fields required to remove adsorbates are usually appreciably lower than those required to remove surface atoms. Once formed, the ions are accelerated and collected by an ion detector. To measure the time elapsed between the formation of the ion and its detection—allowing then the determination of its mass to charge (m/z) ratio by time-of-flight mass spectrometry—pulses are necessary to trigger ionization within a very narrow time interval. These pulses are imposed by nanosecond voltage pulses or thermal pulses using picosecond or femtosecond laser sources. The combination of field emission methods with time-of-flight mass spectrometry bears the generic term of 'atom probe microscopy' (APM).[18,19]

Investigating Nano-structured Catalysts by FIM and APT

To unravel the *local* chemical composition of the imaged surfaces with nanometric or even atomic spatial resolution, accurate determination of the location of ionization is needed. An attractive feature of FIM is that it can be quite easily combined with a time-of-flight mass spectrometer to construct a local chemical probe able to measure the surface composition on some tens of square nanometers while imaging the reaction on the entire surface of the tip apex. The first generations of such atom probe devices were based on the concept of drilling a small aperture, a 'probe hole',[20–22] in the middle of the imaging screen so that ions passing through the hole could be mass separated. Figure 10.2 illustrates this principle. Because of the equivalence of ion trajectories in electrostatic fields, surface atoms which are ionized and removed from the small part of the imaged region aligned over the probe hole will themselves pass through the hole and enter the mass spectrometer.[23] The dark region on the micrograph is therefore the area which is subject to chemical analysis. In this example (see Section 10.3.1.2), the surface composition of a rhodium tip sample is investigated during the ongoing water production in the presence of a reactive H_2/O_2 gas mixture.[16]

A significant breakthrough in the development of atom probes came with the introduction of position sensitive ion detectors[24] (Section 10.2.3.1). The latest generations of atom probes are such that several millions of ion impacts per minute can be processed individually.[19] In this case, a FIM screen is no longer strictly necessary. These atom probes are mainly used for materials studies because they routinely provide three-dimensional maps of individual atoms constituting a sample, with close-to-atomic spatial

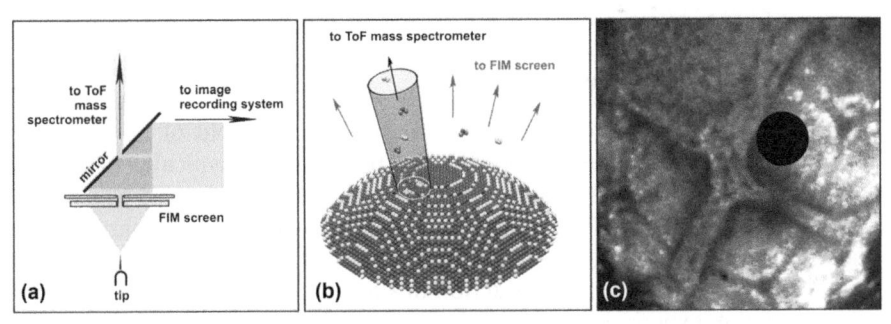

Figure 10.2 (a) Schematic principles for the first design of the atom probe. Gas Ions hitting the FIM screen provide a micrograph of the tip apex. Field desorbed or field evaporated ions passing through the probe aperture are mass separated and identified by their flight time. (b) Scheme of the probe-hole principle: the hole in the screen of the device shown in (a) selects a small area of the tip sample, possibly during a catalytic reaction in progress. (c) Field ion micrograph during the ongoing $H_2 + O_2$ reaction. The micrograph was taken under experimental conditions suitable to investigate the bistability of the reaction ($pO_2 = 1.5 \times 10^{-3}$ Pa, $pH_2 = 2.5 \times 10^{-3}$ Pa, $T = 500$ K). The field ion micrograph suggests the simultaneous presence of nanoscale regions of strongly different local composition on the surface.

resolution. The three-dimensional atomic-scale reconstruction technique has become known as atom probe tomography (APT). This makes possible the establishment of the relationships between the structure of a material, down to the atomic level, with its macroscopic features (ductility, hardness,[25] magnetism,[26] radiation damage,[27] *etc.*). Today, these atom probes are recognized as mainstream microscopy tools and are the subject of a dramatic increase of interest, evidenced by a recent increase in the number of publications from academic and industrial laboratories that now have such facilities. For catalysis studies, these atom probes can unravel subtle variations of the surface and bulk composition of catalyst particles after exposure to reactive environments.

This chapter will first outline the basic principles of APM. Subsequent sections will address selected examples of its application in catalysis studies with an increasing level of complexity, as follows: (i) adsorption studies of single gases on single metal tip samples, (ii) the imaging and the chemical probing of nanosized tips during catalytic reactions; (iii) the study of segregation effects of alloy tips samples after physico-chemical treatments, and (iv) APT studies on single catalyst particles extracted from supported catalysts used, amongst others, for the Fischer–Tropsch reaction. This bottom-up approach will aim to show the value of APM for the development of solid catalysts with an atomic level of precision.

10.2 Imaging and Local Chemical Analysis of Nanosized Crystals Before, During and After Catalysis

10.2.1 Sample Preparation

All APMs require the presence of an intense electric field of the order of 2–50 V nm^{-1}. When the voltage on the tip is raised typically to several kilovolts, these values are achieved by field enhancement at the highly curved surfaces.[5] For this reason, specimens are shaped as a sharp needle tip, the extremity of which is the feature of interest with typical radii of curvature of 5–100 nm. The production of such samples is thus an important aspect of the experimental procedure. Extended and recent reviews on sample preparation are available for a wide variety of materials.[18,28] We will address here those relevant for catalysis studies.

10.2.1.1 *Electropolishing Methods*

Electrochemical polishing was historically the very first method to produce tip samples from wires or 'matchsticks' of bulk materials and is still very widely used. An electrochemical cell is built so as to progressively dissolve the outermost material of the wire in a controlled manner when appropriate conditions are used. The identification of optimum conditions relies on

Investigating Nano-structured Catalysts by FIM and APT 253

empirical investigations. Recommended procedures are listed in most textbooks on APM.[5,23,29] Usually, a positive voltage is applied to the sample material to render it as the cell anode while the counter electrode acts as the cathode. For some materials, alternating voltage is better suited for the development of a sharp needle. A current indicates the occurrence of an oxidation–reduction process involving the oxidation of metal species which are removed as complex ions in the liquid phase from the specimen material. Electropolishing is achieved either in beakers in a pure electrolyte or in biphasic fluids where the top phase contains the electrolyte and the bottom phase is electrochemically inactive. This latter set-up serves to develop necked regions that eventually become the tip apex. For a number of metals, the process can be finalised in a small electrolyte droplet suspended by capillarity inside a gold or platinum loop of 3–4 mm diameter (Figure 10.3a). In the case of platinum group metals (especially iridium,

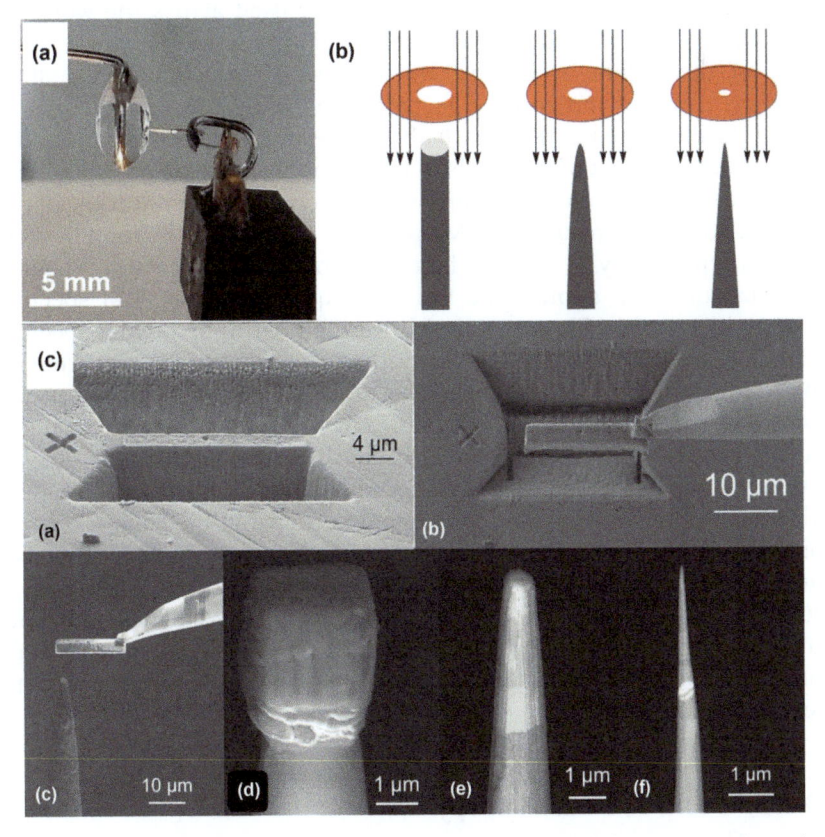

Figure 10.3 (a) Sample preparation by the electropolishing procedure in a suspended electrolyte drop. (b) Schematic principle of the annular ion milling process using circular masks of decreasing diameters. (c) Demonstration of the sample preparation of a targeted region on a flat material using FIB.
(Courtesy of S. V. Lambeets).

rhodium and platinum), electropolishing in molten salt mixtures at ~520 °C gives the best results. However, if this procedure is used, rigorous cleaning of the specimens must be carried out after electropolishing. The evolution of sample preparation is usually followed by optical microscopy. In spite of limited resolving power (~200 nm) well above the typical size of tip apex, it usually suffices to qualitatively assess the sharpness of a specimen and thus its suitability for imaging by APM. Electropolishing systems make use of inexpensive basic materials comprising a AC- and DC power supplies, beakers, melting pots, furnaces, counter electrode material, and an optical microscope.

10.2.1.2 Focused Ion Beam Methods

For materials which are not suited for electropolishing, focused ion beam (FIB) methods are the alternative now widely used across the world. The general concept is to make use of accelerated ions so as to confer a desired shape of the region of interest by erosion. To develop needle shaped geometry, annular ion milling is performed with decreasing inner diameters, as illustrated in Figure 10.3b. The level of control of FIB methods allows a wide number of materials to be prepared as sharp tips. FIB methods have made a significant contribution to widening the application of APM. The first successful results of FIM specimen preparation by the FIB technique was reported by A. R. Waugh *et al.* using a liquid metal ion source.[30] Later, Larson *et al.* applied the FIB technique to prepare FIM specimens from multilayer thin films.[31,32] By fabricating pre-thinned blank specimens using an annular ion beam, the authors succeeded in preparing FIM tips from multi-layer thin films in the direction perpendicular to the film layers. They also succeeded in analysing chemical compositional changes at the interface of the multi-layers with atomic layer resolution. Today, the development of so-called 'lift-out' procedures allows for the production of hybrid samples. In this case, the sample is prepared by using a truncated cone of a conducting metal as a base, with the material of interest being fixed on the flat part of the post. Examples of this procedure are shown in recent papers[33] and one is illustrated by Figure 10.3c. Well-chosen regions of flat materials can now be routinely characterized by APT. Single catalytic particles can also be fixed on the top of sharp posts for the same purpose (Section 10.3.3.2).

10.2.2 Field Ion Microscopy

The Field Ion Microscope (FIM) is an evolution of the Field Electron Emission Microscope (FEM) developed earlier by the same inventor, Erwin W. Müller. The FIM permitted single atoms to be imaged for the first time in October 1955.[7] A historical account of this event has been reported by A. Melmed.[34]

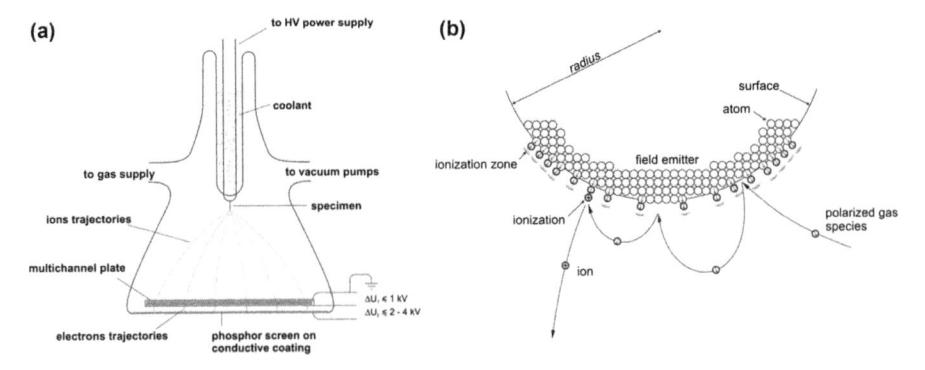

Investigating Nano-structured Catalysts by FIM and APT

Figure 10.4 (a) Schematic diagram of a field ion microscope (b) Ionization of a polarized gas atom at the vicinity of the tip apex.

10.2.2.1 Principles

A schematic diagram of a FIM is presented on Figure 10.4a. The specimen tip is positively biased, cooled to cryogenic temperatures and a low amount of image gas—typically helium or neon at 10^{-5} mbar—is introduced into the system. A highly magnified image of the surface of the tip apex is formed on the screen placed in front of the sample. The field ion microscope is elegant in its simplicity of design and operation. As the voltage on the specimen is increased, the image gas atoms close to the apex of the specimen are polarized and attracted to the specimen by the inhomogeneous high electric field. When the voltage on the specimens is increased to generate a sufficiently high field, typically 30–50 V nm^{-1}, the image gas atoms which come close to the surface are field ionized. The resulting ions are projected towards a screen positioned about 50 mm in front of the specimen. The field ion image is the result of field ionization of gas atoms over individual atoms on the surface of the entire apex region of the specimen. Accordingly, a field ion image such as shown in Figure 10.1c depends on ionization probability of a gas above a curved and charged conducting surface. By further increasing the field on the specimen, the surface atoms themselves can be field ionized and evaporated from the specimen as charged species. This process is termed field evaporation and is useful to clean the tip sample and also to permit the interior of the specimen to be examined. The phenomenon is fundamental to the development of APT. Microchannel plate image intensifiers are positioned in front of the phosphor screen to increase the brightness on the screen by a factor of approximately 10 000.

10.2.2.2 Applications in Catalysis

The aim of the present chapter is to demonstrate the importance of providing real-time information on surface reactions by imaging and chemically monitoring variations in adsorbate coverages and underlying surface structures. FIM and atom-probe techniques are respectively well suited for these goals and both are capable of providing information on the nanostructure of model catalysts. The co-existence of a considerable number of small planes and facets, separated by steps, make field emitter tips ideal to study coupling effects *via* surface diffusion.[35] However, the presence of an imaging field may introduce some complexity that must be considered in 'operando' studies of surface reactions. Processes like field dissociation, charge transfer, *etc.* may lead to misinterpretation if field effects are neglected.[36,37] With the probe-hole method illustrated in Figure 10.2, a local chemical analysis can be made during the ongoing reaction by using field pulses.[21] Their amplitude can be chosen so as to desorb all the gaseous species with one pulse. With this procedure, a fresh surface is restored after each pulse and kinetic studies can be undertaken by a systematic variation of the pulse repetition rate, from 0.1 ms to seconds. Another approach is to use a static imaging voltage and to add pulses of moderate amplitude which do not affect significantly the surface reaction in progress. This so-called 'pulsed field desorption mass spectrometry' (PFDMS) can therefore be considered as a partially destructive method of analysis. Unfortunately, the absence of commercially available designs means that these methods are currently restricted to a limited number of laboratories worldwide.

10.2.3 Atom Probe Tomography (APT)

10.2.3.1 Principles

Atom-probe tomography (APT) uses a position-sensitive detector to determine the original location of atoms field evaporated from the tip. The most widely used position sensitive single ion detectors are currently based on a set of crossed delay lines.[38] When a charged particle hits one line of a conducting material, it causes a variation of the electric potential that propagates towards the extremity of the line. The time difference between the signal detection at the two ends of the line indicates where the charged particle interacted with the line. Multi-hit events, *i.e.* when several ions hit the detector within the time window after a single evaporation pulse, are spatially resolved by crossing two or three delay lines. A microchannel plate electron multiplier is mounted in front of the delay line detector. This converts the initial ion impact into a highly localized, intense shower of electrons, of sufficient amplitude for detection by the delay line electronics system. As illustrated on Figure 10.5, the latest generations of instruments for APT incorporate a local counter-electrode situated in front of the tip. This evolution allowed a considerable reduction of the flight paths and operating voltages without sacrifice of the spectral resolution. It has been termed the Local Electrode Atom Probe (LEAP™) by the inventors and is commercially available. These instruments offer considerable advantages in terms of their

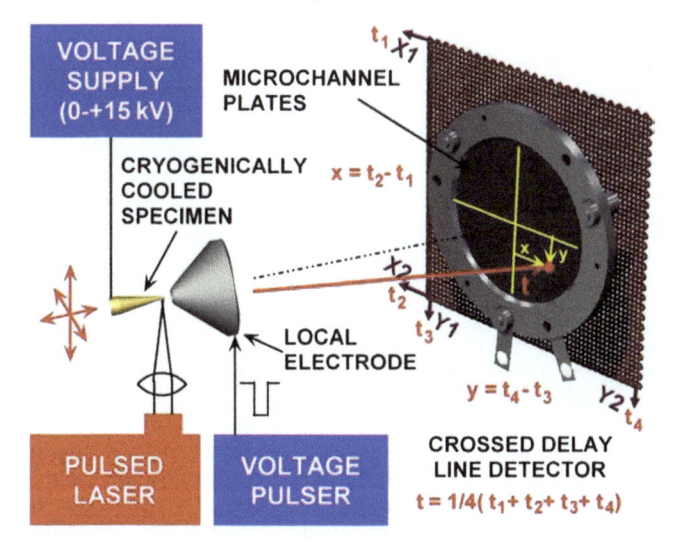

Figure 10.5 Scheme of the atom-probe set-up with a local counter-electrode and a delay line detector.
(Reprinted from ref. 39, copyright (2014), with permission from Elsevier).

wide field of view and high data collection rates (up to 5×10^6 ions per minute), enabling much larger sample volumes to be studied. Equally important is the ability to use laser pulsing to remove ions rather than voltage pulses (Figure 10.5). The use of green or especially UV-wavelength lasers now enables the routine analysis of fragile, poorly conductive materials such as thick oxide layers.

The combination of the flight time and the impact position of the ions on the detector provides the necessary data to reconstruct a three-dimensional map of the probed material. The accuracy of the reconstruction algorithm in APT is not straightforward and directly relies on the understanding and knowledge of two physical mechanisms. The first is the erosion of the typically conical sample by field evaporation and the second is the trajectory of ions from the specimen surface onto the position-sensitive detector. As discussed in the recent review by Vurpillot *et al.*,[40] standard reconstruction algorithms have been mostly based on empirical description of the field evaporation process and from experimental observations by FIM. Examples of atom map reconstructions of materials of interest in heterogeneous catalysis are shown in Figure 10.6. APT reconstructions from the samples exposed to oxidation were generated and focused on the oxide–metal interfaces. In comparison with the aluminium samples, the extent of oxidation is markedly less on Pt-based alloys. This is expected on these noble metal surfaces, and demonstrates the applicability of APT methods for studying relative behaviour trends in a range of metal systems, including those relevant for catalysis.

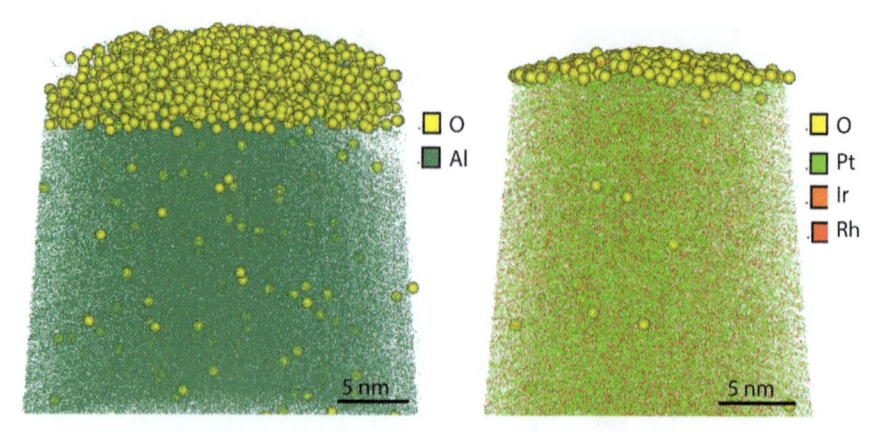

Figure 10.6 Atom maps of Al (left) and Pt-17.3%Rh-14.0%Ir (at%) (right) following oxidation in a separate environmental cell under 20 mbar of O_2 for 30 min at 723 K. The relative thickness of the respective layers demonstrates that Al much more readily oxidizes than the Pt-based alloy. (Reprinted from ref. 17, copyright (2014), with permission from Elsevier).

10.2.3.2 Reaction Cells for Catalysis Studies

For catalysis studies, the topmost atomic layers of tip samples are the region of interest whereas most applications of APT focus on the bulk composition of materials. For this reason, adsorption studies are only relevant if reactive surfaces are not exposed to air contamination during transfers between gas exposures and atom probe analyses. Bagot *et al.* developed an experimental set-up in which an environmental reaction cell is mounted on an atom probe instrument.[41] In this catalytic atom probe (CAP), the reaction cell and gas-dosing rig were designed and integrated into the machine to allow specimens to be exposed to various gas mixtures at pressures of up to one atmosphere and at temperatures of up to 873 K. Sample transfers are then performed under high vacuum (HV) conditions. Among a number of studies on segregation effects produced by physico-chemical treatments (Section 10.3.2), the method has shown its capability to map nitric oxide molecules (NO) selectively adsorbed at the edge of a (012) region of a tip sample after exposure to 10 mbar NO for 15 min at 292 K in the reaction cell.[41] Furthermore, the detector area in this atom probe is comparable with the projected size of specific crystallographic poles, allowing exploration of how different surfaces may react differently to the same exposure conditions.[41,42]

More recently, an integrated environmental cell has been designed and developed for the latest generation of APT LEAP™ instruments, allowing controlled exposure of samples to gases at pressures up to some tens of millibars and at temperatures currently up to ~800 K.[17] This new design for an environmental APT offers a new dimension for studying the influences of high pressure and temperature as well as the effect of surface crystallography and surface curvature on gas–solid interactions. This ability to bridge the pressure gap and the materials gap can be considered as a milestone in the continuous development of new classes of characterization tools for catalytic materials formulations before and after their exposure to conditions approaching those of applied catalysis.

10.3 Case Studies

10.3.1 Catalytic Reactions and Surface Reconstructions on Metals: FIM Studies

The design of nanosized particles for catalysis only makes sense if the equilibrium shape, structure and composition of the catalyst are known at a molecular level. Adsorption can modify not only the local surface energy but also the metal–metal interaction; the mobility of some atoms can therefore lead to new morphologies. The catalytic reaction itself may have similar effects. If a surface reaction is sensitive to the structure,[1] morphological changes may influence the overall catalytic process. These morphological/structural transformations during the reaction and their influences over the local catalytic activity must be assessed experimentally in catalysis studies.

FEM and FIM methods have been used to characterize the tip apex of samples before, during and after gas adsorption and/or catalytic reactions. As previously mentioned, the very apex of the tip closely approaches the structure of one single nanoparticle. Both methods provide a way to investigate the influence of these changes over the catalytic activity. From a quasi-hemispheric sample, reconstructions may be thermally activated,[43] and/or assisted by the addition of reactive gas species at the surface of the catalyst.[44] One single adsorbate species often suffices to increase the mobility of the surface atoms, whereas in some cases, the reaction between two species is required to form the surface compound responsible for surface reconstructions.[45] In most cases, it is a subtle combination of the applied temperature and gas exposure that leads to a variety of reconstructed particle shapes, making theoretical predictions of the latter difficult, even for metals with similar electronic properties.

The following paragraphs describe recent work which gives direct evidence of reconstructions on Pd, Ir, Rh and Pt catalysts after exposure to oxygen gas or during catalytic reactions.

10.3.1.1 O_2/Pd and O_2/Ir

The first cases presented are the oxygen-induced faceting of Pd[46] and Ir,[47] which are good examples of morphological reconstructions due to the adsorption of one single gas species. The procedure used for both samples is similar: a clean hemispherical Pd or Ir field emitter tip is exposed to a chosen dose of O_2 gas at low temperature (55–65 K) under field-free conditions, to produce an oxygen pre-covered sample. This is then annealed for a fixed time (typically 120 s) at a target temperature ranging between 300 and 750 K for Pd, and 300 and 1100 K for Ir, in the absence of an electric field. The sample is then cooled and imaged by FIM using neon as the imaging species. Blank heating experiments in the absence of oxygen are also performed in both cases so as to distinguish oxygen-assisted faceting from thermally induced reconstruction.

In the case of Pd, a (111)-oriented sample has been studied, similar to the one presented in Figure 10.7a. Blank heating experiments do not exhibit any facet development. However, coarsening is observed at a temperature of 500 K and above, and the central pole shrinks when the temperature is increased to 750 K.

When oxygen is dosed on the sample, the development of facets is visible after annealing at 500 K and above. The oxygen dose ranges from 1 to 6 L (1 Langmuir = 1 L = 1.3×10^{-4} Pa s). The formation of dark regions of variable size and their separation by lines of bright spots is consistent with a strong faceting of the crystal (Figure 10.7e). Adjacent facets develop and

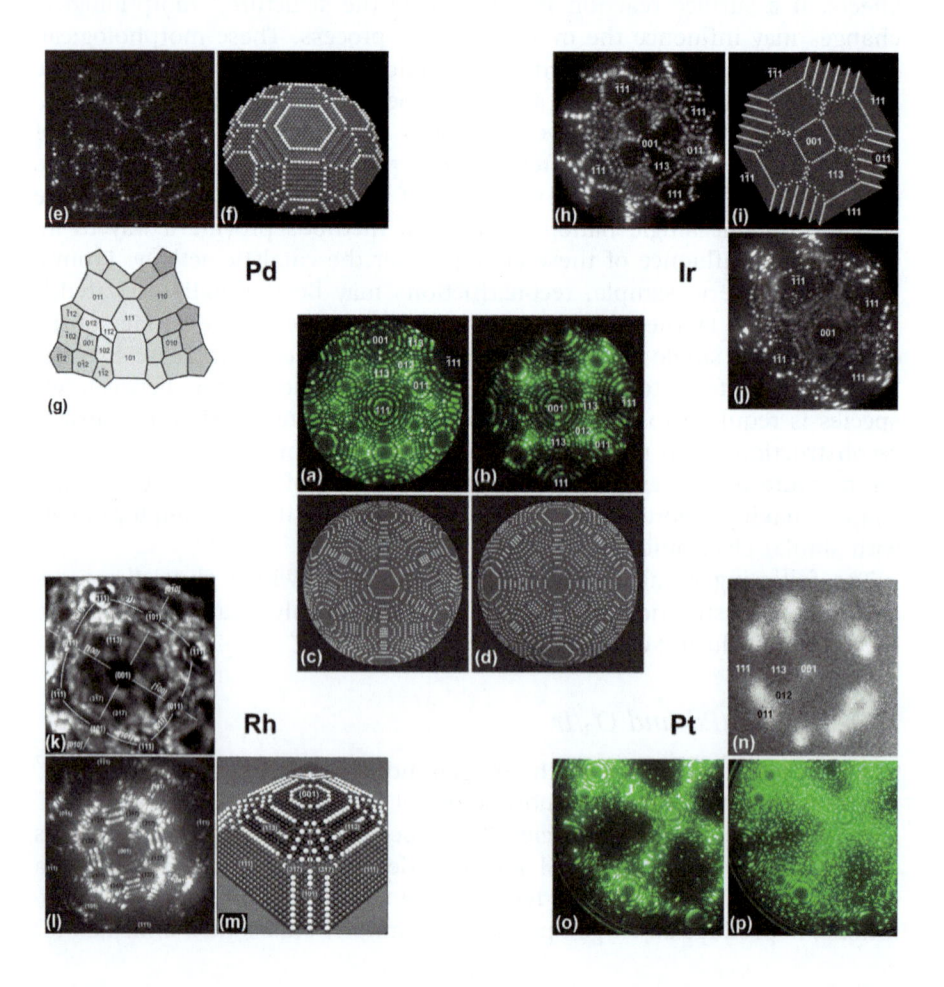

Investigating Nano-structured Catalysts by FIM and APT 261

produce edges which appear brighter on the FIM pattern due to the higher local electric field. The flattening of the facets does not allow atomic resolution of the inner surface areas of these regions. Facet edges are fully developed after annealing at 550–600 K and no further changes are observed at 650 K. For clarity, a ball model is presented in Figure 10.7f to highlight the morphological changes, as compared to the ball model in Figure 10.7c. The faceting occurs as follows: regions around the {001} planes break up into

Figure 10.7 Field ion micrographs of (a) a clean (111)-oriented (three-fold symmetry) and (b) a (001)-oriented (four-fold symmetry) metallic tip with fcc-type crystal lattice (Pt in this case). Imaging conditions $P_{Ne} = 1.0 \times 10^{-3}$ Pa, $T = 55$ K, $F_{tip} \approx 35$ V nm^{-1}. (c) and (d) Ball models of a hemisphere constructed according to an fcc crystal lattice, (111)-, and (001) respectively. An excellent correlation is found between the model catalysts on (a) and (b) and the ball models. (e) Field ion micrograph of a (111)-oriented Pd tip after exposure to 3 L of oxygen and subsequent annealing at 650 K. The development of facets is visible. Imaging conditions: $P_{Ne} = 1.0 \times 10^{-3}$ Pa, $T = 65$ K, $F_{tip} \approx 36$ V nm^{-1}. (f) Ball model of the previous faceted Pd tip apex highlighting the surface reconstruction (as compared to ball model of (c)). (g) Schematic drawing of the Pd tip after oxygen exposure, as well as the Miller indexes of the facets. (h) Field ion micrograph of a (001)-oriented Ir tip after exposure to 3 L of oxygen and subsequent annealing at 700 K. Imaging conditions: $P_{Ne} = 1.0 \times 10^{-3}$ Pa, $T = 55$ K, $F_{tip} \approx 35-36$ V nm^{-1}. (i) Ball model of the previous reconstructed Ir tip apex highlighting the faceting of the sample (as compared to the ball model on (d)). (j) Field ion micrograph of the same Ir tip after annealing at 950 K, without any exposure to oxygen. Faceting along with the expansion of low Miller index planes is observed corresponding to a transition to a polyhedral tip shape. (k) Field ion micrograph of a (001)-oriented Rh tip at $T = 505$ K exposed to an O$_2$–H$_2$ (1:2) gas mixture (total pressure of 4.5×10^{-3} Pa). The imaging species are H$_2$O$^+$ ions resulting from the ionization of the reaction product. (l) Field ion micrograph of a Rh sample after exposure to ~10 000 L of oxygen in the absence of field at $T = 390$ K followed by field evaporation leaving oxygen dissolved into the bulk. The image is taken after a subsequent heating of 30 s duration at $T = 500$ K of the field evaporated crystal. Imaging conditions: $P_{Ne} = 1.0 \times 10^{-3}$ Pa, $T = 63$ K, $F = 35$ V nm^{-1}. (m) Ball model of the previous morphological changes on Rh tip with emphasis on the {137}-type facets. (n) Field emission patterns during catalytic NO$_2$ hydrogenation over Pt (001)-oriented sample during surface explosion due to water production. (001)-center, {011} and {012} facets are active for the reaction, but {113} facets remains inactive. Conditions: $P_{H2} = 1.05 \times 10^{-4}$ Pa, $P_{NO2} = 3.6 \times 10^{-6}$ Pa, $T = 390$ K, $F = 4$ V nm^{-1}. (o) Field ion micrograph of a (001)-oriented Pt tip sample showing dark {113} regions. Imaging conditions: $P_{He} = 2 \times 10^{-3}$ Pa, $T = 29$ K, $F = 37$ V nm^{-1}. (p) Same as in the previous FIM micrograph with an imaging field $F = 43$ V nm^{-1} with resolved {113} facets.
(Panels e–g adapted with permission from ref. 46. Copyright (2010) American Chemical Society. Panels h–j adapted with permission from ref. 47. Copyright (2011) American Chemical Society. Panels n–p adapted from ref. 64. Copyright (2014) with permission from Elsevier).

small {100}, {112} and {012} facets. In addition to this, {011} facets become larger and the (111) center persists. A diagram of the reconstructed Pd tip after oxygen exposure, showing the Miller indices of the facets is presented in Figure 10.7g. The faceting behavior is in accordance with recently calculated equilibrium shapes of Pd crystals in the presence of small amounts of adsorbed oxygen.[48,49]

The overall pattern weakly depends on the oxygen dose (1, 2, 3 and 6 L) which has been calculated to be below the saturation coverage. The major reconstruction observed is thus due to the presence of only sub-monolayer amounts of oxygen which dissociates upon annealing. It is worthy of note that the phenomenology of the faceting process of a quasi-hemispheric tip apex is independent of the crystal orientation.

In the case of Ir, (001)-oriented samples were used to study the morphological changes. Experiments in the absence of oxygen highlight the extension of low Miller index facets at a temperature of 850 K and above. Figure 10.7j depicts an Ir sample after annealing at 950 K where the morphology is dominated by large flat {111} and {001} facets separated by a few straight steps of multiple atomic heights, which corresponds to a transition to a polyhedral shape.[50,51] The exposure of the sample to oxygen also induces reconstructions which differ from the thermally induced reconstruction.

The dosing of 3 L of oxygen and subsequent annealing at 550 K and above forms a center cross-like pattern associated with ⟨011⟩ zone lines. Figure 10.7h shows this structure after annealing at 700 K. The formation of this pattern can be considered as the initial stage of {113} plane development around the central pole. Upon annealing to at least 850 K, {113} facets develop and grow to the detriment of vicinal {012} facets. The final cuboctahedral morphology of the Ir nanosized crystal is represented by a ball model in Figure 10.7i and the shape transformation results from the break-up of all kinked planes to finally expose only facets of {001}, {113}, {111} and {011} type. The 3D ball model also shows the missing row reconstructed {011} facets. Oxygen exposures from 1 to 6 L give very similar results. However, smaller dosing of 0.3 L caused {113} facets to develop upon annealing at 700 K (instead of 850 K) and above, but the tip morphology is not stable at temperatures higher than 950 K.

Note that a careful analysis of the field ion micrograph on Figure 10.7h highlights (1×3) and (1×4) missing row reconstruction of the {011} facets, as has been observed earlier for 2D crystals.[52–54]

These two systems exhibit major reconstruction. Faceting is thermodynamically driven by changes in the surface free energy, which is in turn influenced by adsorption of gas species. In order to be able to observe the faceting, surface diffusion must take place and, in the present cases, adsorbed oxygen lowers the barrier to trigger the mobility of surface atoms within the duration of the experiments. These durations vary from tens of minutes to hours in these reported examples. The systems show that only very small amounts of gas (sub-monolayer) are sufficient to produce drastic morphological changes of the sample.

10.3.1.2 O_2/Rh and $O_2 + H_2/Rh$

Morphological reconstructions over rhodium after exposure to pure oxygen gas imaged by FIM have been presented and compared to the field ion micrographs recorded during the ongoing $O_2 + H_2$ reaction.[55,56] Despite the different experimental conditions used for the two sets of experiments, the similarity of the pictures is quite remarkable. Figure 10.7b shows a clean (001)-oriented sample after field evaporation. Starting from a similar sample, the Rh crystal is exposed to a dose of 10 000 L at 390 K. After this oxygen dosing, the removal of some 30 Rh layers by field evaporation is sufficient to restore the nearly hemispherical sample morphology. Heating this hemispherical sample at $T = 500$ K during 30 s and in the absence of gaseous oxygen, causes the polyhedral reconstruction shown in Figure 10.7l.[57] A clean Rh crystal does not suffer from reshaping at this temperature, and thus, the polyhedral reconstruction is taken as a direct indication of reversible oxygen sub-surface diffusion. It was therefore proposed that the dissolved oxygen atoms diffuse from the bulk to the surface and trigger the reported reconstruction. A much lower dose of 0.5 L of O_2 over a clean Rh tip at $T = 500$ K in the absence of electric field is also sufficient to reach the same result.[58] During the reconstruction process, the only high-index planes present in the polyhedrally reconstructed form are of {137}-type. They appear in pairs on either side of the ⟨100⟩ zones lines, i.e., lines between the (001) central pole and the {011} peripheral facets, and remain stable within a temperature range of 480 K to 530 K. Local reconstructions of the missing-row type are identified in {011} and {113} planes, which are similar to STM results obtained with extended single crystal surfaces.[59,60] A ball model presented in Figure 10.7m illustrates this polyhedral reconstruction. The fact that a local reconstruction of the missing-row type can be clearly identified on field emitter tips is interesting from the viewpoint that planes do not exist independently but may communicate with each other *via* diffusing adsorbates. In comparison to Rh, Pd and Ir morphological reconstructions are due to annealing of samples with pre-adsorbed oxygen and lead to a faceting with mainly low Miller index facets. In the case of Rh, sub-surface oxygen can also act in the reconstruction process. Higher Miller index facets may extend under defined conditions and can be observed during the ongoing catalytic hydrogenation of oxygen.

It is also interesting to notice that in the context of adsorbate-induced restructuring, the $O_2 + H_2$ reaction on Rh is observed to occur on a tip reshaped to a polyhedron.[61] Figure 10.7k shows a field ion micrograph captured during $O_2 + H_2$ reaction over Rh at $T = 505$ K. The H_2 pressure is kept sufficiently high to prevent surface oxidation.[62] Despite a lower resolution as compared to Figure 10.7l, it is evident that the tip has undergone a transformation to coarser and polyhedral structure. The missing-row type reconstructions of the {011} and {113} facets are not visible due to a poorer resolution under these conditions, but the extension of the {137} facets is clearly visible as dark regions of both sides of the ⟨100⟩ zone lines. 1D-atom probe studies indicate that the imaging species are H_2O^+ ions resulting from the ionization of the reaction product.[62] Reconstructions observed under reaction conditions (Figure 10.7k) are undoubtedly assisted by oxygen. A morphological change to such a polyhedral shape is not observed in experiments with pure hydrogen, or in the presence of an electric field alone, or in its absence under the same experimental conditions.

At higher temperatures (up to 600 K) the extension of {113} and {011} facets occurs to the detriment of their vicinal orientations. The origin of the contrast of around the dark extended areas is due to the extension and the flattening of the facets leading to a polyhedral morphology: the displacement of rhodium atoms—probably in the form of mobile Rh_xO_y species—to sites of higher stability caused the appearance of flat areas where the electric field decreases.

The onset of oxidation during the $O_2 + H_2$ reaction shows a strong sensitivity to the surface structure at temperatures of 505 K and lower, and occurs first along the ⟨001⟩ zone lines.[35,63] This effect vanishes with increasing temperatures. A cross comparison of the respective influences of the temperature and the tip morphologies indicates that reconstructions have only a limited influence on the reactivity at temperatures above ∼500 K.

10.3.1.3 O_2/Pt and $NO_2 + H_2$/Pt

This section reports on the occurrence of periodic oscillations during the $NO_2 + H_2$ interaction over Pt tip samples,[64] focussing on the surface transformation from a freshly developed quasi-hemispherical Pt tip to a supposedly polyhedral shape. The transition from the first to the second is observed through the evolution of the surface reactivity.

The Pt catalyst is imaged with high resolution by FIM before (Figure 10.7b) and after the reaction. The reaction itself and related patterns during the ongoing reaction are followed by FEM. The main observations during the ongoing process are as follows: the reaction between NO_2 and H_2 on Pt samples shows self-sustained kinetic instabilities that are expressed as peaks of brightness around the {011} facets. Their ignition is synchronized within the time resolution of the video recording device (40 ms), suggesting the presence of a strong diffusional coupling. During the first minutes of the interaction, while the tip apex is freshly developed and exhibits a quasi-hemispherical shape, only {012} and {011} facets are active for the given reaction. After a time delay of approximately 15 min, facets located along the ⟨001⟩ zone lines become active, as well as the topmost (001)-center. The activity of the (001) facet results from the progressive reconstruction of the Pt tip sample. Finally, during the whole process, the {113} and {111} facets remain totally inactive for the reaction.[8] The activity/inactivity of these facets can be seen on the field emission pattern during surface explosions. A snapshot taken during one of these surface explosions is presented in Figure 10.7n.

The enlargement of major facets occurs to the detriment of surrounding high Miller-index planes, as is the case for rhodium. However, the authors suggest that the lack of brightness is due to the presence of chemisorbed oxygen atoms which inhibit local field emission. Surface oxidation of platinum with the occurrence of sub-surface diffusion of oxygen atoms and the formation of an oxide layer is rather unlikely at a low temperature of 390 K. The oxidation of platinum nanoparticles in the presence of NO_x is only observed at higher temperature (~ 573 K).[65] During the reaction, dissociative adsorption of NO_2 gas takes place leading to NO(ads) and O(ads) species. Adsorbed oxygen atoms are active species which react with fast diffusing H(ads) atoms to form water. From the reported field emission patterns, the surface explosions indicate the location of an ongoing catalytic conversion. This is supported by the fact that adsorbed oxygen is present and substantially increases the work function between surface explosions.[66] The local current density of emitted electrons is lowered accordingly. The reaction of O(ads) with H(ads) causes the local work function to decrease within some 200 ms. This phenomenon is translated to a sudden peak of the local brightness which images surface regions that are transiently metallic. Subsequent adsorption of NO_2 and its dissociation over these areas causes the field emission patterns to darken. Particular attention is paid to the {113} facets which do not show brightness peaks. The imaged area of these regions also extends with time. This can be attributed either to an important extension of these regions which are less reactive towards the $NO_2 + H_2$ reaction, or to the occurrence of a more complex adsorption process which acts on the local field emission.

To further assess the morphological shape of the Pt tip after reaction, low temperature FIM observations were carried out using helium imaging gas. Images are presented in Figure 10.7o and p and lead to conclusions different from those drawn for Rh. The Pt tip sample was first exposed to pure oxygen ($P_{O2} = 2 \times 10^{-3}$ Pa at 300 K during 180 s, in the absence of field), and then imaged at 29 K. Under these conditions, crystal planes lying along the $\langle 100 \rangle$ lines are resolved with atomic resolution. Facets along the $[\bar{1}10]$ line appear dark although they were visible at higher temperatures (not represented). The experiment was conducted to avoid field evaporation as much as possible. In Figure 10.7p, the dark regions centered by the {113} facets are now resolved by field ionization of helium atoms in the presence of a higher imaging field. A careful analysis of the sample excludes the occurrence of field evaporation between the two micrographs. This observation differs from the rhodium case where field evaporation of several tens of atomic layers—counted on the bright regions—are necessary to restore the original field evaporation end form, i.e., a quasi-hemispherical shape. In the case of Pt, the difference between the two field ion micrographs has its origin in the imaging process. The presence of an O(ads) layer over the Pt{113} and vicinal facets surface is proposed, which can drastically decrease the ionization probability over the sample. Pt undergoes only limited structural changes during the ongoing $NO_2 + H_2$ reaction at 390 K. These have important consequences for the reactivity of the (001) plane which becomes catalytically active as the reconstruction proceeds.

10.3.1.4 NO/Pt and NO/Pt-17.4at%Rh

Another example of surface reconstruction can be found after the dosing of nitric oxide on Pt and Pt-17.4at%Rh field emitter tips at a temperature of 573 K.[67,68] The main observations in the Pt case are the presence of brighter spots on the (001)-center, which have been identified as platinum oxides species by APT experiments, and a roughening of the surface around the {012} facets. Thus roughening is likely due to the presence of O-species on an atomically rough region.[69] In the $NO_2 + H_2$ case (Section 10.3.1.3), {113} facets show an apparent extension after O_2 dosing. Changing the oxidizing species to NO gas causes the restructuring to occur on the {012} facets. This difference is not yet fully understood.

In the case of Pt-17.4at%Rh alloy, the same dosing procedure (NO at 10 mbar at 573 K during 15 min) leads to drastic changes in the field ion pattern. As can be seen in Figure 10.8a, dark regions where the atomic structure is not resolved appear in the field ion micrograph. This extension of facets occurs on the {012} regions, as well as the (001) center, as in the case of NO dosing on pure Pt. It was noticeable that the Pt-17.4at%Rh alloy appeared much more susceptible to restructuring under NO than Pt. Field evaporation of a few atomic layers was necessary to recover the initial hemispherical sample.

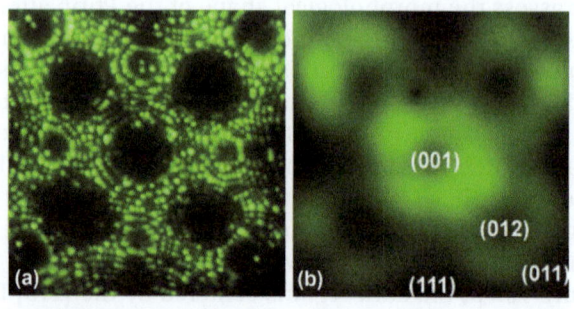

Figure 10.8 (a) Field ion micrograph of a (001)-oriented Pt-17.4at%Rh tip sample after exposure to NO at 10 mbar at 573 K for 15 min in an integrated catalytic cell showing dark regions on {012} facets. (b) Field emission pattern of the same Pt-17.4at%Rh tip after the same treatment showing the presence of dark circles at the {012} facets and a bright ring around the (001) center.
(Reprinted from ref. 67, copyright (2007), with permission from Elsevier).

A field emission image of the same (001)-oriented Pt-17.4at%Rh sample (Figure 10.8b) shows a good correlation with the corresponding field ion micrograph and highlights the presence of dark circles at the {012} facets, as well as one bright ring around the (001) center. These rings are an indication of the edge of a facet or a region attacked by oxygen.

These two examples show that similar systems may behave either in a similar or highly contrasting manner. O_2 dosing on Pt leads to the formation of weakly adsorbed species on {113} facets with a strong influence on the field emission pattern. In contrast NO dosing on Pt leads to the formation of {012}-facet roughening that require field evaporation to recover the original surface. The effect is even more drastic for a Pt-17.4at%Rh alloy where extension of dark regions around {012} facets are visible. However, in the case of alloys, surface segregation must also be considered. Thermal treatment as well as adsorption of chemical species may induce segregation of one of the metals, which may in turn influence the local catalytic activity and the occurrence of morphological reconstruction. The next section is dedicated to 3D atom probe examples of surface segregation in binary and ternary alloys.

10.3.2 Surface Enrichment of Platinum Group Metal-based Alloys: APT Studies

As has been indicated in the previous section, the analysis of alloys is quite challenging by FIM and FEM because the surface composition of alloys usually deviates from the bulk composition. Synergistic effects on alloy catalysts may drastically influence their activity compared to their constituting pure metals. APT allows the study of this behavior with subnanometer resolution. This section reports different behavior of platinum group metals (PGM) alloys subject to various physico-chemical treatments. As briefly described in Section 10.2.3.4, dedicated environmental cells have been developed for atom probe instruments. In order of increasing complexity, the study of binary and then ternary alloys will be presented with a focus on PGM-based alloys for automotive applications.

10.3.2.1 Binary Alloys

10.3.2.1.1 Pd-38at%Au: An Example of Chemically Driven Surface Enrichment. Gold-based catalysts have shown unprecedented catalytic activity for some reactions (CO oxidation or H_2O_2 production[70,71]) and could be considered to improve NO_x selective catalytic reduction (SCR) of exhaust gases. The alloy studied is composed of 62 at% Pd and 38 at% Au.[72] Atom probe analysis of a clean sample heated at 573 K for 20 min under UHV conditions reveals a homogeneous composition within the bulk and at the surface. This is highlighted is the cumulative 'ladder' diagram in Figure 10.9a where the slope remains constant. This diagram plots the

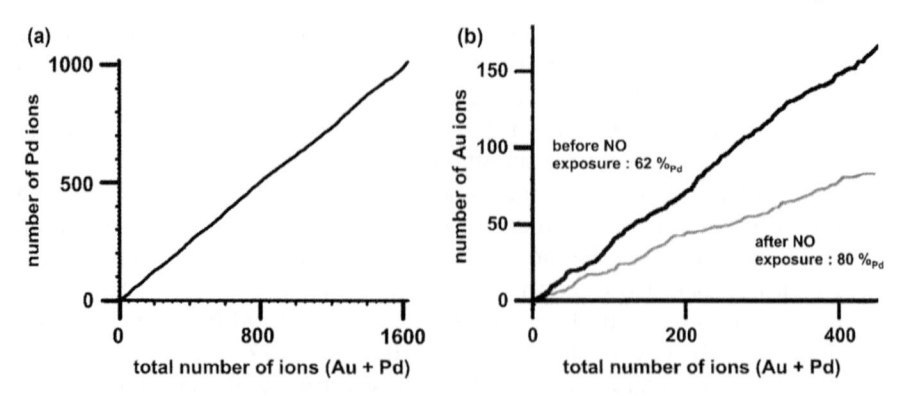

Figure 10.9 (a) Cumulative ladder diagram of a clean Pd-38at%Au alloy. (b) Comparison of the previous diagram after exposure to NO gas at 573 K highlighting surface segregation of Pd, with a concentration of 80 at%. (Reprinted from ref. 72, copyright (2009), with permission from Elsevier).

number of collected ions for a single element (Pd in this case) as a function of the total number of ions collected, so by extension, as a function of the depth that is probed. A straight line in the cumulative diagram denotes a homogeneous phase and its slope is representative of the alloy composition (Figure 10.9a). Any changes in the depth profile are reflected by local changes of this slope. A further exposure of the sample to NO gas (6×10^3 Pa for 15 min at 573 K) and subsequent analysis by APT leads to a drastic change in the slope of the ladder diagram (see Figure 10.9b): Pd reaches a concentration of 80 at% as compared to 62 at% before NO exposure. The same treatment at room temperature does not lead to any surface segregation: the Pd surface enrichment is thus thermally activated. 1D atom probe experiments have shown that NO dissociates on pure Pd field emitter tips, but not on pure Au field emitter tips.[73] The formation of the Pd–O bond and the absence of Au–O bond[74] induces a chemically driven surface enrichment of the Pd atoms by O species. Interestingly, annealing experiments in the absence of oxidizing species lead to the opposite trend, *i.e.*, gold is enriched at the surface. These opposing influences open a way to adjust the surface composition of a catalyst with atomically precise control.

10.3.2.1.2 Pt–Rh Alloys: Towards the Preparation of Pure Pt@Rh Core-Shell Structures

Part 1: From FIM Morphological Reconstruction to Surface Enrichment: NO-adsorption Studies

Pt-Rh alloys are of particular interest because of their use in automotive catalytic converters. Structural changes of Pt-17.4at%Rh samples observed by FIM were described in the previous section. These observations can be complemented by chemical analyses of the surface by APT.[41,67] Surface enrichment in Rh is expected in Pt-17.4at%Rh alloy after exposure to oxidizing gases such as NO and O_2, owing to the favorable Rh–O bond strength over Pt–O.[75] The focus will be on {111}, {001}, {011} and {012} crystallographic orientations using the smaller field-of-view CAP. When Pt-17.4at%Rh is exposed to NO gas during 15 min at a pressure of 10 mbar at increasing temperatures, a general observation is that the number of NO species detected by mass spectrometry decreases as the number of atomic oxygen species increases. At 573 K, few oxygen species are left on the Pt-17.4at%Rh surface except on the {111} regions which desorb significant amounts of RhO and PtO species. On {111} regions, the exposure of NO at increasing temperatures also leads to an increasing segregation of Rh in the very first atomic layer (up to 42 at% at 573 K). The second atomic layer suffers from depletion in Rh species, suggesting the occurrence of diffusion from the bulk to the surface. However, the Rh enrichment exceeds the Rh depletion of the underlying atomic layer. To understand this, an analysis on a wider total area of the surface is necessary. The atom maps established on the {001} regions show a more complex behavior. Rh enrichment is observed after NO exposure below 523 K. This Rh segregation turns into Rh depletion at 573 K and above, with less than 4 at% of Rh at the surface. A careful analysis of the corresponding atom map shows that Rh is present only at the edges of the facet. A plausible explanation is the surface diffusion of Rh away from {001} facets, which explains the depletion of {001} regions as well as the greater enrichment of the {111} regions. Accordingly, the reported surface enrichments result from a complex combination of atom displacements along directions parallel and perpendicular to the surface. The {011} facet is equally complex: surface segregation is observed up to 423 K, beyond which a gradual reduction in surface Rh content with further temperature rises is noted, leading to severe Rh depletion (less than 3 at% at 673 K). Regarding the {012} facet, the combined first and second layers show Rh enrichment up to around 473 K – the trends are less clear here probably because of the inherent roughness of the {012} facets. Increasing the temperature induces a gradual loss of Rh from the first layer, down to a minimum value at 673 K. These results are consistent with the FIM micrographs of the sample showing flat {001} and {012} surfaces.

Similar trends are obtained in the case of O_2 and N_2O exposure. Rh enrichment increases on the {111} facets up to a maximum of 30–35 at%, while Rh depletion is observed on the {001} surface after exposure at 573 K. These experiments indicate that the presence of free oxygen atoms drives surface segregation in Pt-17.4at%Rh. A greater apparent activity is observed in the case of NO exposure as compared to O_2 on the same surface, which may be due to a higher NO sticking coefficient.

In blank experiments, heating the sample at 823 K under UHV conditions, or exposing the sample to N_2 or C_2H_2 (at respectively room temperature and 423 K) did not induce any segregation (similar results were also reported in ref. 76 and 77). A model for such oxidation-driven surface segregation on Pt alloys has also been proposed.[42]

Part 2: APT Analysis During Exposure to Multiple Gas Mixtures

The aim of these experiments is to mimic the operating conditions of a catalytic converter where the catalyst is exposed to a wide range of gases at the same time. As an example, the NO + CO system will be presented. The major challenge in the conversion of exhaust gas remains the reduction of NO_x under oxidizing conditions. As a first step towards a better understanding of this process, $NO + O_2$ reactions will be discussed.[67]

When Pt-17.4 at% Rh alloy is exposed to CO and NO at the same time, for 15 min at a total pressure of 20 mbar, it appears that CO and NO can co-adsorb without interference at 373 K on {111} surfaces. However, increasing the temperature does not lead to Rh surface segregation at 473 K as was the case with pure NO gas. In contrast, depletion of Rh has been observed at 573 K, an observation not seen on any {111} surfaces subjected to a purely oxidising environment. This observation can be explained by the reduction by CO(ads) of the oxide layer resulting from NO dissociation. The top Rh layer can thus be removed forming $Rh(CO)_x$.[78] No oxide can be detected in the analysis of {001} facets. If NO exposure is carried out prior to CO exposure, no oxides are seen as CO likely strips them away. Segregation of Rh at the surface however is still observed, although less pronounced. The effect of CO is thus to prevent the build-up of oxide layers and to reduce the extent of Rh segregation seen under NO exposure alone. This is in good agreement with work from Medvedev *et al.* on reconstructions of Rh tip samples in the presence of O_2 and CO gases.[58]

To investigate the possibility that the alloy dissociates NO under oxidizing conditions, the same Pt-Rh alloy was exposed to pure O_2 gas (at 10 mbar during 15 min), and then exposed to pure NO gas in the same conditions. Greater quantities of oxygen species can be detected as compared to all other previous experiments. On Pt–Rh {111}, the segregation–depletion trend of Rh remains.

On the Pt–Rh {001} region at 473 K, Rh segregation in the first atomic layer and Rh depletion in the second atomic layer is observed. By increasing the temperature to 573 K, oxide species are less abundant on {001} regions as compared to the {111} regions and there is an accumulation at the step edges rather than on central flat region. Apart from these oxides, Rh depletion is observed. Both NO and O_2 are able to drive surface segregation. In the presence of pre-oxidized surfaces, no intact NO molecules were detected confirming the ability of oxidized Pt-17.4 at% Rh to dissociate nitric oxide.

Part 3: From Atom Probe Analysis to Nano-engineering of Core–Shell Structures

The oxidation behavior of Pt-22at%Rh has been studied by atom probe at 873, 973 and 1073 K[79] using the modern LEAP instruments. Strong oxidation of the specimens is observed after treatments for 5 h in 1 bar of oxygen. TEM has been used to observe some specimens before and after oxidation. Figure 10.10a shows TEM images of APT samples after oxidation at three

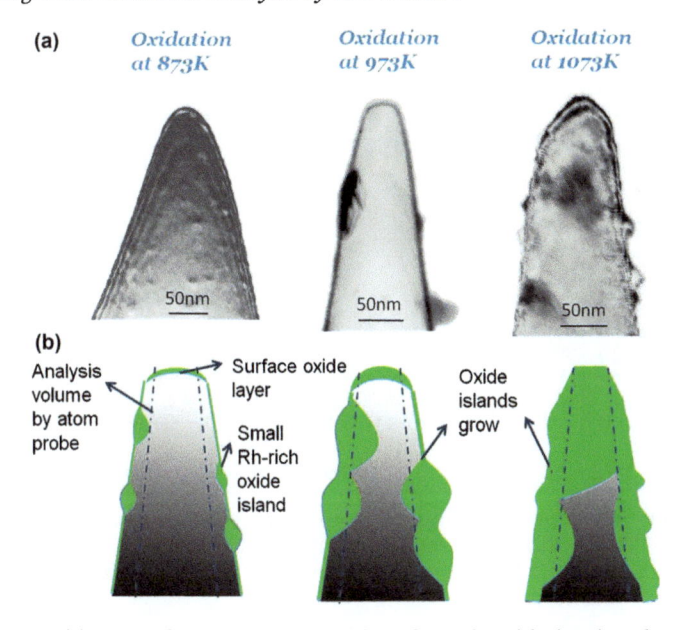

Investigating Nano-structured Catalysts by FIM and APT 271

Figure 10.10 (a) TEM of Pt-22at%Rh samples after 5 h oxidation in 1 bar of oxygen at different temperatures. (b) Schematic representation of the oxidation process.
(Reprinted from ref. 79, copyright (2011), with permission from Elsevier).

different temperatures and Figure 10.10b represents a schematic of the oxidation process. Due to the wide but limited acceptance angle of the APT method, only the central part of the specimen is analyzed, as is indicated by the dashed lines in Figure 10.10b. We can observe the formation of the oxide phase *via* the nucleation of small islands at 873 K. A thin layer of oxide is visible by atom probe. Increasing the temperature to 973 K leads to the formation of oxide crystals, with an approximate size of 30 nm. The inward growth of some of those crystals is captured by APT analysis. Finally, an even greater level of oxidation reflected by a thicker oxide layer is obtained at 1073 K.

To study the composition of the sample, the treatment at 873 K has been analyzed in more detail. The proximity histogram is represented in Figure 10.11. In order to plot this histogram, an iso-concentration surface is defined, which corresponds to the oxide/metal interface (35 at% O in this case) as marked at a distance of '0 nm'. The proximity histogram allows analysis of the concentration of species perpendicular to the iso-concentration surface. In Figure 10.11, Pt and Rh are plotted independently of their chemical environment (so Rh corresponds to pure Rh as well as Rh_xO_y species). Three distinct regions can be observed: (1) a Rh-rich oxide region, (2) a Rh-depleted region (transition depth of ~ 3 nm) and (3) the bulk composition of the alloy. Similar histograms can be plotted for the oxidation of Pt-17.4at%Rh at lower temperatures under NO, O_2 and N_2O gas exposure.

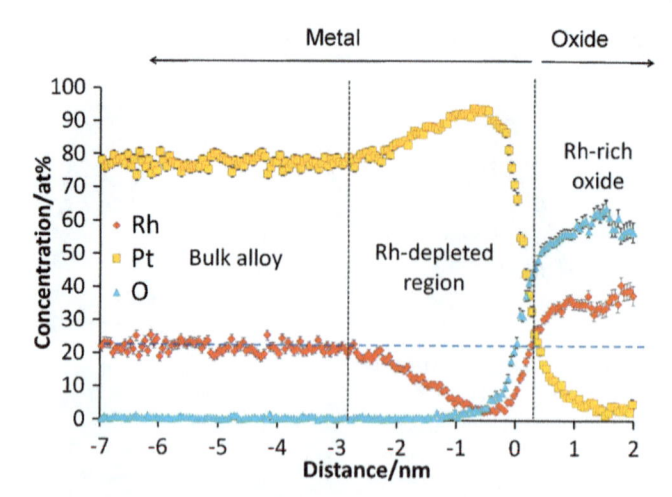

Figure 10.11 Proximity histogram presenting three different compositions within the oxidized Pt-22at%Rh sample.
(Reprinted from ref. 79, copyright (2011), with permission from Elsevier).

The composition of the oxide phase varies only slightly with temperature: the Pt concentration decreases as the temperature of oxidation increases. The measured stoichiometry is consistent with M_2O_3 (M = Rh + Pt), and thus corresponds to Rh_2O_3 in agreement with previous studies.[75,80–86]

The concentration of Rh in the oxide layer remains consistent as the oxidation time increases (37–39 at%). However, the thickness of the oxidation layer, which is also related to the depth of the Rh depleted layer, increases with time. Growth of Rh_2O_3 oxide is limited by the diffusion rate of Rh in the alloy, and the Rh depleted zone therefore provides a measure of the Rh diffusion coefficient in Pt, which is close to other reported values.[87] Also, the analysis at different temperatures allows determination of an activation energy of 236 ± 41 kJ mol^{-1}, in good agreement with values for Rh diffusion into Pt crystals planes.[88,89]

Further experiments consisted of a cycle of oxidation/reduction (oxidation at 1073 K for 1 h in 1 bar of O_2 and then reduction at 673 K for 1.5 h in 1 bar of H_2). After this subsequent reduction, an almost pure 2 nm thick layer of Rh is observed (more than 99 at%). This suggests a more straightforward method to produce a core–shell structure from a homogeneous Pt-22at%Rh alloy. The reduction temperature and time need to be carefully chosen in order to get complete reduction of the surface together with the absence of diffusion of Rh atoms back into the bulk.

As a conclusion, the extent of segregation of Pt-Rh alloys is strongly dependent of the temperature, treatment and crystallographic plane. It is then possible to control the extent of the chemically driven segregation by careful control of the exposure temperature and alloy composition.

Investigating Nano-structured Catalysts by FIM and APT

Also, subsequent reduction of the oxide phase left a thin and almost pure layer of Rh, suggesting a rather simple route for engineering the formation of a core–shell structure. The addition of a third metal to the alloy will be discussed in Section 10.3.3.2.

10.3.2.1.3 Pd-6.4at%Rh: Formation of Nano-islands.

Palladium-rhodium alloys are used in catalytic converters for automotive pollution control. The use of these alloys may present a good opportunity to generate two separate active phases and to precisely adjust the load of PGM in the catalytic converter. The time-dependent oxidation of Pd-6.4at%Rh was studied at 873 K after exposure at 1 bar of oxygen for 10, 30 or 180 min.[90] After 10 min of treatment, two distinct regions of oxidation are observed: one at the apex of the tip, and the other is composed of oxide islands at the side of the specimen. Figure 10.10 shows the region of the sample analyzed by the atom probe. The oxide present at the side of the tip results from the migration of species from the specimen shank. The analysis of this oxide has been performed with a ladder diagram (Figure 10.12a) and proximity histogram (Figure 10.12b). With a bulk content of 6.4 at% Rh the surface is slightly Rh-depleted (4.1 at%) and the underlying region is slightly Rh-enriched (7.8 at%). The proximity histogram highlights the fact that the oxide at the surface has a stoichiometry close to PdO, which is consistent with previous studies.[91–93]

The location of oxide on the shank of the sample is shown schematically in Figure 10.13. Treatment for 30 min or longer leads to severe oxidation. After 30 min, PdO remains the dominant oxide. However, the presence of Rh-rich oxide is noticed close to the apex, with a stoichiometry for a mixed oxide phase of $(Rh_1Pd_1)O_2$ which is quite unexpected at this temperature (as compared to 1075–1125 K in other works[91–93]). Such Rh-rich islands are formed from within the PdO phase. Increasing the oxidation time to 180 min leads to formation of the Pd-rich sub-oxide Pd_2O. The analysis of a sample oxidized at 873 K for 20 min and then subsequently reduced at 673 K for

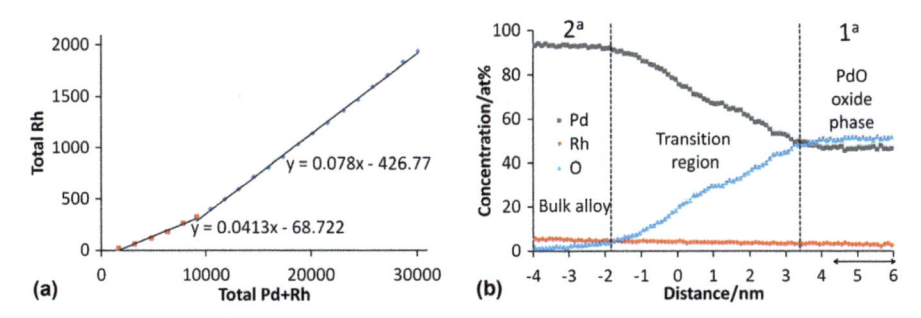

Figure 10.12 (a) Ladder diagram of the Pd-6.4at%Rh alloy after 10 min of oxidation at 873 K. (b) Proximity histogram at the oxide/metal interface. (Reprinted with permission from ref. 90. Copyright (2012) American Chemical Society).

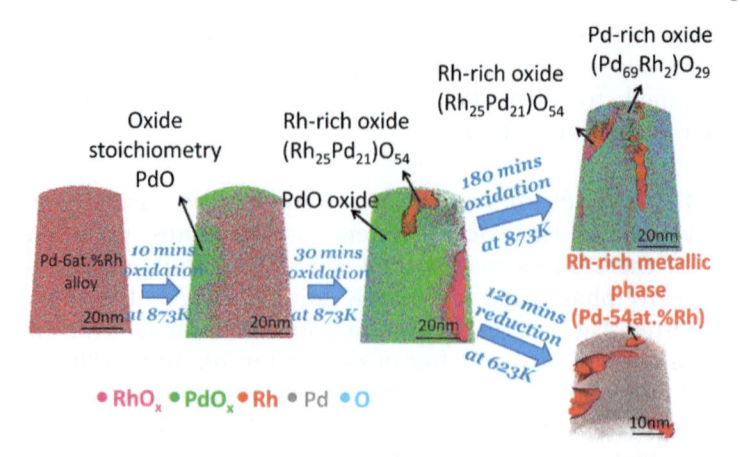

Figure 10.13 Time-dependent behavior of the oxidation process in a Pd-6.4at%Rh alloy with preferential oxidation from the apex and the shank with PdO stoichiometry at 10 min, presence of mixed oxide phase (Rh$_1$Pd$_1$)O$_2$ after 30 min, appearance of Pd-rich sub-oxide (Pd$_2$O) at 180 min, and subsequent reduction leading to subsequent reduction leads to the formation of Rh-rich island on the sample.
(Reprinted with permission from ref. 90. Copyright (2012) American Chemical Society).

120 min in 1 bar of hydrogen efficiently removes the oxygen from the sample, but does not significantly affect the distribution of Pd and Rh species. Separate Pd-rich oxide and Rh-rich oxide regions formed during the oxidation process generate Pd-rich and Rh-rich metallic regions after reduction, which can be seen as the formation of islands of a catalytic active phase (Figure 10.13).

10.3.2.1.4 Pt-31at%Pd Alloy: Formation of Core–Shell Structures from Pt@Pd to Pd@Pt.

Pt-31at%Pd alloy is used for oxygen reduction and methanol oxidation at low temperature in fuel cell technology. The use of core–shell structures enhances the catalytic activity due to the synergy between the core and shell layers.[94,95] The oxidation behavior of Pt-31at%Pd between 673 and 1073 K has been studied[96] (5h of treatment under 1 bar of oxygen gas). From these experiments, three different types of behavior have been observed. The oxidation pathway is very strongly dependent on the exposure temperature. At low temperatures (673–773 K), a thin oxide layer some 1–2 nm thick enriched in Pd is observed, with approximately 50% of the metal surface species being Pd. This behavior is expected as the Pd oxide is thermodynamically more stable than Pt oxide.[97,98] A more severe oxidation occurs at 873 K, at least 30 nm thick, where the surface is maximally enriched in Pd (45.8 ± 0.4 at%) and heavily depleted in Pt (1.2 ± 0.7 at%). This corresponds to a thick and stable oxide layer with stoichiometry close to PdO. Further increase of temperature above 873 K results in the absence of oxide species, and the enrichment in Pt at the surface, with an increasing Pt content as the temperature of oxidation increases (97.8 ± 0.7 at% at 1073 K). At high temperature, the formation of volatile oxides may explain the absence of solid oxide.[99,100] Furthermore, the evaporation rate of PdO is much higher than that of PtO; the surface may then become deficient in Pd and enriched in Pt,[101] which could explain the presence of an almost pure Pt layer. This study shows that it is possible to fine-tune the surface composition of Pt–Pd nanoparticles to be either Pd-rich or Pt-rich, simply by adjusting the oxidation temperature.

10.3.2.1.5 Pt-10.1at%Ir.

Another Pt-based alloy studied is Pt-10.1at%Ir.[17] Iridium is known to lower the light-off temperature for NO reduction, and has demonstrated activity for NO reduction with propene.[102] This alloy is also used in spark-plug applications where it is subjected to harshly oxidizing conditions. Oxidation studies involving APT analysis have been performed at 723 K under 20 mbar of oxygen for 30 min. The proximity histogram in Figure 10.14 shows a depth of oxidation of approximately

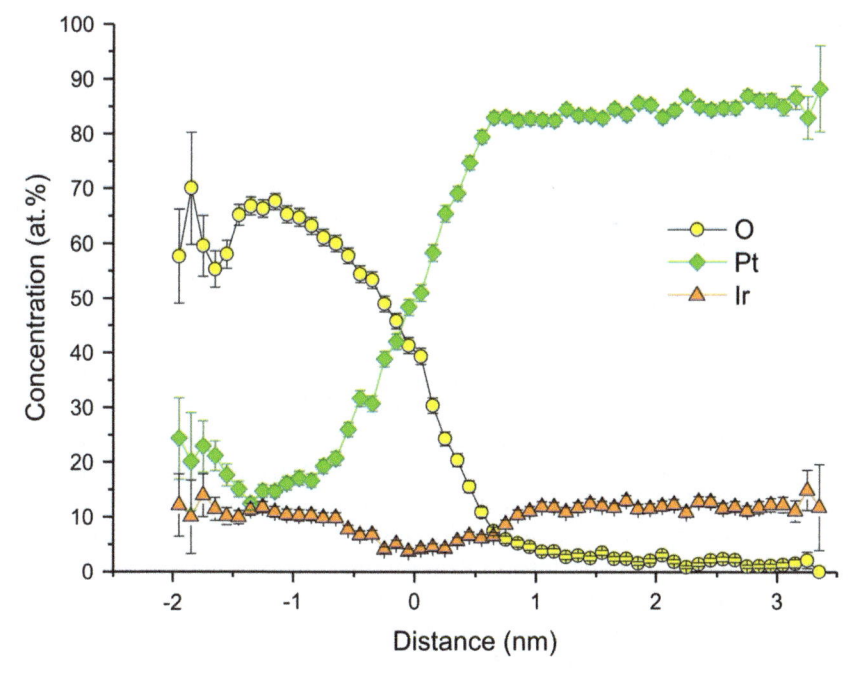

Figure 10.14 APT analysis of an oxidized Pt-10.1at%Ir alloy (723 K, 20 mbar oxygen, 30 min). The proximity histogram shows an O-rich surface where Pt concentration is drastically reduced, compared to the bulk. The existence of an Ir-depleted sub-surface region indicates a process of Ir diffusion and surface segregation.
(Reprinted from ref. 17, Copyright (2014), with permission from Elsevier).

3 nm. As can be seen, the dominant species at the surface of the specimen is oxygen (more than 50 at%). In addition to this, the ratio of metallic elements deviates significantly from the bulk composition: Pt with an initial bulk content of 90 at%, drops to ~65 at% within the 2–3 first atomic layers (ratio of the Pt/Ir metallic species, without consideration of the oxygen concentration). Below these layers, a region of 2 nm marks the transition to the bulk value of the Pt/Ir ratio. More precisely, the Pt/Ir decreases down to a minimum of ~50% and then rises up to 90%. This transition region is also characterized by the decrease of the oxygen concentration and a slight Ir depletion that indicates the diffusion of Ir from the sub-surface layers towards the oxygen-rich surface. Annealing a similar alloy under UHV conditions induces Pt segregation:[103] Ir enrichment is thus a chemically induced process.

10.3.2.1.6 Pt-8.9at%Ru. Ruthenium is known to exhibit high selectivity for conversion of NO into N_2.[104,105] A Pt-8.9at%Ru alloy was oxidized under 1 bar of oxygen gas for 5 h, over a range of temperatures.[106] A rather low level of oxidation is observed at 773 K, with the presence of both PtO_x and RuO_x species observed. The oxide layer is enriched in Ru (the Ru/(Pt+Ru) ratio increases to 0.22). Beneath the oxide layer, the Ru content is accordingly depleted for approximately 3 nm. At 873 K, the oxide concentration remains low but in this case, only PtO_x species were detected. There is, however, a Ru-depleted region of approximately 4 nm beneath the oxide layer, suggesting that more Ru has diffused to the surface oxide, but became volatile at this temperature once RuO_x was formed.[107,108] A further increase in temperature to 973 K shows that the surface is Pt-rich, together with the absence of oxide species. A 5 nm Ru-depleted region is also present which is in agreement with the formation of volatile Ru-oxides.

10.3.2.1.7 Conclusion on Binary Alloys. The most important overall conclusion from these studies of binary alloys is that there is a marked variations in susceptibility to oxidation and in the atomic-scale structures formed, depending on alloy composition and environmental exposure conditions. Chemically induced diffusion leads to the segregation of one of the metals in the presence of reactive gas. This may lead to an enrichment of the surface in one species, for example in Pd-38at%Au and Pt-10.1at%Ir alloys, or to the formation of a core–shell structure with a nearly single element surface, such as in the Pt-22at%Rh alloy after oxidation and subsequent reduction (Figure 10.15a). For Pt-17.4at%Rh, lower oxidation temperatures highlight the presence of two different diffusion processes: perpendicular to the surface (which corresponds to exchanges between the bulk and the surface) and lateral (surface diffusion between different crystallographic surfaces). In some cases, core–shell structures can be finely controlled by varying the oxidation temperature. Pt-8.9at%Ru forms a Ru-rich oxide at low temperatures and a Pt-rich shell at high

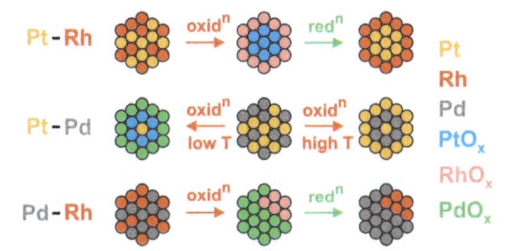

Investigating Nano-structured Catalysts by FIM and APT

Figure 10.15 Representation of segregation behavior that is strongly alloy dependent with (a) formation of metallic Pt-shell after oxidation/reduction of Pt-22at%Rh alloy, (b) switching the surface enrichment by controlling the temperature of a Pt-31at%Pd alloy, and (c) generating two separate phases of active catalyst in a Pd-6.4at%Rh alloy. (Reprinted from ref. 96, Copyright (2013), with permission from Elsevier).

temperatures. In the case of Pt-31at%Pd alloys, a nearly pure Pd-shell oxide is formed at low temperatures, while a nearly pure Pt-shell is formed at high temperatures (Figure 10.15b). The last case of Pd-6.4at%Rh exhibits the unique formation of Rh islands after an oxidation/reduction cycle, with the presence of two catalytically active phases at the surface (Figure 10.15c). A diverse range of structures may thus be formed, depending on parameters such as the temperature, the extent of exposure to various gases, the use of subsequent reduction treatments, the presence of defects in the sample such as grain boundaries, and the chemical nature of the alloy. The addition of Ir or Ru to Pt–Rh to form ternary alloys creates further scope for significantly altering the oxidation behavior, and this will be discussed in the next section.

10.3.2.2 Ternary Alloys

10.3.2.2.1 Pt-17.3%Rh-14.0%Ir (at%). More complex alloys allow a catalyst to be tailored towards higher efficiency, as well as a greater resistance to poisoning. In the present case, the addition of Ir to a Pt–Rh alloy has been investigated. The system studied is Pt-17.3at%Rh-14.0at%Ir.[109] The following experiments have been performed using a smaller field-of-view atom probe, and the discussion will be focused on the {111} and {001} facets because they show the greatest differences from the behavior of binary Pt-17.4at%Rh alloys. The specimens were subjected to an NO exposure at 10 mbar for 15 min, at a temperature range of 295–773 K. In the case of Pt–Rh–Ir {111}, the dissociation of NO is observed across the full range of temperature. The formation of atomic oxygen leads to the field evaporation of oxide species during APT analysis. Oxides of Pt, Rh and Ir are observed at the surface of the sample. A gradual enrichment of Rh, as well as a smaller enrichment of Ir is observed as the temperature of NO exposure is increased. The segregation of Rh is in accord with experiments on the {111} planes of binary Pt–Rh alloys, but to a slightly lesser extent suggesting competition between Rh and Ir. On the (001) crystallographic pole, the three oxides are again observed at 473–573 K in the two first atomic layers. Rh-enrichment is also observed at 573 K, but this changes to severe depletion at higher temperatures. The Ir species follow a similar trend, with Ir-enrichment turning to Ir-depletion as oxidation temperature rises. At 698 K, the depletion is such that the surface layer is almost completely composed of Pt. It has been concluded that NO interacts with Ir in a similar manner to its interaction with Rh, but interestingly the addition of Ir to the binary Pt–Rh alloy stabilizes any Rh-enriched layers on {001} surfaces, raising the temperature necessary to initiate depletion from 523 K to 573 K.

Another set of experiments was performed more recently with this alloy, using a LEAP.[17] The oxidation reactions were performed at 723 K, under 20 mbar of oxygen for 30 min. The proximity histogram (not represented) shows an O-rich surface as in the case of a binary Pt-10.1at%Ir alloy. Rh appears to be the metal which exhibits the strongest surface segregation. Beneath the Rh-rich region is a Rh-depleted region, as in the case of Pt–Rh alloys. Due to the significant surface enrichment of Rh atoms, Pt and Ir undergo a severe depletion within the first atomic layer. The three metals recover their bulk concentration 2 nm below the surface.

10.3.2.2.2 Pt-23.9%Rh-9.6%Ru (at%). Another ternary alloy which has been studied is Pt-23.9at%Rh-9.6at%Ru.[109] The specimen was subjected to the same oxidizing conditions in the CAP as the Pt–Rh–Ir alloy described above (NO adsorption 10 mbar for 15 min). In the Pt–Rh–Ru {111} case, NO dissociates at temperatures as low as 295 K. As in the Pt-17.3at%Rh-14.0at%Ir alloy, the three oxides are found in increasing amounts as the exposure temperature increases. However, there is no indication of any segregation, and the surface chemistry is not altered over the temperature range studied (295–773 K). The addition of Ru thus appears to stabilize the {111} planes. The {001} regions present an interesting behavior. As in the {111} case, three oxide species are observed and their content increases with increasing temperature. From 373 K to 523 K, a slight Rh-enrichment is identified, to an excess of ≤6 at% above the bulk Rh concentration. Above 548 K, Rh depletion is observed, and thereafter shows the same trend as in the Pt-17.4at%Rh case. Regarding the Ru species, no significant changes are observed up to 423 K. At higher temperatures, the surface concentration of Ru decreases to 2 at% at 673 K. A plausible explanation would be the lateral diffusion of Ru towards other crystallographic regions. However, Ru is known to be prone to removal under oxidizing conditions,[110,111] and the loss of Ru is observed at lower temperature than Rh depletion. Overall, it is concluded that Ru lessens the extent of Rh enrichment and depletion, and helps to maintain a consistent Rh composition across certain temperature ranges.

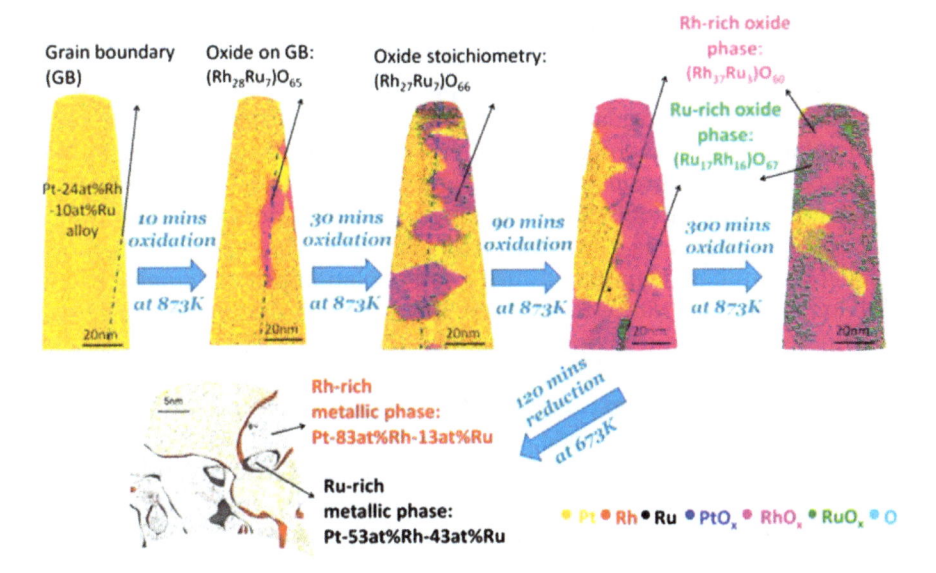

Investigating Nano-structured Catalysts by FIM and APT 279

Figure 10.16 Time-dependent behavior of the oxidation process in a Pt-23.9at%Rh-9.7at%Ru alloy exposed to 1 bar of oxygen at 873 K showing preferential oxidation at a grain boundary (10 min); oxidation of the surface and the bulk leading to an approximate MO_2 stoichiometry (30 min); and separation of Rh-rich and Ru-rich oxides (90 and 300 min) which remain separate phases after subsequent reduction.
(Reprinted with permission from ref. 106. Copyright (2012) American Chemical Society).

To investigate the time dependent evolution of oxide formation of Pt-23.9at%Rh-9.7at%Ru, the temperature of 873 K was selected[106] (an intermediate value between Ru-rich and Pt-rich surface as highlighted in the Pt-8.9at%Ru alloy). In this case, the whole sample was analyzed by LEAP rather than examining individual crystal planes using the CAP. The treatment consists of 1 bar of oxygen gas for either 10, 30, 90 or 300 min. Atom maps resulting from these experiments are shown in Figure 10.16.

From the atom map after 10 min of treatment, the extent of oxidation is more pronounced in the bulk rather than at the surface. The oxidation is concentrated along a grain boundary (highlighted with the dashed line). Such regions are known to be more prone to oxidation than grain interiors in a wide range of materials. The grain boundary region appears enriched in both Rh and Ru. In the regions immediately adjacent to the grain boundary, Rh and Ru are correspondingly depleted. This is due to the high diffusion coefficient of species at grain boundaries, and explains why the initial oxide formation appears at this particular feature. If no grain boundary is present, the oxide occurs only at the surface as a thin layer after 10 min of treatment. After 30 min, the apex of the sample presents a 10 nm thick oxide layer, and the oxide continues to grow along the grain boundary. The analysis reveals the presence of separate oxide regions rather than a uniform phase.

The mixed oxide (enriched in both Rh and Ru) has a stoichiometry close to MO_2. After 90 min of oxidation, distinctly spatially separated Rh-rich oxide and Ru-rich oxide regions become apparent. Treatment for 300 min increases the level of oxidation with more extensive formation of separate oxide phases with stoichiometries approximating to RuO_2 and Rh_2O_3. The analysis of a sample oxidized at 873 K for 1 h and subsequently reduced at 673 K for 2 h in 1 bar of hydrogen highlights the presence of Rh-rich and Ru-rich metallic phases, as also shown in Figure 10.16. The significant chemical changes and phase separation during the oxidation process survive the reduction treatment. Nano-island structures with two separate active phases may therefore be induced by an oxidation/reduction cycle, as was the case for the Pd-6.4at%Rh alloy.

A comparison between Pt-22at%Rh, Pt-8.9at%Ru and Pt-23.9at%Rh-9.7at%Ru alloys can be made. A cycle of oxidation (1073 K) + reduction (673 K) induces the formation of a Rh-shell in the Pt-22at%Rh alloy. In the case of Pt-8.9at%Ru, oxidation treatments at temperatures below 873 K generate the formation of very thin Ru-rich oxides, while above 873 K, only a Pt-rich surface is present with little oxides apparent. The oxidation of Pt-23.9at%Rh-9.7at%Ru at 873 K produces Rh-rich oxides and Ru-rich oxides as distinct separate phases. Subsequent reduction at 673 K retains the separation as Rh-rich and Ru-rich metallic regions. Thus the ternary alloys exhibit drastically different behavior, which cannot be predicted by respective trends in the binary alloys. This opens up a wide range of new opportunities for nanoscale engineering of catalyst surface compositions.

10.3.3 APT Studies of Catalytic Nanoparticles

The previous section has demonstrated that APT is a powerful technique to investigate the composition of alloy catalysts after exposure to reactive gases. These types of studies are therefore providing a unique tool to study model catalysts using the sharp sample tip as a particle approximation. The next step towards a better understanding of catalytic nanoparticles lies in the study of actual catalyst formulations used in practice. The size of catalytic nanoparticles is usually less than 10 nm[112] and thus smaller than the apex of field emitter tips (20 nm $< R_c <$ 100 nm). However, conventional tip samples used in atom probe experiments can act as a support for these smaller nanoparticles.

10.3.3.1 Methods of Atom Probe Sample Preparation for Nanoparticles

A detailed description of sample preparation for the APT analysis of nanoparticles is beyond the scope of this chapter, and thus only a brief explanation is presented. Such developments are very recent and only two methods have been reported: an electrophoretic method and a combination of chemical vapor deposition (CVD) and focused ion beam (FIB).

Investigating Nano-structured Catalysts by FIM and APT 281

10.3.3.1.1 Electrophoresis Method. The concept of this method[113] is to use a pre-sharpened needle-shaped substrate consisting of a Pt (or Pt-Rh) atom probe samples produced by electropolishing in a molten salt mixture. On this substrate, nanoparticles are deposited by electrophoresis. To perform this deposition, a liquid droplet containing a suspension of nanoparticles is placed in a gold loop and a voltage bias is applied between the loop and the substrate tip. On dipping the tip into the loop, nanoparticles are drawn through the potential gradient by electrophoresis towards the apex region of the specimen where they are subsequently immobilized. A schematic representation can be found in Figure 10.17a, which is similar to the electropolishing procedure described in Section 10.2.1.1. Samples are then inspected by transmission electron microscopy (TEM) to confirm the deposition of a suitable layer of nanoparticles. Figure 10.17b shows a clean Pt-22at%Rh preformed tip, rather blunt, before deposition and Figure 10.17c shows the same tip after deposition of Ag@Pd nanoparticles[113] (the notation 'Ag@Pd' denotes a core made of silver with a shell made of palladium).

Factors affecting the deposition process include the particle type (colloidal solution of isolated particles[113,114] or a suspension of crushed, catalyst support material containing nanoparticles[14]), concentration, size and charge polarity, along with the specimen sharpness, loop voltage and duration of immersion. An important factor in the success of APT study is the number of deposited nanoparticles: agglomeration of nanoparticles makes the atom probe analysis difficult and reconstructed atom maps susceptible to particle overlapping from trajectory aberrations. In the case of supported samples, the nanoparticles are relatively well separated, which is not the case for a colloidal solution. A good dispersion of supported nanoparticles is required and can be obtained *via* dissolution and immersion in an ultrasonic bath. Typical deposition conditions are 5–20 V DC during $\sim$10–20 s.

Figure 10.17 (a) Schematic representation of the electrophoresis method where a Pt-22at%Rh pre-formed tip is dipped into a liquid droplet containing a suspension of nanoparticles. (b) TEM of a Pt-22at%Rh specimen before deposition. (c) TEM of the same specimen after deposition of Ag@Pd nanoparticles.
(Panel (a) adapted with permission from ref. 14. Copyright (2014) American Chemical Society; panels (b) and (c) adapted by permission from Macmillan Publishers Ltd: Nature Nanotechnology, ref. 113, copyright 2011).

10.3.3.1.2 Preparing an APT Tip from a Catalyst Powder: CVD-FIB Method.
This method has been successfully applied to produce atom probe specimens from powders,[115–117] and more recently, from CoCuMn particles.[118] The alloy is synthesized in the absence of a support and is produced in a porous form having a very high specific area of some 170 m^2 g^{-1}. The first step of specimen preparation require the use of electron beam assisted chemical vapor deposition (eBCVD) of methylcyclopentadienyl (trimethyl) platinum(IV) in order to fill the pores in the catalyst structure. This fully dense sample is then shaped by focused ion beam (FIB) milling with an accelerating voltage of 10 kV. A final clean-up of the sample is made by FIB with a lower voltage of 5 kV.

10.3.3.2 Analysis of Unsupported Nanoparticles

APT has been successfully used to image and characterize the core–shell structure of nanoparticles in cases where diffraction techniques and TEM were unsuccessful, either due to the ultra-thin shell broadening the X-ray or energy-loss peaks, or due to the lack of atomic-species contrast. This section reports on the study of two nanoparticle systems used for specific catalytic reactions: Ag@Pd core–shell nanoparticles for hydrogen production from formic acid, and CoCuMn nanoparticles for CO hydrogenation. APT results provide a better understanding of the activity and the efficiency of such catalysts for these particular reactions.

10.3.3.2.1 Ag@Pd Used in Hydrogen Production.
Ag–Pd core–shell nanocatalysts have been used for production of hydrogen from formic acid at ambient temperature. The formation of Ag–Pd core–shell structures is achieved by wet chemical synthesis. Details of the synthesis can be found elsewhere.[113] The aim is to enhance the catalytic activity towards formic acid decomposition at room temperature, essential for small mobile fuel cell devices, through the retention of Pd atoms at the surface. The activity of these can be enhanced *via* electronic interactions with a suitable core in a core–shell bimetallic particle.[119,120] APT has been used to confirm the core–shell configuration and to study the thickness of the Pd-shell as a function of the Ag : Pd ratio used for the wet-chemistry synthesis route.

The samples were produced by the electrophoresis method on Pt-22at%Rh support substrates of 80–100 nm diameter (dipping conditions: 5–15 V for 10 s). The 3D reconstruction of each individual particle is sliced for atomic depth profiling. Figure 10.18a presents an atom map of a single Ag@Pd nanoparticle, where the core–shell structure can be clearly observed. The inner core is composed of Ag (gray dots) and the shell is composed of Pd (yellow spheres). The interface between the two phases is very sharp. The thickness of the shell regions can be analyzed. In the case of a 1 : 1 Ag : Pd ratio, Pd shells are restricted to thicknesses of 1 to 2 atomic layers, as seen in Figure 10.18. In the case of a 1 : 3 Ag : Pd ratio a broader shell of 5–10 atomic layers is measured (not shown). Higher catalytic activity of the 1 : 1 ratio

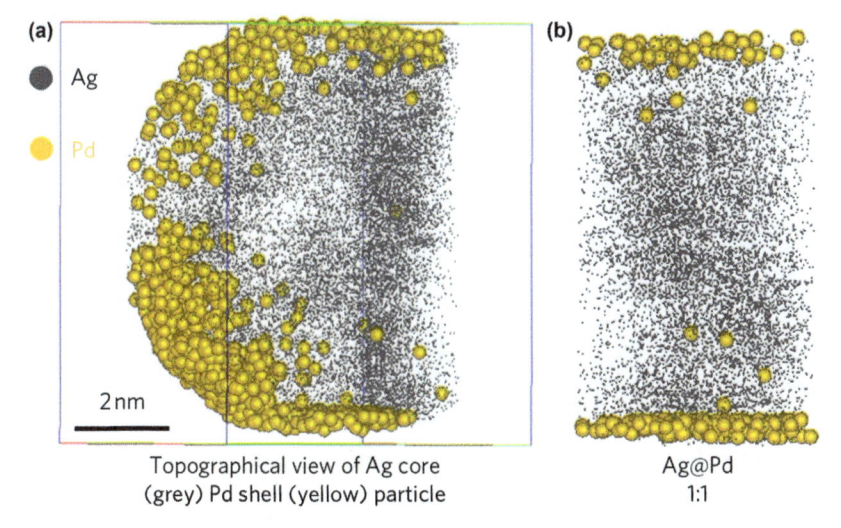

Investigating Nano-structured Catalysts by FIM and APT 283

(a) Topographical view of Ag core (grey) Pd shell (yellow) particle

(b) Ag@Pd 1:1

Figure 10.18 (a) Atom map of a single Ag@Pd nanoparticle core–shell structure with sharp interface (Ag = gray dots; Pd = yellow spheres). (b) Slice through a 1:1 Ag:Pd ratio showing a shell thickness of one to two atomic layers. (Adapted by permission from Macmillan Publishers Ltd: Nature Nanotechnology, ref. 113, copyright (2011)).

suggests that the electronic modification of Pd due to underlying Ag has a rather short range. This discovery is a breakthrough in the understanding of atomically precise tailored catalysts.

10.3.3.2.2 CoCuMn Used in Fischer–Tropsch Reaction. CoCuMn nanoparticles have been used during catalytic CO hydrogenation in order to produce long-chain terminal alcohols. The synthesis of the core–shell structure of the catalyst has been achieved using the co-precipitation of Co-, Cu- and Mn-oxalates followed by thermal decomposition. Details of the synthesis can be found elsewhere.[121,122] An intimate mixing of the metals seems to be the key to improving the selective formation of long-chain hydrocarbons with terminal functionalization. APT has been used to demonstrate the presence of a metallic phase with the three metals being present in the same particles.[118]

A passivated catalyst powder sample has been conditioned to form a nanosized tip using the CVD-FIB method. The 3D reconstruction of a 10 nm thick slice through a single grain of catalyst is presented in Figure 10.19a where Co atoms are represented as blue spheres, Cu as orange spheres, Mn as green spheres and O as white spheres. A core–shell chemical structure can be observed with major amounts of Co forming the core, with all three elements present in a Cu-dominated shell of about 2 nm thickness. An enlarged view is presented Figure 10.19b showing the distribution of oxygen throughout the core–shell interface. Only small amounts of oxygen are detected, proving the presence of a largely metallic CoCuMn phase.

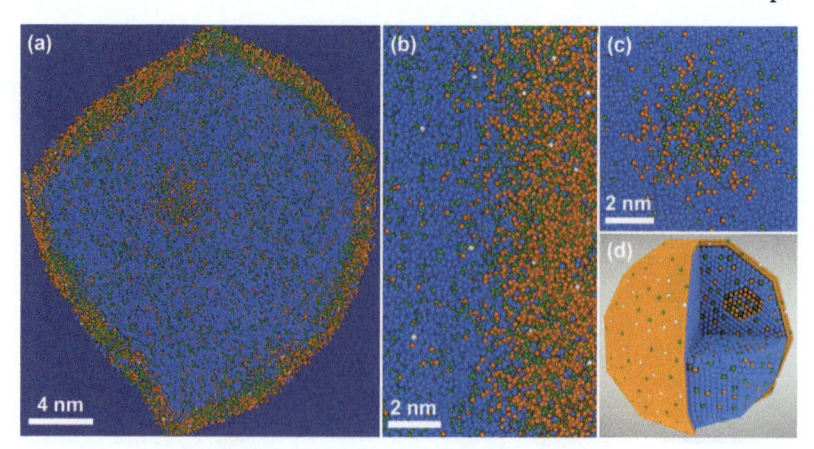

Figure 10.19 (a) Atom map of a single grain of catalyst showing a core–shell structure with Co-rich core and Cu-dominated CoCuMn mixed shell (Co = blue spheres; Cu = orange spheres; Mn = green spheres; O = white spheres). (b) Enlarged view of the core–shell interface. (c) Enlarged view of intra-core cluster with high concentration of Cu and Mn. (d) 3D sectional view of the element distribution.
(Adapted with permission from ref. 118. Copyright (2013) American Chemical Society).

Precipitates of 5 nm thickness and with high concentrations of Cu and Mn are also found inside the core structure (Figure 10.19c). A 3D sectional view of the element distribution is represented in Figure 10.19d to highlight the presence of the core–shell structure and intra-core clusters.

This atom probe study provides a direct proof of the occurrence of a core–shell structure of a CoCuMn catalyst in an atomically quantified manner. It proves the validity of the concept of establishing relationships between the structure of a catalyst at the atomic scale and its activity at the laboratory scale.

10.3.3.3 Atom Probe Studies of Supported Nanocatalysts: Towards a Better Understanding of Metal–Support Interaction

We now turn the discussion to the characterization of commercially available Pt-based nanoparticles supported on carbon black, with a high specific area.[14] The internal structure, composition and the distribution of trace elements will be presented for three chosen catalysts: supported Pt, Pt-25.3at%Co alloy and an isolated Ir@Pt core–shell structure. The interest in investigating such catalysts is two-fold. On the one hand, APT is used as a powerful quality control tool, allowing the identification of differences in the final product as compared to the targeted specifications. These results can be used to improve preparation methods of catalysts. On the other hand, these studies provide new ways to unravel the relationships between the structure and the activity taking into account the presence and the influence of the support.

Various tip substrates may be used for the deposition of these nanoparticles. W, Al and Pt were tested, but only Pt and its alloys turn out to be suitable for atom probe analysis because of an apparently stronger interaction with the above-mentioned nanoparticles. A Pt-22at%Rh alloy was chosen for the deposition of Pt-based particles to aid distinguishing the origin of the Pt atoms. The onset of Rh detection indicates that the analysis has reached the tip substrate. Regarding the 3D data reconstructions, several factors increase the difficulty of analysis, such as a lack of crystallographic information and non-uniform evaporation rate across the surface. Standard procedures of reconstruction are not suitable and a cross-comparison of the sizes and shapes of nanoparticles in APT reconstructions with TEM images was found to be the best way of optimizing the reconstruction.

10.3.3.3.1 Pt-Supported Nanoparticles. The synthesis of the sample is made *via* a wet chemical method using surfactants as stabilizers; the latter induce the presence of a thick organic layer around the nanoparticles.[123] The particles show a mean diameter of ~ 7 nm on top of the porous C-support, with smaller sizes of 3–4 nm in deeper pores. The 3D atom map presented in Figure 10.20a shows a rather complex morphology with the presence of Pt (red spheres), C (black spheres), Na (blue spheres) and complex ions C_xH_y (green spheres). Top views of the upper 5 nm layer (see gray arrow) are presented in Figure 10.20b and c to clarify the position of Pt on the C-support. C_xH_y species correspond to fragments of the carbon support, surfactants from the synthesis or methanol molecules from the solvent itself. The main impurities present consist of sodium ions, which are located spatially with the C_xH_y species. This suggests that they are introduced during the synthesis. Na species are known to act adversely on catalytic activity in some cases.[124] Cycles of washing are usually made to remove this impurity, but atom probe experiments show the relatively high retention of Na on the sample.

The variations in particles sizes within the analyzed volume are made visible with a better spatial view than in TEM experiments. An enlarged view of the black rectangular region of Figure 10.20b is presented on Figure 10.20d showing four Pt nanoparticles with a size of 3–4 nm.

10.3.3.3.2 Analysis of the Inhomogeneity of Pt-25.3at%Co Nanoparticles. Synthesis of Pt-Co nanoparticles has also been performed *via* wet chemical methods, leading to a nominal atomic composition of 74.7 at% of Pt and 25.3 at% of Co (determined from ICP analysis), with an average particle diameter of 5.6 ± 0.2 nm. The sizes of Pt-25.3at%Co particles on the 3D atom map reconstruction shown in Figure 10.20 matches TEM experiments. Furthermore, analysis of isolated particles indicates the presence of a homogeneously alloyed internal structure, which is consistent with STEM/EDX experiments. The analysis by APT is restricted to well-defined

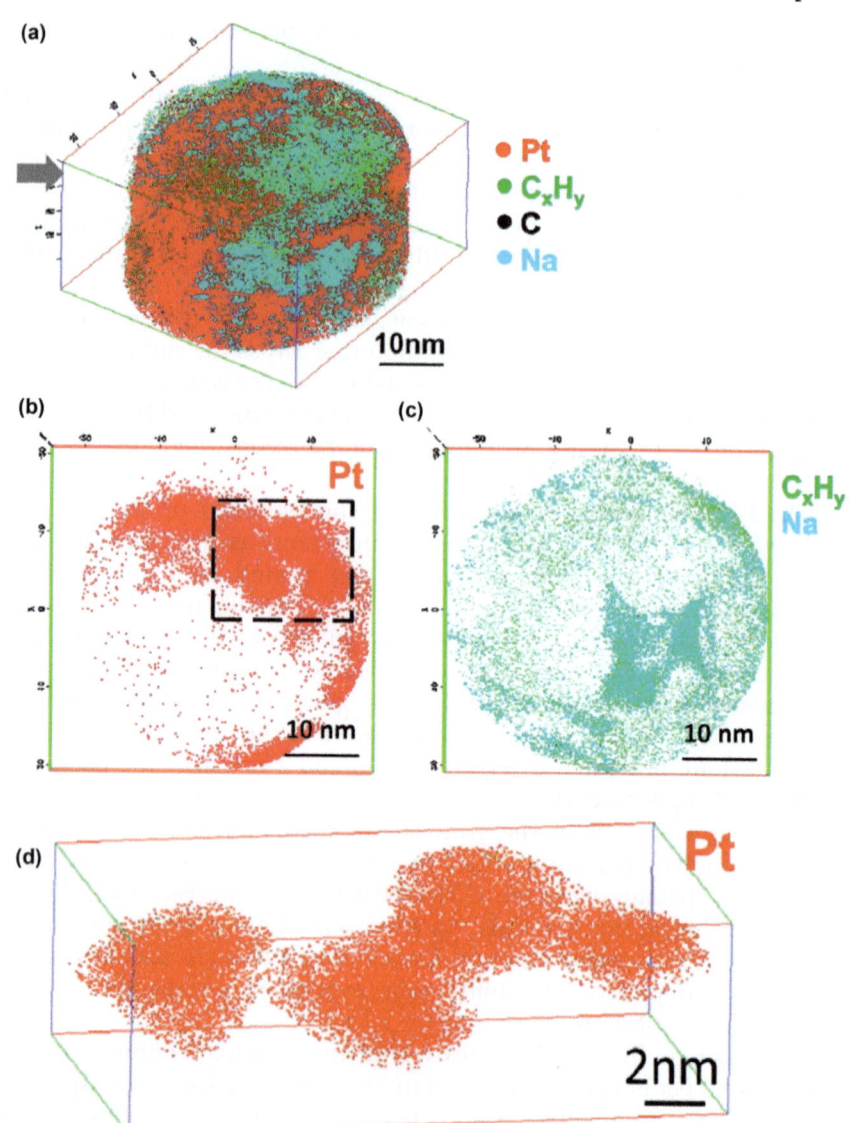

Figure 10.20 (a) 3D atom map presenting a complex morphology of Pt nanoparticles within a small region of a carbon-supported catalyst. (b) Top view of the upper 5 nm layer for Pt atoms only. (c) Same as in (b) for C_xH_y and Na species. Na impurities are only present with C_xH_y. (d) Reconstruction of four Pt nanoparticles from the black box on (b) with a 3–4 nm size. (Adapted with permission from ref. 14. Copyright (2014) American Chemical Society).

and well-separated nanoparticles, as shown in Figure 10.21a. In Figure 10.21b, the Co concentration is plotted as function of the particle size (expressed as the number of atoms on the lower x axis, and as

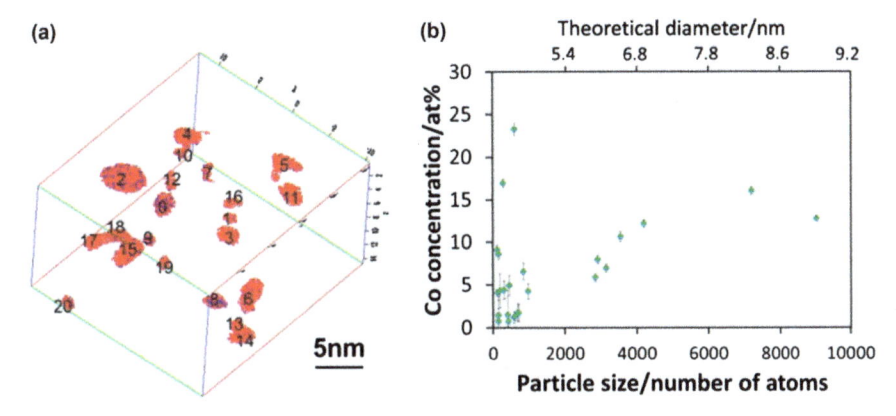

Figure 10.21 (a) Reconstruction of well-defined and well-separated Pt-25.3at%Co nanoparticles. (b) Analysis of the size-dependence variation in Co-content.
(Adapted with permission from ref. 14. Copyright (2014) American Chemical Society).

theoretical spherical diameter on the upper x axis). Within the average composition of 93.7 ± 5.8 at% Pt and 6.3 ± 5.8 at% Co, a clear and marked size-dependent variation in composition can be observed. The variations of Co content are also confirmed by STEM/EDX experiments. These results suggest that the impregnation method could be improved to get a narrower compositional distribution.

Pt-Co alloys can be magnetic and this feature is dependent on the composition (at ambient temperatures, the Curie transition point lies at about 12 at% Co). The electrophoresis method used in this case may thus promote the deposition of particles with lower Co concentrations. Furthermore, the field present in the sample preparation may induce the agglomeration in solution of particles with higher Co concentration. Nevertheless, particles deposited and analyzed by APT present a broad size and compositional distribution and the APT analysis remains of interest, although not fully representative of the total composition of the material.

10.3.3.3.3 Analysis of the Intermixing in Ir@Pt Nanoparticles. The synthesis of Ir@Pt core–shell structures is based on an industrial method in which platinum is deposited selectively over 4 nm Ir nanoparticles.[125] The characterization of the thickness of the shell and the composition of this alloy remains challenging with high-resolution TEM due to the similar atomic weight of the two elements. Within the scope of this study, only isolated Ir@Pt particles were deposited using the electrophoresis method. Figure 10.22a shows a typical 3D atom map for this system. Different types of structure may be found: some clearly shows signs of a core–shell form, with Pt enrichment in the outer layers, but others consist mainly of Pt

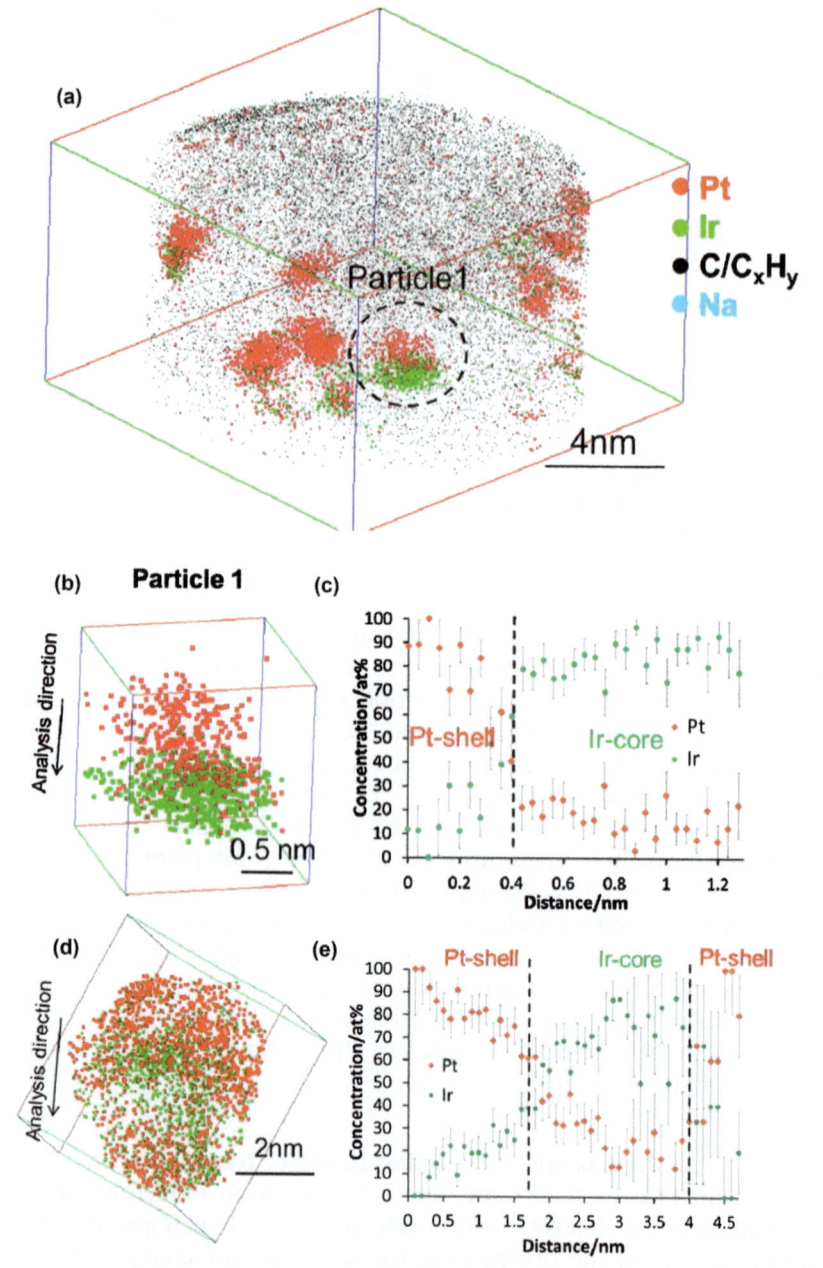

Figure 10.22 (a) 3D reconstruction of Ir@Pt nanoparticles with Pt as red spheres and Ir as green spheres. (b) Enlargement of the particle 1 from the previous atom map. (c) 1D profile of the previous particle presenting a duplex structure rather than a complete core–shell structure. (d) 3D reconstruction of another nanoparticle presenting a core–shell structure. (e) 1D profile of the previous nanoparticle.
(Adapted with permission from ref. 14. Copyright (2014) American Chemical Society).

(not shown). The particle labeled 1 in Figure 10.22a is enlarged in Figure 10.22b, together with its 1D profile in Figure 10.21c. This particle presents a duplex structure consisting of a ~0.4 nm thick Pt-rich layer in direct contact with a ~0.9 nm thick Ir-rich region. The composition profile has been obtained along the analysis direction shown in Figure 10.22b. The composition of the shell and core are: Pt-rich: 81 ± 5 at% Pt/19 ± 2 at% Ir and Ir-rich: 22 ± 5 at% Pt/78 ± 9 at% Ir. Figure 10.21d and e provide a second example from a different data set. In this case, the particle has a more pronounced Pt(shell)–Ir(core) structure with an overall diameter of 4.0 ± 0.2 nm. The compositions are as follows: Pt-shell: 75 ± 11 at% Pt/25 ± 5 at% Ir and Ir-core: 30 ± 3 at% Pt/70 ± 5 at% Ir. On average, the outer shell has a Pt content of 80 ± 8 at% with the inner core containing 77 ± 11 at% Ir. This set of results provides a clear indication of variations in both core and shell contents between different particles of comparable size. Also, significant intermixing of the Ir-core and Pt-shell has taken place, rather than a sharp separation as was the case in Ag@Pd nanoparticles.

Because of the poor contrast and the same fcc crystal structure of the two elements concerned, this level of information is not possible from high-resolution STEM techniques. APT thus provides a unique tool to study this type of catalyst with unprecedented spatial resolution.

10.4 Conclusions

The use of samples prepared as nanosized tips is particularly suited to the study of catalytic reactions occurring on metallic nanoparticles. In the context of automotive pollution control, the design of nanoparticles is particularly important to get better thermal stability, lower light-off temperatures, greater resistance to poisoning and more efficiency under lean-burn engine conditions. Atomically precise design of such particles can be achieved if the behavior of both the catalytic reaction and the catalyst is determined. In its working state, the texture of the catalyst, the morphology of the particles, and their local chemical composition are some of the chief factors that have to be considered when designing a catalyst that must exhibit selectivity, activity and resistance to ageing. FEM and FIM can be used to study the reaction itself, FIM is used to observe the structural and morphological changes of the sample before, during and after gas adsorption and/or reaction, making it possible to assess synergistic influences between structure and reactivity. The dosing of oxygen on Pt, Rh, Pd and Ir samples has been shown to give rise to a wide diversity of behavior; from local surface reconstructions to drastic faceting. These studies highlight the morphological reconstructions of the catalyst and their effects on catalytic activity. However, the main limitation of these two methods is the lack of chemical analysis. APM methods are complementary to study the modification of the chemical nature of the surface and near-surface regions, as well as the synergies between the bulk and the surface. In the early development of atom probe, the analysis was limited to a narrow field of view, restricting the discussion to different crystallographic orientations. More recently, the development of the laser-pulsed LEAP has increased the potential to analyze fragile or complex samples such as those containing nanoparticles, as well as widening the size of the analysis regions and the collection rates. These new features have opened a global investigation of changes occurring at the surface and in the bulk of catalytic samples, with atomic spatial resolution. Dedicated environmental cells have been developed to avoid contamination during the sample transfers between reaction and analysis chamber. As in FIM studies, different alloys show different response towards similar treatment, including segregation of one metal at the surface by chemically driven surface enrichment, formation of core–shell structures, as well as formation of separate active phases in the form of nano-islands. These features are highly dependent on the composition of the alloy, the temperature and duration of treatments, the crystallographic orientation, and the presence of surface or bulk defects. As an example, the behavior of a ternary alloy cannot be predicted from that of its respective binary alloys. A systematic study has to be undertaken for each alloy composition. The next step towards the understanding of catalytic behavior lies in the study of applied formulation of catalysts. In that case, the tip sample becomes the support for the deposition of catalyst particles. Supported on carbon, dispersed in colloidal solution, or even powders of catalytic nanoparticles have been successfully analyzed by APT revealing core–shell structures and mixed metallic phases. APT can also be used as a quality control tool, which could serves as feedback to adjust and improve the synthesis of nanoparticles at the atomic scale.

Looking forward, a new generation of ambient-pressure atom probes is also being developed. Tools of this nature will provide the highest impact studies to address the complexities of catalytic systems/particle upon exposure to reactive environments.

Acknowledgements

C.B. thanks the Fonds de la Recherche Scientifique (F.R.S.-FNRS) for financial support (PhD grant). The Wallonia-Brussels Federation is gratefully acknowledged for supporting this research (Action de Recherches Concertées n° AUWB 2010–2015/ULB15), as well as the Foundation Wiener-Anspach. C.B. and T.V.d.B. also thank Professors Yannick De Decker and Norbert Kruse for fruitful discussions. P.A.J.B. acknowledges financial support from the UK Engineering and Physical Sciences Research Council (EPSRC, grant EP/077664/1) and from The Queen's College Oxford. Provision of samples and support from Johnson Matthey plc is also gratefully acknowledged.

References

1. G. A. Somorjai and Y. Li, *Introduction to Surface Chemistry and Catalysis*, Wiley & Sons, New-Jersey, USA, 2nd edn, 2010.
2. J. W. Gibbs, The Collected Works, *Thermodynamics*, Longmans, New York, 1931, vol. 1, p. 320.
3. G. Wulff, *Z. Kristallogr.*, 1901, **34**, 449.
4. L. D. Marks, *Surf. Sci.*, 1985, **150**, 358.
5. E. W. Müller and T. T. Tsong, *Field ion microscopy, principles and applications*, Elsevier, New York, USA, 1969.
6. E. W. Müller, *Z. Phys.*, 1951, **31**, 136.
7. E. W. Müller, *Z. Naturforsch.*, 1956, **11a**, 88.
8. C. Barroo, S. V. Lambeets, F. Devred, T. D. Chau, N. Kruse, Y. De Decker and T. Visart de Bocarmé, *New J. Chem.*, 2014, **38**, 2090.
9. C. Barroo, Y. De Decker, T. Visart de Bocarmé and N. Kruse, *J. Phys. Chem. C*, 2014, **118**, 6839.
10. J.-S. McEwen, P. Gaspard, T. Visart de Bocarmé and N. Kruse, *Proc. Natl. Acad. Sci. U. S. A.*, 2009, **106**, 3006.
11. V. Gorodetskii, J. Lauterbach, H.-H. Rotermund, J. H. Bock and G. Ertl, *Nature*, 1994, **370**, 276.
12. V. Gorodetskii, W. Drachsel and J. H. Block, *Catal. Lett.*, 1993, **19**, 223.
13. Y. Suchorski, R. Imbihl and V. K. Medvedev, *Surf. Sci.*, 1998, **401**, 392.
14. T. Li, P. A. J. Bagot, E. Christian, B. R. C. Theobald, J. D. B. Sharman, D. Ozkaya, M. P. Moody, S. C. E. Tsang and G. D. W. Smith, *ACS Catal.*, 2014, **4**, 695.
15. R. Imbihl, *New J. Phys.*, 2003, **5**, 62.
16. N. Kruse and T. Visart de Bocarmé, in *Handbook of Heterogeneous Catalysis*, ed. G. Ertl, H. Knözinger, J. Weitkamp and F. Schüth, Wiley-VCH, 2nd ed, 2008, p. 870.
17. S. Dumpala, S. R. Broderick, P. A. J. Bagot and K. Rajan, *Ultramicroscopy*, 2014, **141**, 16.
18. B. Gault, M. P. Moody, J. M. Cairney and S. P. Ringer, *Atom Probe Microscopy*, Springer-Verlag, New York, 2012.
19. T. F. Kelly and D. J. Larson, *Annu. Rev. Mater. Res.*, 2012, **42**, 1.
20. E. W. Müller, J. A. Panitz and S. B. McLane, *Rev. Sci. Instrum.*, 1968, **39**, 83.
21. N. Kruse, *Ultramicroscopy*, 2001, **89**, 51.
22. T. T. Tsong, *Atom-Probe Field Ion Microscopy: Field Ion Emission, and Surfaces and Interfaces at Atomic Resolution*, Cambridge University Press, UK, 2005.
23. M. K. Miller and G. D. W. Smith, *Atom Probe Microanalysis: Principles and Applications to Materials Problems*, Materials Research Society, Pittsburgh USA, 1989.
24. A. Cerezo, T. J. Godfrey and G. D. W. Smith, *Rev. Sci. Instrum.*, 1988, **59**, 862.

25. Y. Aruga, D. W. Saxey, E. A. Marquis, A. Cerezo and G. D. W. Smith, *Ultramicroscopy*, 2011, **111**, 725.
26. X. W. Zhou, H. N. G. Wadley, R. A. Johnson, D. J. Larson, N. Tabat, A. Cerezo, A. K. Petford-Long, G. D. W. Smith, P. H. Clifton, R. L. Martens and T. F. Kelly, *Acta Mater.*, 2001, **49**, 4005.
27. E. A. Marquis, J. M. Hyde, D. W. Saxey, S. Lozano-Perez, V. de Castro, D. Hudson, C. A. Williams, S. Humphry-Baker and G. D. W. Smith, *Mater. Today*, 2009, **12**, 30.
28. M. K. Miller, K. F. Russell, K. Thompson, R. Alvis and D. J. Larson, *Microsc. Microanal.*, 2007, **13**, 428.
29. M. K. Miller, *Atom Probe Tomography: Analysis at the Atomic Level*, Springer, New-York, 2000.
30. A. R. Waugh, S. M. Payne, G. M. Worrall and G. D. W. Smith, *J. Phys. (Paris)*, 1984, **45**(C-9), 207.
31. D. J. Larson, D. T. Foord, A. K. Petford-Long, T. C. Anthony, I. M. Rozdilsky, A. Cerezo and G. D. W. Smith, *Ultramicroscopy*, 1998, **75**, 147.
32. D. J. Larson, A. K. Petford-Long, A. Cerezo and G. D. W. Smith, *Acta Mater.*, 1999, **47**, 4019.
33. M. K. Miller and K. F. Russell, *Ultramicroscopy*, 2007, **107**, 761.
34. A. J. Melmed, *Appl. Surf. Sci.*, 1996, **94/95**, 17.
35. J.-S. McEwen, P. Gaspard, T. Visart de Bocarmé and N. Kruse, *J. Phys. Chem. C*, 2009, **113**, 17045.
36. J.-S. McEwen, A. G. C. Rosa, P. Gaspard, T. Visart de Bocarmé and N. Kruse, *Catal. Today*, 2010, **154**, 75.
37. J. H. Block, H. J. Kreuzer and L. C. Wang, *Surf. Sci.*, 1991, **246**, 125.
38. T. F. Kelly and M. K. Miller, *Rev. Sci. Instrum.*, 2007, **78**, 031101.
39. M. K. Miller and R. G. Forbes, *Mater. Charact.*, 2009, **60**, 461.
40. F. Vurpillot, B. Gault, B. P. Geiser and D. J. Larson, *Ultramicroscopy*, 2013, **132**, 19.
41. P. A. J. Bagot, T. Visart de Bocarmé, A. Cerezo and G. D. W. Smith, *Surf. Sci.*, 2006, **600**, 3028.
42. P. A. J. Bagot, H. J. Kreuzer, A. Cerezo and G. D. W. Smith, *Surf. Sci.*, 2011, **605**, 1544.
43. A. Vázquez, J. M. Domíguez and S. Fuentes, *Appl. Surf. Sci.*, 1990, **44**, 331.
44. K. Tanaka, *Surf. Sci.*, 1996, **357–358**, 721.
45. K. Bobrov and L. Guillemot, *Surf. Sci.*, 2007, **601**, 3268.
46. R. Bryl, T. Olewicz, T. Visart de Bocarmé and N. Kruse, *J. Phys. Chem. C*, 2010, **114**, 2220.
47. R. Bryl, T. Olewicz, T. Visart de Bocarmé and N. Kruse, *J. Phys. Chem. C*, 2011, **115**, 2761.
48. J. Rogal, K. Reuter and M. Scheffler, *Phys. Rev. B: Condens. Matter Mater. Phys.*, 2004, **69**, 075421.
49. F. Mittendorfer, N. Seriani, O. Dubay and G. Kresse, *Phys. Rev. B: Condens. Matter Mater. Phys.*, 2007, **76**, 233413.
50. M. Drechsler and J. F. Nicholas, *J. Phys. Chem. Solids*, 1967, **28**, 2609.

51. H. Sang and W. A. Miller, *Surf. Sci.*, 1971, **28**, 349.
52. H. Bu, M. Shi and J. W. Rabalais, *Surf. Sci.*, 1990, **236**, 135.
53. J. L. Taylor, D. E. Ibbotson and W. H. Weinberg, *Surf. Sci.*, 1979, **79**, 349.
54. I. Ermanoski, C. Kim, S. P. Kelty and T. E. Madey, *Surf. Sci.*, 2005, **596**, 89.
55. T. Visart de Bocarmé, T. Bär and N. Kruse, *Surf. Sci.*, 2000, **454–456**, 320.
56. T. Visart de Bocarmé, T. Bär and N. Kruse, *Ultramicroscopy*, 2001, **89**, 75.
57. C. Voss and N. Kruse, *Surf. Sci.*, 1998, **409**, 252.
58. V. K. Medvedev, Yu. Suchorski, C. Voss, T. Visart de Bocarmé, T. Bär and N. Kruse, *Langmuir*, 1998, **14**, 6151.
59. F. M. Leibsle, P. W. Murray, S. M. Francis, G. Thornton and M. Bowker, *Nature*, 1993, **363**, 706.
60. V. R. Dhanak, K. C. Prince, R. Rosei, P. W. Murray, F. M. Leibsle, M. Bowker and G. Thornton, *Phys. Rev. B: Condens. Matter Mater. Phys.*, 1994, **49**, 5585.
61. T. Visart de Bocarmé and N. Kruse, *Top. Catal.*, 2001, **14**, 35.
62. T. Visart de Bocarmé, G. Beketov and N. Kruse, *Surf. Interface Anal.*, 2004, **36**, 522.
63. J.-S. McEwen, P. Gaspard, T. Visart de Bocarmé and N. Kruse, *Surf. Sci.*, 2010, **604**, 1353.
64. C. Barroo, N. Gilis, S. V. Lambeets, F. Devred and T. Visart de Bocarmé, *Appl. Surf. Sci.*, 2014, **304**, 2.
65. M. Yu. Smirnov, E. I. Vovk, A. V. Kalinkin, A. V. Pashis and V. I. Bukhtiyarov, *Kinet. Catal.*, 2012, **53**, 117.
66. P. T. Dawson and Y. K. Peng, *Surf. Sci.*, 1980, **92**, 1.
67. P. A. J. Bagot, A. Cerezo and G. D. W. Smith, *Surf. Sci.*, 2007, **601**, 2245.
68. P. A. J. Bagot, A. Cerezo, G. D. W. Smith, T. Visart de Bocarmé and T. J. Godfrey, *Surf. Interface Anal.*, 2007, **39**, 172.
69. M. K. Miller, A. Cerezo, M. G. Hetherington and G. D. W. Smith, *Atom Probe Field Ion Microscopy*, Oxford University Press, Oxford UK, 1996.
70. M. Haruta, *Catal. Today*, 1997, **36**, 153.
71. G. J. Hutchings, *Gold Bull.*, 2004, **37**, 3.
72. T. Visart de Bocarmé, M. Moors, N. Kruse, I. S. Atanasov, M. Hou, A. Cerezo and G. D. W. Smith, *Ultramicroscopy*, 2009, **109**, 619.
73. G. Maire, L. Hilaire, P. Legare, F. G. Gault and A. O'Cinneide, *J. Catal.*, 1976, **44**, 293.
74. L. Hilaire, P. Legare, Y. Holl and G. Maire, *Surf. Sci.*, 1981, **103**, 125.
75. B. Moest, S. Helfensteyn, P. Deurinck, M. Nelis, A. W. Denier van der Gon, H. H. Brongersma, C. Creemers and B. E. Nieuwenhuys, *Surf. Sci.*, 2003, **563**, 177.
76. J. Florencio, D. M. Ren and T. T. Tsong, *Surf. Sci.*, 1996, **345**, L29.
77. T. T. Tsong, D. M. Ren and M. Ahmad, *Phys. Rev. B: Condens. Matter Mater. Phys.*, 1988, **38**, 7428.
78. N. Kruse and A. Gaussmann, *Surf. Sci.*, 1992, **266**, 51.

79. T. Li, E. A. Marquis, P. A. J. Bagot, S. C. Tsang and G. D. W. Smith, *Catal. Today*, 2011, **175**, 552.
80. M. Rubel, M. Pszonicka, M. F. Ebel, A. Jablonski and W. Palczewska, *J. Less-Common Met.*, 1986, **125**, 7.
81. D. B. Beck, C. L. DiMaggio and G. B. Fisher, *Surf. Sci.*, 1993, **297**, 293.
82. D. B. Beck, C. L. DiMaggio and G. B. Fisher, *Surf. Sci.*, 1993, **297**, 303.
83. T. Wang and L. D. Schmidt, *J. Catal.*, 1981, **71**, 411.
84. R. M. Wolf, J. Siera, F. C. M. J. M. van Delft and B. E. Nieuwenhuys, *Faraday Discuss. Chem. Soc.*, 1989, **87**, 275.
85. W. B. Williamson, H. S. Gandhi, P. Wynblatt, T. J. Truex and R. C. Ku, *AIChE Symp. Ser.*, 1980, **76**, 212.
86. A. R. McCabe and G. D. W. Smith, *Platinum Met. Rev.*, 1983, **27**, 19.
87. W. F. Gale and T. C. Totemeier, *Smithells Metals Reference Book*, Elsevier, Oxford, 8th edn, 2004.
88. Y. L. He, J. K. Zuo and G. C. Wang, *Surf. Sci.*, 1991, **255**, 269.
89. H. T. Wu and T. T. Tsong, *Surf. Sci.*, 1994, **318**, 358.
90. T. Li, P. A. J. Bagot, E. A. Marquis, S. C. E. Tsang and G. D. W. Smith, *J. Phys. Chem. C*, 2012, **116**, 4760.
91. G. W. Graham, T. J. Potter, W. H. Weber and H. S. Gandhi, *Oxid. Met.*, 1988, **29**, 487.
92. R. J. Baird, G. W. Graham and W. H. Weber, *Oxid. Met.*, 1988, **29**, 435.
93. D. Wang, T. B. Flanagan, R. Balasubramaniam and Y. Sakamoto, *Scr. Mater.*, 2000, **43**, 685.
94. J. A. Rodriguez and D. W. Goodman, *Science*, 1992, **257**, 897.
95. J. G. Chen, C. A. Menning and M. B. Zellner, *Surf. Sci. Rep.*, 2008, **63**, 201.
96. T. Li, P. A. J. Bagot, E. A. Marquis, S. C. Edman Tsang and G. D. W. Smith, *Ultramicroscopy*, 2013, **132**, 205.
97. L. Brewer, *Chem. Rev.*, 1953, **52**, 1.
98. W. M. Haynes and D. Lide, *Handbook of Chemistry and Physics*, CRC-Press, Boca Raton, 91st edn, 2010.
99. M. Chen and L. D. Schmidt, *J. Catal.*, 1979, **56**, 198.
100. G. Samsonov, *The Oxide Handbook*, Plenum Press, New York, 1973.
101. W. L. Phillips Jr, *Trans. Am. Soc. Met.*, 1964, **57**, 33.
102. E. Iojoiu, P. Gélin, H. Praliaud and M. Primet, *Appl. Catal., A*, 2004, **263**, 39.
103. M. Ahmad and T. T. Tsong, *J. Chem. Phys.*, 1985, **83**, 388.
104. M. Shelef and H. S. Gandhi, *Ind. Eng. Chem. Res.*, 1972, **11**, 393.
105. H. S. Gandhi, G. W. Graham and R. W. McCabe, *J. Catal.*, 2003, **216**, 433.
106. T. Li, P. A. J. Bagot, E. A. Marquis, S. C. E. Tsang and G. D. W. Smith, *J. Phys. Chem. C*, 2012, **116**, 17633.
107. W. E. Bell and M. J. Tagami, *Phys. Chem.*, 1963, **67**, 2432.
108. H. J. Jehn, *J. Alloys Compd.*, 1984, **100**, 321.
109. P. A. J. Bagot, A. Cerezo and G. D. W. Smith, *Surf. Sci.*, 2008, **602**, 1381.

110. J. Hrbek, D. G. van Campen and I. J. Malik, *J. Vac. Sci. Technol., A*, 1995, **13**, 1409.
111. H. S. Gandhi, H. K. Stepien and M. Shelef, *Mater. Res. Bull.*, 1975, **10**, 837.
112. *Chemisorption and Reactivity on Supported Clusters and Thin Films*, ed. R. M. Lambert and G. Pacchioni, Springer, New York, 1997.
113. K. Tedsree, T. Li, S. Jones, C. W. A. Chan, K. M. K. Yu, P. A. J. Bagot, E. A. Marquis, G. D. W. Smith and S. C. Edman Tsang, *Nat. Nanotechnol.*, 2011, **6**, 302.
114. K. M. K. Yu, W. Tong, A. West, K. Cheung, T. Li, G. Smith, Y. Guo and S. C. Edman Tsang, *Nat. Commun.*, 2012, **3**, 1230.
115. R. Larde, J. Bran, M. Jean and J. M. Le Breton, *Powder Technol.*, 2011, **208**, 260.
116. F. Wu, P. Bellon, M. L. Lau, E. J. Lavernia, T. A. Lusby and A. J. Melmed, *Mater. Sci. Eng., A*, 2002, **327**, 20.
117. F. Wu, D. Isheim, P. Bellon and D. N. Seidman, *Acta Mater.*, 2006, **54**, 2605.
118. Y. Xiang, V. Chitry, P. Liddicoat, P. Felfer, J. Cairney, S. Ringer and N. Kruse, *J. Am. Chem. Soc.*, 2013, **135**, 7114.
119. M. Ojeda and E. Iglesia, *Angew. Chem., Int. Ed.*, 2009, **48**, 4800.
120. A. M. Lossack, D. M. Bartels and E. Roduner, *Res. Chem. Intermed.*, 2001, **27**, 475.
121. P. Buess, R. F. I. Caers, A. Frennet, E. Ghenne, C. Hubert and N. Kruse, *U.S. Patent* 6362239 B 1, 2002.
122. A. Frennet, C. Hubert, E. Ghenne, V. Chitry and N. Kruse, *Stud. Surf. Sci. Catal.*, 2000, **130**, 3699.
123. L. Keck, J. S. Buchanan and G. A. Hards, *U.S. Patent* 5068161, 1991.
124. C. Kwak, T.-J. Park and D. J. Suh, *Appl. Catal., A*, 2005, **278**, 181.
125. E. Gyenge, M. Atwan and D. Northwood, *J. Electrochem. Soc.*, 2006, **153**, A150.

Subject Index

acetylene cyclotrimerization, 46–48
active sites, 1
adsorption-driven surface-limited reactions, 148–149
advanced fuel cell electrocatalysts, 161–163
Ag–Pd core–shell structures, 282–283
ALD. *See* atomic layer deposition
aldehydes, catalytic selective hydrogenation, 136–138
alkanes
    hydrogenolysis of, 17–18
    metathesis, 18
alkylidynes, hydroxyl group reactions, 8–10
α,β-unsaturated ketones/aldehydes, 98–100
anisotropic infiltration, 174–176
APT. *See* atom probe tomography
atomic layer deposition (ALD)
    adsorption-driven surface-limited reactions, 148–149
    catalysis
        early work preparation, 179–181
        literature overview, 179
        ordered mesoporous materials, 185–189
        photo-active nanoparticles, mesoporous films, 189–191
        photocatalysis, 184
        supported noble metal catalysts, 181–183
        synthesis of catalytic membranes, 184–185
    definition, 167
    description, 168–171
    displacement-driven surface-limited reactions
        hydrogen adsorption, 155–156
        ML core–shell type electrocatalysts, 150–154
        Pt monolayer core–shell type electrocatalysts, 154–155
    methodology, 146–147
    nanoporous materials
        anisotropic infiltration, 174–176
        high surface area powder particles, 173–177
        isotropic infiltration, 174
        nanometer-sized mesopores, 171–173
    supported catalysts preparation, 177–179
    surface limited reactions, 147
atom probe tomography (APT)
    catalytic nanoparticles
        atom probe sample preparation, 281–282

Subject Index

metal–support
interaction, 285–287
unsupported
nanoparticles, 282–284
definition, 251–252
platinum group metal-based
alloys
binary alloys, 267–276
ternary alloys, 277–280
principles, 252, 256–258
reaction cells, catalysis studies,
258–259

bimetallic catalysts
conventional preparation
procedures, 56
definition, 55
effects in heterogeneous
catalysts, 56
ensemble effects, 56
geometric and electronic
effects, 56
heterometallic catalysts
precursors
bridging ligands,
59–61
ions pairs, 58–59
mixed-metal precursors,
61–63
M–M′ metal–metal bond,
57–58
other bimetallic
combinations, 61
nanoparticle
experimental
methodology, 67–68
key concepts, 68–70
ligand-stabilized
nanoparticles, 77–82
mixed-metal clusters,
70–77
binary alloys, APT studies
chemically driven surface
enrichment, 267–268
core–shell structures
formation, 274–275

nano-islands formation,
273–274
Pt-10.1at%Ir, 275–276
Pt-8.9at%Ru, 276
pure Pt@Rh core–shell
structures preparation,
268–273

calcined zeolite, 39
carbon–carbon coupling reactions
Sonogashira cross-coupling
reaction, 139–140
Ullmann-type homo-coupling
reaction, 138–139
carbon monoxide oxidation, 93–96
catalysis
applications, surface
organometallic chemistry
alkane metathesis, 18
deperoxidation reactions,
20–21
epoxidation reactions,
20–21
ethylene to propene by
tungsten hydride,
18–19
hydrogenolysis of
alkanes, 17–18
olefin metathesis, 19
trimerization of ethylene,
19–20
atomic layer deposition
early work preparation,
179–181
literature overview, 179
ordered mesoporous
materials, 185–189
photo-active nanoparticles,
mesoporous films,
189–191
photocatalysis, 184
supported noble metal
catalysts, 181–183
synthesis of catalytic
membranes, 184–185
clusters in, 65–66

catalysis *(continued)*
   gold nanoparticle-based catalysts
      α,β-unsaturated ketones/aldehydes, 98–100
      carbon monoxide oxidation, 93–96
      description, 91–93
      styrene epoxidation, 96–98
   oxidation reactions
      CO to CO_2, 131–134
      styrene, 134–135
      sulfides, 135–136
   selective hydrogenation
      aldehydes, 136–138
      ketones, 136–138
      nitrophenol, 138
catalytic cycle, 46–48
catalytic membranes synthesis, 184–185
catalytic nanoparticles
   atom probe sample preparation
      CVB-FIB method, 282
      electrophoresis method, 281
   metal–support interaction
      inhomogeneity of Pt-25.3at%Co, 285–287
      intermixing in Ir@Pt, 287–289
      Pt-supported nanoparticles, 285
   unsupported nanoparticles
      Ag–Pd core–shell structures, 282–283
      Fischer–Tropsch reaction, 283–284
CeO_2-based catalysts, 213–216
chemically driven surface enrichment, 267–268
chemical vapor deposition (CVD) method, 30

cluster
   bimetallic nanoparticle catalysts
      experimental methodology, 67–68
      key concepts, 68–70
      ligand-stabilized nanoparticles, 77–82
      mixed-metal clusters, 70–77
   in catalysis, 65–66
   definition, 64
   surface reactions, 66
   synthesis, 65
cluster–surface analogy, 65
colloids, 64
crystal structures, gold nanoclusters, 128–129
Cu under-potentially pre-deposited monolayer, 150–154

Deacon process, 198–199
density functional theory (DFT), 110–112
deperoxidation reactions, 20–21
direct methanol fuel cells (DMFCs), 239
displacement-driven surface-limited reactions
   hydrogen adsorption, 155–156
   ML core–shell type electrocatalysts, 150–154
   Pt monolayer core–shell type electrocatalysts, 154–155
DMFCs. *See* direct methanol fuel cells
durability, 146

electrocatalysis
   electrochemical reduction, 242–243
   formic acid oxidation reaction
      intermetallic NP catalyst, 240–242
      Pt-based alloy NP catalysts, 239–240

Subject Index

oxygen reduction reaction
  Pt-based alloy NP catalysts, 234–237
  Pt-based core shell NPs, 237–238
electrocatalysts
  advanced fuel cell, 161–163
  conventional chemical methods, 156
  oxygen reduction reaction, 145
electrochemical deposition
  of monolayers
    adsorption-driven surface-limited reactions, 148–149
    displacement-driven surface-limited reactions, 149–156
  strategies, 147
electrochemical polishing, 252–254
electrochemical reduction, 242–243
electro-deposition techniques
  advantages of, 163
  bimetallic Pd alloys, 157–161
  Pd nanostructures, 157–161
  Pd/WNi refractory alloys, 161–163
electrophoresis method, 281
electrospinning
  CeO_2-based catalysts, 213–216
  metal oxide fibers, 208–211
  RuO_2-based fibers, 211–213
electrospun nanofibers, 217
ensemble effect, 56
epoxidation
  catalysis applications, 20–21
  styrene, 96–98

FAOR. *See* formic acid oxidation reaction
FAU zeolite, 33
FEM. *See* field emission microscopy
Fermi hole effect, 112
FIB. *See* focused ion beam
field emission microscopy (FEM)
  applications, 250
  description, 249–250

field ion microscopy (FIM)
  applications, 250, 255–256
  catalytic reactions and surface reconstructions
    NO_2+H_2/Pt, 264–266
    NO/Pt, 266–267
    NO/Pt-17.4at%Rh, 266–267
    O_2+H_2/Rh, 263–264
    O_2/Ir, 259–263
    O_2/Pd, 259–263
    O_2/Pt, 264–266
    O_2/Rh, 263–264
  description, 249–250
  features, 251
  principles, 255
FIM. *See* field ion microscopy
Fischer–Tropsch reaction, 283–284
Fischer–Tropsch synthesis, 17
focused ion beam (FIB), 254
formic acid oxidation reaction (FAOR)
  intermetallic NP catalyst, 240–242
  Pt-based alloy NP catalysts, 239–240

gold nanoclusters
  carbon–carbon coupling reactions
    Sonogashira cross-coupling reaction, 139–140
    Ullmann-type homo-coupling reaction, 138–139
  catalytic oxidation
    CO to CO_2, 131–134
    styrene, 134–135
    sulfides, 135–136
  catalytic selective hydrogenation
    aldehydes, 136–138
    ketones, 136–138
    nitrophenol, 138

gold nanoclusters (*continued*)
   crystal structures, 128–129
   reversible conversion, 130–131
   size focusing methodology, 124–128
   thermal stability, 129–130

gold nanoparticle-based catalysts
   catalysis
      α,β-unsaturated ketones/aldehydes, 98–100
      carbon monoxide oxidation, 93–96
      description, 91–93
      styrene epoxidation, 96–98
   characterization, 90–91
   correlation of magnetic structure–catalysis, 114–116
   description, 87–88
   electronic structure
      photoemission spectroscopy, 105–108
      X-ray absorption near edge structure, 102–105
   future perspectives, 116–117
   magnetic structure
      density functional theory (DFT) calculations, 110–112
      electronic configuration, 108
      Fermi hole effect, 112
      oscillator strengths, 113
      spin arrangements, 109
      spin–orbit coupling, 112
   synthesis, 88–91

grafted organometallic complexes
   description, 3
   hydroxyl group reactions
      alkylidynes, 8–10
      consequences, 6
      metal alkyl complexes, 6–8
      metal alkylidynes, 8–10
   reactions, 2

HAADF. *See* high-angle annular dark-field

Haber–Bosch process
   ammonia activation, 21–22
   nitrogen activation, 21–22

HCl oxidation reactions
   Deacon process, 198–199
   electrospinning
      CeO_2-based catalysts, 213–216
      metal oxide fibres, 208–211
      RuO_2-based fibers, 211–213
   single crystalline RuO_2 films
      atomic-scale properties, 204–205
      learnings from, 205–208
      synthesis, 202–203

heterometallic catalysts precursors
   bridging ligands, 59–61
   ions pairs, 58–59
   mixed-metal precursors, 61–63
   M–M' metal–metal bond, 57–58
   other bimetallic combinations, 61

high-angle annular dark-field (HAADF), 37–38, 152–153
high surface area powder particles, 173–177
HSSZ-53 zeolite, 35, 37
hydrogen adsorption, 155–156
hydrogenolysis of alkanes, 17–18
hydroxyl group reactions
   grafted organometallic complexes
      alkylidynes, 8–10
      consequences, 6
      metal alkyl complexes, 6–8
      metal alkylidynes, 8–10
   organometallic compounds
      α-H abstraction, 11
      β-H abstraction, 11–12
      further reactions, 10–11

Subject Index

γ-H abstraction, 12
hydrogen reactions, 12–14
other molecules reactions, 14–15

impregnation, 146
intermetallic nanoparticle catalyst, 240–242
ion-exchange method, 29–30
isotropic infiltration, 174

ketones, catalytic selective hydrogenation, 136–138

ligand-bridged heterometallic catalysts precursors, 59–61
ligand effects, 237
ligand-stabilized nanoparticles, 77–82

mesoporous films, photo-active nanoparticles, 189–191
metal alkyl complexes, hydroxyl group reactions, 6–8
metal alkylidynes, hydroxyl group reactions, 8–10
metal complexes reactivity, 41–46
metal nanoparticles, electrocatalysis
electrochemical reduction, 242–243
formic acid oxidation reaction
intermetallic catalyst, 240–242
Pt-based alloy catalysts, 239–240
oxygen reduction reaction
Pt-based alloy catalysts, 234–237
Pt-based core shell NPs, 237–238
metal oxide fibres, 208–211
metathesis
alkane, 18
olefin, 19
mixed-metal clusters, 70–77

mixed-metal precursors, 61–63
M–M' metal–metal bond, 57–58
model catalysis
electrospinning
CeO_2-based catalysts, 213–216
metal oxide fibres, 208–211
RuO_2-based fibers, 211–213
reasons for, 199–202
single crystalline RuO_2 films
atomic-scale properties, 204–205
learnings from, 205–208
synthesis, 202–203
monodisperse nanoparticles synthesis
activation for catalysis, 232
anisotropic growth, reverse micelle, 230–232
kinetic control, 229–230
nanoparticle formation, 226–228
shape control, 229–232
size control, 228
monolayer core–shell type electrocatalysts, 150–154

naked supported nanoparticles, 77–82
nano-islands formation, 273–274
nanometer-sized mesopores, 171–173
nanoparticle
bimetallic catalysts
experimental methodology, 67–68
key concepts, 68–70
ligand-stabilized nanoparticles, 77–82
mixed-metal clusters, 70–77
monodisperse synthesis
activation for catalysis, 232
anisotropic growth, reverse micelle, 230–232

nanoparticle (*continued*)
    formation, 226–228
        kinetic control, 229–230
        shape control, 229–232
        size control, 228
nanoparticle gold catalysts
    catalysis
        α,β-unsaturated ketones/aldehydes, 98–100
        carbon monoxide oxidation, 93–96
        description, 91–93
        styrene epoxidation, 96–98
    characterization, 90–91
    correlation of magnetic structure–catalysis, 114–116
    description, 87–88
    electronic structure
        photoemission spectroscopy, 105–108
        X-ray absorption near edge structure, 102–105
    future perspectives, 116–117
    magnetic structure
        DFT calculations, 110–112
        electronic configuration, 108
        Fermi hole effect, 112
        oscillator strengths, 113
        spin arrangements, 109
        spin–orbit coupling, 112
    synthesis, 88–91
nanoporous materials, atomic layer deposition
    anisotropic infiltration, 174–176
    high surface area powder particles, 173–177
    isotropic infiltration, 174
    nanometer-sized mesopores, 171–173
nanosized crystals
    field ion microscopy
        applications, 255–256
        principles, 255

    sample preparation
        electrochemical polishing, 252–254
        focused ion beam method, 254
nitrophenol, catalytic selective hydrogenation, 138

olefin metathesis, 19
ordered mesoporous materials, 185–189
organometallic hydroxyl group reactions
    α-H abstraction, 11
    β-H abstraction, 11–12
    further reactions, 10–11
    γ-H abstraction, 12
    hydrogen reactions, 12–14
    other molecules reactions, 14–15
ORR. *See* oxygen reduction reaction
oxygen reduction reaction (ORR)
    electrocatalysis
        Pt-based alloy NP catalysts, 234–237
        Pt-based core shell NPs, 237–238
    electrocatalysts, 145
    platiunum monolayer
        electro-deposited bimetallic Pd alloys, 157–161
        electro-deposited Pd nanostructures, 157–161
        electro-deposited Pd/WNi refractory alloys, 161–163

Pb under-potentially pre-deposited monolayer, 154–155
PEMFCs. *See* polymer electrolyte membrane fuel cells
photo-active nanoparticles, mesoporous films, 189–191
photocatalysis, 184
photoemission spectroscopy, gold nanoparticles, 105–108

Subject Index 303

platinum group metal-based alloys
  binary alloys
    chemically driven surface enrichment, 267–268
    core–shell structures formation, 274–275
    nano-islands formation, 273–274
    Pt-10.1at%Ir, 275–276
    Pt-8.9at%Ru, 276
    pure Pt@Rh core–shell structures preparation, 268–273
  ternary alloys
    Pt-17.3%Rh-14.0%Ir, 277–278
    Pt-23.9%Rh-9.6%Ru, 278–280
platiunum monolayer
  core–shell type electrocatalysts, 154–155
  electro-deposited bimetallic Pd alloys, 157–161
  electro-deposited Pd nanostructures, 157–161
  electro-deposited Pd/WNi refractory alloys, 161–163
polymer electrolyte membrane fuel cells (PEMFCs), 144–145
precursors
  heterometallic catalysts
    bridging ligands, 59–61
    ions pairs, 58–59
    mixed-metal precursors, 61–63
    M-M' metal–metal bond, 57–58
    other bimetallic combinations, 61
  metal complexes, 38–41
Pt-based alloy nanoparticle catalysts
  formic acid oxidation reaction, 239–240
  oxygen reduction reaction, 234–237

Pt-based core shell nanoparticles, 237–238
Pt-supported nanoparticles, 285
reduction/co-reduction, 146
RuO_2-based Deacon catalysts, 213–216
RuO_2-based fibers, 211–213

Schrock catalyst, 9
silica gel, isotropic infiltration, 174
single crystalline RuO_2 films
  atomic-scale properties, 204–205
  learnings from, 205–208
  synthesis, 202–203
size focusing methodology, 124–128
SLRR. *See* surface limited redox replacement
SLRs. *See* surface limited reactions
sol-gel method, 63
Sonogashira cross-coupling reaction, 139–140
spin–orbit coupling, gold nanoparticles, 112
SSZ-53 zeolite, 35, 37
strain effects, 237
styrene
  catalytic oxidation, 134–135
  epoxidation, 96–98
sulfides, catalytic oxidation, 135–136
surface limited reactions (SLRs), 147
surface limited redox replacement (SLRR)
  definition, 147
  ML core–shell type electrocatalysts, 150–154
  Pt monolayer core–shell type electrocatalysts, 154–155
surface organometallic chemistry
  ammonia activation, 21–22
  catalysis applications
    alkane metathesis, 18
    deperoxidation reactions, 20–21
    epoxidation reactions, 20–21
    ethylene to propene by tungsten hydride, 18–19

surface organometallic chemistry (*continued*)
  hydrogenolysis of alkanes, 17–18
  olefin metathesis, 19
  trimerization of ethylene, 19–20
 cationic complexes formation, 15–16
 concept, 2
 grafted organometallic hydroxyl group reactions
  alkylidynes, 8–10
  consequences, 6
  metal alkyl complexes, 6–8
  metal alkylidynes, 8–10
 grafting reactions, organometallic complexes, 2
 grafting sites, 3–5
 nitrogen activation, 21–22
 organometallic hydroxyl group reactions
  α-H abstraction, 11
  β-H abstraction, 11–12
  further reactions, 10–11
  γ-H abstraction, 12
  hydrogen reactions, 12–14
  other molecules reactions, 14–15
 zeolite-supported molecular metal complex catalysts, 31

ternary alloys, APT studies
 Pt-17.3%Rh-14.0%Ir, 277–278
 Pt-23.9%Rh-9.6%Ru, 278–280
thin-film rotating-disc electrode (RDE) method, 145
thiolate-protected gold nanoclusters, 129–130
trimerization of ethylene, 19–20

Ullmann-type homo-coupling reaction, 138–139
under-potentially pre-deposited (UPD)
 Cu monolayer, 150–154
 definition, 149
 Pb monolayer, 154–155
unsupported catalytic nanoparticles
 Ag–Pd core–shell structures, 282–283
 Fischer–Tropsch reaction, 283–284
UPD. *See* under-potentially pre-deposited

valence band study, photoemission spectroscopy, 105–108

XANES. *See* X-ray absorption near edge structure
X-ray absorption near edge structure (XANES), 102–105

zeolite-supported molecular metal complex catalysts
 characterization techniques, 31–33
 molecular chemistry
  catalytic cycle, 46–48
  supported metal complexes, 41–46
 structural uniformity
  description, 28–29
  precursor metal complexes, 38–41
  spectroscopic characterization techniques, 41
  type of zeolite materials, 33–38
 synthesis
  chemical vapor deposition method, 30
  ion-exchange method, 29–30
  surface organometallic chemistry approach, 31
Zeotile-4, anisotropic infiltration, 174–176